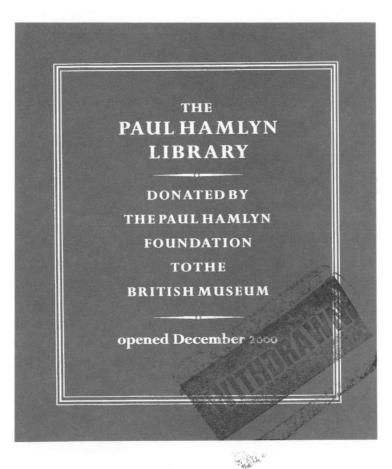

The Harwin chronology of
Inventions Innovations Discoveries
A chronology from pre-history to the present day

The Harwin chronology of

Inventions
Innovations
Discoveries

from pre-history to the present day

Kevin Desmond

HARWIN

Constable *London*

First published in Great Britain 1987
by Constable and Company Limited
10 Orange Street London WC2H 7EG
Copyright © 1986 by Kevin Desmond

ISBN 0 09 466150 2

Designed by Ivor Kamlish FSIAD

Printed in Great Britain by
St Edmundsbury Press
Bury St Edmunds, Suffolk

This chronology is dedicated to GOD Creator of all things

God hath made man upright; but they have sought out many inventions.

Ecclesiastes vii 9, *c.* 250 BC

Countless ages will beget many new inventions, but my own is mine.

Pedanius Dioscorides,
fl. first Century AD

Tell me when anything was ever made.

Leonardo da Vinci, *c.* 1515

'Tis frivolous to fix pedantically the date of a particular invention. They have all been invented over and over fifty times. Man is the arch machine of which all these shifts drawn from himself are toy models.

Ralph Waldo Emerson,
Conduct of Life: Fate, 1860

Invention and innovation are often preceded by a flash of thought, but it requires infinite patience, concentration and hard work to develop that thought to the point where the idea can be said to *be* an invention or innovation. Often this point is only the beginning of more intensive labour to educate contemporaries in the use and application of the idea. When reading this book it must be remembered that invention and innovation can only be judged in retrospect. It is history that decides whether the 'mad inventor' is a genius or not.

The Harwin Group of companies are used to innovation, and sometimes invention, in the esoteric world of connectors and components for the electronics industry, which is why we are happy and excited to sponsor Kevin Desmond and Constable in this innovative presentation of man's development.

Damon de Laszlo
Chairman, Harwin, Portsmouth, Hants.

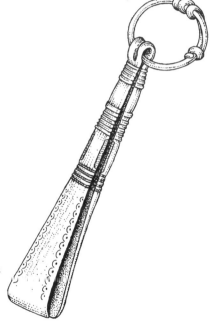

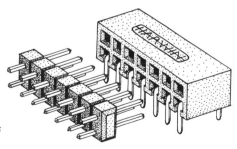

'Ancient and Modern': Anglo-Saxon bronze tweezers (5th-7th century) alongside Harwin electronic connectors (20th century)

General thanks are due to the public relations departments of the many companies whose innovations appear in this book – too numerous to single out; thanks also to those editors who published my 'appeals' in their magazines – and to the choice few who responded to them. I would also have been lost but for the patient help of a whole chain of librarians – particularly those in my own locality of North-West London.

Last but not least, special gratitude must go to Andrew Macdonald who has done such fine additional illustrations; to Keith Macqueen for a mammoth indexing task; to Imogen Olsen; to my editors Miles Huddleston and Prudence Fay; to Ivor Kamlish for the typography and design; but above all to my long-suffering wife, Alex.

K.D. 1986

Acknowledgements

The scientific and industrial progress recorded in this book has been achieved in an extraordinarily short period in history. Scientists tell us that the earth is about 4½ billion years old. Imagine this period represented by a 24-hour earth clock. The first faint traces of life appeared at about 2 p.m. (1400 hours). The dinosaur showed up at about 11 p.m. (2300 hours). And the human species? Man finally made the scene at two seconds before midnight. The entire last 6000 years of human endeavour have occurred in the final one-tenth of a second.

Within that tenth of a second, it took man 1750 years from the year 1 AD to double his technological knowledge. A mere 150 years later, by the year 1900, he had doubled his knowledge again. The next doubling took only 50 years (1900–50); the next, ten years (1950–60); and man's knowledge has quadrupled during each of the last two decades.

Another way of estimating the same progress is to state that there have been somewhere between 40,000 and 60,000 generations of *Homo sapiens* on this planet. In Great Britain, where the Industrial Revolution was born, only seven generations have been involved with that revolution – just a blip in history.

If most inventors' lives were to be averaged out, one would find that they spent 23 out of their 70 full years fast asleep; that they spent six years eating and two years keeping themselves clean. Given that no fewer than twenty of the years remaining to them are merely formative – devoted to the learning of established science – this leaves them just twenty years of their entire life to accomplish the practical work of inventing or discovering something totally new. No wonder the era of the individual inventor is no longer with us and has been replaced by ever-growing Research and Development armies of university-linked, multi-national corporations.

One should not leave out the fact that *only half* of the world's total accumulated population throughout history was given the opportunity to invent or discover. The women who feature in this chronology may be counted on the fingers of both hands. Only now perhaps is a 6000-year-old prejudice being overturned: that woman's only role has been to conceive and bring up the latest generation to both wield and suffer from the newest weaponry which man has been able to devise.

A major part of our recent and astonishing development of technological knowledge has been due to the use of an essentially simple tool – the computer. It has been projected that, by the end of this millennium, artificially intelligent computers will have outpaced human intelligence. Yet can this be possible? The human brain can store about $2\frac{1}{2}$ million times more information than today's most advanced computer. The human brain can hold some 10 million million 'bits' of information, and all tucked away in a skull that measures about one-twentieth of a cubic foot. The human brain weighs about 3 lb. A modern mini-computer – a 1000 million bit machine – takes up about 40 cubic feet and holds about 20 lb of memory units. To attempt to equal – let alone surpass – *cerebrum humanum* would probably require the long-term pooling of all the available computer scientists, psychologists, linguists and brain surgeons in the USSR, USA, Europe and Japan.

Writing in severely impoverished conditions in 1919, William Friese-Greene, English inventor of the cinema some 40 years earlier, recalled how '. . . the expense and enthusiasm of the Inception and for proving the possibilities of Cinematography, ended my business career. One man cannot compete against the prejudice and jealousy of an Invention which is not believed at the time. So the Invention goes through the stages: 1st: Rubbish, impossible! 2nd: Impossible! 3rd: Possible and probable. 4th: Yes, possible, but is there any money in

it? 5th: Yes there is money in it (that is the stage of worry). 6th: Plenty of money in it. 7th: Oh I have heard of this Invention long ago! So the ebb and flow of Invention, like the ebb and flow of Reason, will always flow from generation to generation . . .'

Almost endorsing Friese-Greene's analysis, there is now such a library of science-fiction books and films that whenever an item moves out of the realm of fiction and into that of fact – in other words, is *really* invented or discovered – people are so well prepared for it that the reality no longer awes or impresses them. They too have heard of it long ago.

Although three separate topics appear in the title of this book, there are no clear-cut divisions between them. The process is too fluid for that. An 'invention' often appears as a more noticeable step forward than an 'innovation', the latter being defined as a new assembly of existing inventions. 'Discoveries' are usually made either by the invention of new devices or by an innovative assembly of existing equipment with which to observe, measure and demonstrate. Once made, such 'discoveries' in their turn may be utilised as just one part in the innovative assembly that makes up a distinguishable 'invention'.

Books about the history of invention usually choose to list them in alphabetical order or grouped by subjects, appending a chronology of no more than a couple of dozen pages at the back. They also usually give much space to illustrating and explaining how the inventions worked and work. This in turn, for reasons of space, obliges them to limit the nature of the inventions listed to practical day-to-day objects.

In this book I have taken a different approach. As almost 4000 entries have been compiled, it has not been possible to carry out conclusive verifications on the originality or 'first-ness' of each and every entry, but the new chronological approach has enabled me to include a wider range of topics – both the purely scientific and the more whimsical and forgotten. This book therefore does not seek to rival the more conventional approach, but rather to act as a companion to it.

It has been estimated that there is now an accumulation of 25 million patent documents worldwide, including microfiche and ultrafiche systems. Well over 4½ million of these are official US patents, with a total of over 2 million for both the UK and France, not to mention totals of 500,000 to over a million for other industrial nations. Even as you read them, such statistics are already out of date.

Currently anything from 600,000 to three-quarters of a million *new* patent specifications worldwide are accepted each year. In recent years, from 28,000 to 38,000 patents have been granted annually by the UK Patents Office alone – which means that over 500 new patents are laid out for public inspection at Southampton Buildings, London, each week.

As was seen at the start of this Introduction, the rate at which man is inventing and discovering has speeded up tremendously in the past few decades. (Future books of this kind will show whether this frenetic trend has been maintained.) The entries which follow, therefore, can only represent a minute and symbolic fraction of man's astounding technological ingenuity.

Despite that, listing inventions and discoveries in chronological order allows the reader to chart the pattern of technological progress in a slightly different manner: to see, for instance, how certain basic inventions were the seed for a vast plantation of subsequent applications. Witness as such the toothed wheel, the steam engine, steel, the internal-combustion engine, plastic, the thermionic valve, the transistor and the micro-processor.

A year-by-year account will also enable the family historian, for example, to see exactly what were the innovations that would have excited each generation of his ancestors, to find out what was being discovered in the year when he was born, and finally, using the list of inventions still to be realised, to glimpse what may become the latest reality for his children, and their children.

So what direction will man's ingenuity take from now on?

In the course of the past forty years, half a person's lifetime, we have, through invention and innovation, built up an arsenal of weaponry capable of destroying the earth several times over. We have also so 'organised' our trade and industrial environment that the vulnerable wildlife kingdom, which has taken millions of years to evolve, could be virtually wiped out by the end of the century.

And yet in that same short period, we have come up with both cures and preventions for many illnesses hitherto considered fatal; we can replace many parts of the human anatomy that are damaged or in danger of breakdown; we can in a few minutes perform mathematical computations which would have once taken teams of workers years to complete; we have at last begun to quantify the real links between the innermost secrets of the elusive atom

and the outmost mysteries of the universe; finally, we have so improved our telecommunication systems that the world, and even the ionosphere above it, acts as a 'global village', able to show immediate international concern for any disaster which may occur in it.

In the 1890s, Charles Duell, the then US Commissioner of Patents and Trademarks, advocated the abolition of his office on the grounds that everything worthwhile had already been invented. He encountered polite derision.

It would be about as impossible to use a laser beam to change the flight-path of a pre-historic flint spearhead without shattering it, as to change either the pace or the direction of invention without killing it dead.

At the last count, it may only be our deep instinct for survival coupled with that remarkable ingenuity which is evidenced in the following pages, that can give us cause to look bravely ahead towards an even more dazzling, but perhaps less destructive, scientific and technological future.

Kevin Desmond, London
7 April 1985

15,000 million years ago
Formation of the universe ('Big
Bang').

4600 million years ago
Formation of our solar system.

Pre-dates (estimated)

3,000,000 years ago
Flint-knapping.

1,800,000 years ago
Stone tool kits (deliberate
fracture)
Koobi Fora, Rift Valley East Africa.

1,500,000 years ago
Homo erectus.

250,000 years ago
Homo sapiens.

230,000 years ago
Hand-axes, cleavers and flake
tools (stone)
*'Acheulean' industry from the
Kapthurin beds near Lake Baringo,
Kenya, Africa.*

107,000 years ago
Engraving (on bone)
Peche de l'Aze, Dordogne, France.

45,000 years ago
Endscrapers, burins, backed
knives (stone and bone)
*Central and Eastern Europe. Used to
scrape hides clean of hair and fat.*

27,000 years ago
Engraving (on stone)
*La Ferrassie, Périgord, France. Large
blocks of stone engraved with animal
figures and symbols.*

27,000 years ago
Sun-dried clay artefacts.

27,000 years ago
Paintings of animal figures on
rock slabs
*at the Apollo II shelter, Namibia,
Africa.*

The Aurignacian 'Laussel Venus'

22,000
Sculpture (in stone)
the Aurignacian 'Laussel Venus'.

17,000
Sewing needles (in bone)
*Pleistocene peoples of Europe and
North Asia.*

15,000
Cave painting
*Lascaux, France. Depicting reindeer
and bulls.*

13,000
Cave mural
*Altamira, Spain. Depicting bison
herds.*

12,000
**Harpoons, spearthrowers and
fishing hooks**
Man-made fire.

11,000
**Domesticated animal
husbandry**
Thailand, South-east Asia.
Fired clay objects (ceramic)
animal and human figurines.
Stone lamps (animal-fat fuel).

10,000
**Mortar and pestle and digging
stick.**
The zero (mathematics)
Hindu priests.

9–8000
**Domestication of crops (wheat
and barley)**
The Near East.
Flint arrowheads and bows.

8350–7350
Walled town
Jericho, ancient city of Israel.

8050
Domesticated goat
Asiab, Iran.

8000
Quern (grain-crusher)
Ships (reed-boats and shaped
dug-outs).

7700
Domesticated dog
North Yorkshire, England.

7200
Domesticated sheep
Thessaly, Greece.

7000
Coiled pottery
Kurdistan, Iran.
Domesticated peas and beans
Domesticated pigs and cattle
Thessaly, Greece.

6500
Potter's wheel
Asia Minor.

6000
Beer (from malted grain)
Mesopotamia, South-west Asia.
Rice cultivation
Thailand.
Rotary quern
Mesopotamia.
Villages (4–5 hectares)
*Mesopotamia. With irrigation and
religious temples.*

5000
Drilling holes by tying rods
together with thongs
Flaked-stone hoe blade
(agricultural)
Hassana, Mesopotamia.
Pottery repair technique
Nineveh, Northern Mesopotamia.
Woven cloth

4500
Copper-working

*earliest smelting sites, Rudna Glava,
Yugoslavia.*
Stone door sockets (prototype
hinges)
Hassana, Mesopotamia.

4000
Bevel-rimmed open bowl (mass-
produced)
Mesopotamia.
Copper axes
Plough (copper parts)
Reed ropes

3875
Harp (musical instrument)
*attributed to Jubal, son of Lamech
and Adah (see Genesis) – 'father of all
such as handle the harp and organ'.*

3800
Maps (clay tablets)
Sumer.

3761
Jewish Year 1.

3500
Cylinder seal
Mesopotamia.
Gold (mined)
Mesopotamia.
Pictographic statement of
accounts
*at the temple of Uruk, a city-state of
Sumer.*
Rotary grain grinder
Mesopotamia.
Wheeled carts
*Sumer and Syria. Found in tombs at
Kish and Ur with copper-nail rim
studs.*

3200
Ship (rectangular sail)
*depicted on pottery vessel from the
Egyptian Naqada II period.*

3150
Irrigation
River Nile, Upper Egypt. Evidenced

on a ceremonial macehead from the
reign of 'King Scorpion'.

3100
Irrigation (large-scale)
'King Menes' *in the Nile Valley,
Egypt.*
Pictographic inscription
*the slate Palette of Narmer, King of.
Upper and Lower Egypt.*

3000
Boat (timber-built)
*River Nile, Egypt, using ash, meru or
'wood from the Lebanon'.*
Warp-weighted looms.

2950
Unwritten papyrus
*found in a tomb at Saqqara, Egypt,
from the reign of King Den.*

2950–2750
Dam (masonry walls)
Garawi Valley, Egypt.

2800
Bronze working
South-west Asia.

2761
Jewish Year 1000.

2650
Stepped pyramid
Imhotep, *Egyptian architect and
vizier for King Djoser at Saqqara.*

2640
Silk production
Si-Ling Chi, *legitimate wife of
Prince Hoang-ti of China. For her
achievement, Si-Ling Chi was deified.*

2600
Monumental hieroglyphic texts
Egyptian tombs.
**Scribe's outfit (palette, bag for
powdered pigments and reed-
holder)**
Egypt.

Ink (ordinary)
*Egypt and China. Lampblack ground
with a solution of glue or gums.*

2590
Embalmed person
Egyptian Queen Hetepheres, *mother
of Cheops. Only her internal organs
have survived.*

2575
Smooth-sided pyramid
King Sneferu *at North Dahshur,
Egypt.*

2568 (?)
Ark (animal rescue ship)
Noah, *as directed by God. Constructed
of gopher wood and used during the
Great Flood. The Ark was later
'beached' on the side of a mountain
called Ararat in East Turkey.*

2500
Bitumen as glue
Mesopotamia.
Cotton (spun and woven)
India.
Gaming boards
Ur, city-state of Sumer.
**Plough (ox-pulled and man-
guided)**
Egypt.
Stone-working drill
Egypt.

From the Tomb of Rekhmire at Thebes

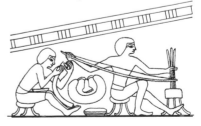

2450
**Senet (board game similar to
draughts)**
Egypt.

Sistrum

2400
Engraved seals
India.
History in hieroglyphs
the 'Palermo Stone'.
Mummy of 'Waty'
Found in the tomb of Nefer, Royal Court singer at Saqqara, Egypt
Sistrum (musical rattle)
Egypt.
Written papyrus
the Abusir Papyrus, from the reign of the Fifth Dynastic King Djedkare Isesi.

2380
'Wisdom Text'
the instructions and precepts of Ptahhotep, vizier of King Djedkare Isesi of Egypt.

2350
Lightweight, horse-drawn chariot (spoked wheels)
the Hurrians of Mesopotamia.

2300
Artificial stone (glazed)
Telltaya, Mesopotamia.

2200
Rough engraving of city layout
Babylon.

2100
Ur-Nammu Ziggurat (building)
Mesopotamia.
Workable road network
Ur-Nammu of Mesopotamia.

2000
Chariot wheels (copper-tyred as opposed to studded)
Susa, capital of Elam.
Snow skis
Northern Norway. Depicted in a rock carving known as 'the Rødøy Man'.

1900–1800
Gynaecological/Veterinarian papyrus
the Illahun Papyrus.

1800
Bath tub
Mari, Babylonia.
Geometrical problem texts (on clay)
Tell Harmal (Baghdad). Includes a demonstration of triangular laws.

1761
Jewish Year 2000.

1700
'Hounds and Jackals' (sophisticated board game)
Egypt.
The Phaistos Disk
Crete.
Royal bath tub
found in the Queen's bathroom at the palace of King Minos in Knossos, Crete.

1650–1600
Mathematical papyrus

the 'Rhind Papyrus', a copy of a much older document.

1600
Alphabet (Sinaitic)
Canaanite scholars.

1500
Contraceptive
description of vaginal plug of lint, ground acacia branches and honey in the Egyptian Ebers Papyrus.

Egyptian war chariot

Glass manufacture
Mesopotamia.
Glazing of pottery
Nuzi, Mesopotamia.
Horse bit (bronze with cheek pieces)
Egypt.
Iron smelting
the Hittites of Anatolia.

Bronze sword hilt

Phoenician alphabet
(30-letter system).
Plank-built boats
North Ferriby, beside the River Humber, Yorkshire, England.
Ritual astromical stone clock
Stonehenge, Salisbury Plain, England.

1470
Polychrome glass vessels
during the reign of King Tuthmosis III of Egypt.

1417–1379
The Colossae of Memnon
King Amenhotep III of Egypt.
Following damage by an earthquake, whenever the sun heated the air inside the statue, it expanded and rushed through the cracks to make a 'whistling' sound.

1400
Accurate plan of a city
Nippur, Mesopotamia.
Horse-drawn chariots (paired-draught animals)
both Shang Dynasty in China and under Thotmes III and IV in Egypt.
Shaduf (well-sweep with counterpoise)
Egypt.

1320
Papyrus map
sketch of an Egyptian gold mine.

1300
Complete alphabet
Ugarit, Syria. 32 letters.
Mobile prefabricated temple
Bezaleel ben Uri (Tribe of Judah), together with Aholiab ben Ahisamach (Tribe of Dan) for Moses. This included the tabernacle, ark, mercy seat, candlestick, priests' garments and all else as specified in Exodus.
Musical notation (on clay tablets)
Ugarit, Syria.

1270
Encyclopaedia
Abulfaraj (Bar-Hebraeus), *Syrian
historian. As master of Syriac, Arabic
and Greek, and equally learned in
philosophy, theology and medicine,
Abulfaraj was admirably suited for the
task.*
Sonic destructor
Joshua, *son of Nun, during his
conquest of Canaan. The blowing of
trumpets, shaped like ram's horns, with
mass shouting is reported to have
destroyed the walls of Jericho.*

1250
**The Wooden Horse of Troy
(disguised troop carrier)**
*mythical artifice by which the Greeks
are said to have got possession of Troy,
with assistance from the captive,
Sinon.*

1200
Horse-bit (jointed mouthpiece)
Egypt.

1150
Chess (board game)
*popularly attributed to Palamedes,
'the handy or contriving one', a
mythical figure, son of Nauplius, and
also erroneously credited with the
invention of lighthouses, measures,
scales, the discus, dice, the alphabet
and the art of regulating sentinels.*

1100
Bell (tintinabulum)
Babylonian Palace of Nimrod.
Dictionary
Pa-out-She, *Chinese scholar. 40,000
characters, mostly hieroglyphic or
zodiacal.*
**Phoenician alphabet (22-letter
system)**
Byblos and Ras Shamra, the Lebanon.

1000
Protogeometric style of pottery
Eastern Mediterranean.

Wooden toy tiger (movable jaw)
Thebes, Upper Egypt.

950
Solomon's Temple
Hiram of Tyre, *Phoenician architect.*

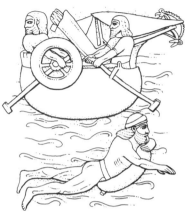

Assyrian army buoyancy aids

880
Inflatable swimming aid
King Assur-nasir-apli II of Assyria.
To enable his troops to cross a river.

870
Canal-building
King Assur-nasir-apli II of Assyria.
*From off the Upper Zab river to water
the orchards and botanic gardens of his
'creation' at Kalhu.*

865
Siege engines
Assyrian generals.

850
Battering rams
Assyrian generals.
**Biremes (galleys with two banks
of oars)**
Assyrian boatbuilders.
**Damp-proofing (bitumen
process)**
Assyrian architects.

Stirrup (horses)
Assyrian blacksmiths.

775
Fully phonetic alphabetic script
Greece. Written from left to right.

770
Gold coins (uninscribed)
Chou Dynasty of China.

761
Jewish Year 3000.

750
Triremes (galleys with three banks of oars)
Greek boatbuilders.

730
The Sundial of Ahaz
Ahaz, *son of Jotham and eleventh King of Judah.*
Written poetry
'The Iliad' by Homer *and 'Works and Days' by* Hesiod, *both Greek poets.*

Greek warrior's helmet, 750 BC

720
Aramaic/Assyrian alphabet.

700
Aqueduct
King Sennacherib of Assyria, *during the rebuilding of Nineveh. Its length was 30 miles.*
Demotic script
Egypt.

673
Lyre (seven-stringed)
Terpander, *possible founder of the first Greek school of music at Sparta.*

650
Glass-making texts (on clay tablets)
Nineveh, Assyria.

640
Roof tiles
on the Temple of Hera, Olympia, Greece.

605
Circus Maximus
Lucius Tarquinius Priscus, *King of Rome.*

600
City sewer (large-scale)
Cloaca Maxima in Rome.
Device-struck coin (electrum alloy)
Ionia, West Turkey. The inscription on the coin reads: 'I am the signet of Phanes'.
Electrical properties (of rubbed amber)
Thales of Miletus, *Greek-Ionian philosopher.*

592
Anchor fluke
Anacharsis, *witty Scythian prince, who travelled widely in search of knowledge.*

570
Geographical/celestial charts
Anaximander of Miletus, *Greek-Ionian philosopher. Also credited with the invention of the gnomon.*
The Hanging Gardens of Babylon
Nebuchadnezzar, *King of Babylon.*

552
Celestial and terrestrial spheres
Anaximander of Miletus.

550
Measurement dial
Anaximander of Miletus.
Silver and gold coinage (inscribed)
Croesus, *King of Lydia, Eastern Mediterranean. Also erroneously attributed to Pheidon, King of Argos and Midas, King of Phrygia.*

540
Harmonic strings
Pythagoras, *wandering Greek philosopher, after hearing four blacksmiths working with hammers in harmony whose weights he found to be six, eight, nine and twelve.*

530
Library
Pisistratus, *'tyrant' of Athens.*

520
Tube structure of the ear
Alcmaeon, *Greek physician and philosopher of Croton. Later known as the Eustachian tubes.*

505
Democracy (practical system)
Cleisthenes, *statesman of Athens, Greece.*
Observatory
erected on top of the temple of Belus at Babylon, possibly used by Naburiannu, the astronomer.

500
Telegraphy (vocal)
King Darius I, *surnamed Hystaspis, of the Persians. He used shouting from hilltops to convey his orders. Mountain horns, trumpets and drums were also used to communicate across distances. So were beacon fires, smoke signals and reflecting mirrors.*

480
Floating bridge(double line of boats)
Xerxes I, *King of Persia. Enabled his army to cross the Hellespont in seven days and nights.*

477
Mnemonics
Simonides of Kos, *called The Younger, a Greek lyric poet working in Athens. He won some 56 poetical contests using his system.*

460
Medical diagnosis
Hippocrates of Kos, *physician.*

447
Chryselephantine statues
Phideas, *Greek sculptor. His statues of Athena, Parthenos and Zeus ranked among the Seven Wonders of the World.*

420
Regal portrait on a coin
Kharai, *King of Lycia.*

Silver tetradrachma coin, used in Athens, 430 BC

400
Automaton flying dove
Archytas of Tarentum, *Greek general and mathematician.*
Gastraphetes ('stomach-bow')
Greek military engineers in Sicily.
Ionian-Greek alphabet
24 letters.
Kite
China.
'Knives and Hoes' (copper coinage)
China.
Pulley
Archytas of Tarentum.
True arch (architectural device)
Palera, Greece.

350
Mausoleum
Mausolus, *King of Caria and husband of Artemisia, who erected a magnificent monument to his memory at Halicarnassus.*
Three conic sections
Menaechmus, *Greek mathematician.*

334
Private library
Aristotle, *Greek philosopher, scientist and physician. He made use of this to educate Alexander, the son of Philip of Macedon – and then at his school, the 'Lyceum' in Athens.*

320
Euclid's Elements
Collated by Euclid, *Greek mathematician.*

312
Aqueduct (long-distance)
Appius Claudius Caecus. *He also constructed the Appian Way.*

300
Aryan alphabet
India.
Canals and rivers (controlled by a series of gates)
China.

Crossbow (individual weapon)
China.
Disc coins (with holes)
China.

300–200
Trip-hammer mill
China.
Ultra-straight road network
Roman engineers throughout Italy.

285
Lighthouse
'The Pharos', near Alexandria, Egypt, was 550 ft high and visible for 42 miles.

280
The Colossus of Rhodes
Chares, *sculptor and engineer. This gigantic figure of Apollo, 700 cubits high, its legs astride the harbour for ships to pass in and out, was destroyed by an earthquake in 224 BC.*

260
Archimedean screw (mechanical lifting device)
Archimedes, *son of the astronomer Pheidias, whilst studying in Alexandria.*

Archimedean screw

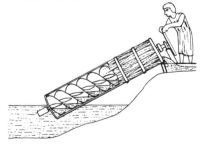

Bibliographic system
Callimachus, *Greek poet and scholar. His 'Pinakes' (tablets), running to 120 books, were an elaborate catalogue of the Great Library at Alexandria.*

250
Hydraulis (water-powered organ)
Ktesibios, or Ctesibius, *physicist and former barber of Alexandria.*
Fire-engine
Ktesibios.
Quinqueremes (war galleys)
several Eastern Mediterranean city-states.
The thirty systematic dynastic divisions
Manetho, *Egyptian priest in Alexandria. Later proved incorrect.*
Toothed wheels
attributed to Ktesibios.

240
Lithotomy ('cutting for the stone')
Ammonius, *Greek surgeon.*

225
Armillery sphere
Eratosthenes, *Greek astronomer, geographer and mathematician working at the Library in Alexandria.*

221
Gunpowder
Alchemists and engineers *in Chin Dynasty, China.*

220
Ellipse and hyperbola
Apollonius of Perga, *Greek mathematician.*
Hydrostatics (fundamental law)
Archimedes of Syracuse *on the island of Sicily to check the extent to which the crown of his distant kinsman, King Hieron II, was pure gold.*
World map
Eratosthenes, *superintendent of Ptolemy Euergetes's library at Alexandria. He also invented the 'Cribrum Arithmeticum'.*

215
Leverage (large-scale)
Archimedes of Syracuse. *For*

defensive war machines to ward off the Roman invasion as led by Marcellus. Catapults and grappling irons, based on lever devices, inspired fear and awe among the Romans for almost 3 years.

214
Planetarium (table size)
Archimedes of Syracuse.

213
Solar energy mirrors
Archimedes of Syracuse. *A barrage of reflectors along the harbour walls which not only dazzled Marcellus's approaching fleet but set fire to the sails and timber work.*

206
Reconnaissance kite
General Han-Sin, *to ascertain the distance between his camp and the palace Wei-Yang-Kong, in China.*

200
Glass-blowing iron
Babylonian glassworkers.

198
Vellum books
Attalus I, *King of Pergamum, for his great library which, under the direction of Crates of Mallus, grew to 200,000 volumes, many of them parchment.*

196
The Rosetta Stone
bilingual text in three scripts: hieroglyphs, demotic and Greek.

193
Warehouse (multiple barrel-vault design)
Porticus Aemilia (the Aemilia gate).

190
Hexameter (verse line of six metrical feet)
Quintus Ennius, *Roman poet.*

Pyrphoros (flame-thrower)
Pausistratus, *a Rhodian admiral.*

142
Stone bridge
Roman engineers, *across the River Tiber, Rome.*

130
Astrolabe
Hipparchus, *Greek astronomer of Rhodes.*
Trigonometry
Hipparchus.

121
Concrete building
The Temple of Concord, Rome.

100
Draw-loom for figured weaves
China.
Sails and rigging (mat-and-batten principle)
China.

Silk-reeling machinery
China.
Wheel-bearings
four-wheeled wagon found at Dejbjerg, Jutland.

85
Seed-drill plough (with hopper)
China.

50
Caesarian Cipher
Gaius Julius Caesar. *Also used by the Emperor Augustus.*

46
Julian Calendar
Sosigenes of Alexandria, *astronomer and mathematician, employed by Julius Caesar to devise a more accurate calendar.*

40
Rotary winnowing machine (with crank handle)
China.

17
Lithotomy (small apparatus)
'Celsus', alias Aulus Cornelius, *Latin physician.*

20
Trip hammer mill (water-powered)
China.

31
Metallurgical blowing engine (water-powered)
Tu Shih, *Prefect of Nanyang, China.*

60
Miniature reading lens
the Emperor Nero, *who used the large transparent gemstone in one of his rings.*

83
Lodestone spoon rotating bronze plate (magnetism)
China.

c. 100
Whirling Aeoliphile (steam-powered)
Hero, *Greek mathematician of Alexandria. Among his other inventions were an olive oil beam press, a grape screw press, a screw-cutting machine, a gastraphetes (crossbow), water- and wind-powered musical organs, automatically opening doors and even a Holy Water slot machine. 80 such gadgets were described in his*

Hero's whirling Aeoliphile

tome, 'Pneumatica'.
Earthen lamp
Epictetus, *former Roman slave and Stoic philosopher of Nikopolis in Epirus. The invention was sold after his death for 3000 drachmas.*

105
Paper-making (mulberry-based)
T'sai Lun, *eunuch director of the Imperial workshops of the Chinese Han Court and later deified as the god of paper-makers.*

124
Pantheon (large circular temple)
Architects and engineers of the Roman Emperor Hadrian.

132
Earthquake alarm
Chang Heng, *Chinese Imperial astronomer, mathematician, cartographer and poet. One of the wonders of the Imperial Observatory, it was 8 ft high and incorporated a free-swinging pendulum.*

150
Map of the world (latitude and longitude grid system)
Claudius Ptolemaeus, *Greek astronomer and geographer.*

170
Edge-runner mill
China.

180
Gimbals
China.
Rotary fan for ventilation
China.
Zoetrope (lamp cover revolved by ascending hot air)
China.

189
Square-pallet chain-pump
China.

200
Artery (as blood carrier)
Galen of Pergamum, *overturning the belief that they were merely air tubes.*

224–241
Circular-plan city
King Ardashir I *at (present-day) Firuzabad, south-west Iran.*

231
Wheelbarrow
China.

239
Jewish Year 4000.

320
Helicopter top (spun by pulling a cord)
China.

340
Wagon-mill (grinding during travel)
China.

366
Prayer beads
attributed to Saint Augustine of Hippo.

370
De Rebus Bellicis
anonymous author. *Containing working drawings for defence machines against the Vandals by Valentinian I, Roman Emperor, and his brother, Valens. Includes an ox-powered paddleboat.*

372
Public hospital
Saint Basil, *Bishop of Caesarea in Cappadocia.*

385
Jewish calendar
Rabbi Hillel II.

400
Edge-runner mill (water-powered drive)
China.
Ship (watertight compartments)
China.

410
Commemorative metal crucifix
On a rocky mount of the open court between the basilica and rotunda of the Holy Sepulchre in Jerusalem. Seven years later, Christian Roman Emperor Theodosius II encased the crucifix with

gold and precious jewels. Soon after, crucifixes started to appear in and on churches.

450
Whisky
attributed to Saint Patrick, *although 'Aqua Vitae' had been brewed by the Romans 500 years before.*

512
Arabic alphabet: 28 letters.

520
Decimal system
Aryabhata and Varamihara, *Indian mathematicians.*

532
Anno Domini (dating system)
Dionysius Exiguus, *Scythian abbot of a monastery near Rome. Not generally employed for several centuries.*
Single-shell dome
Anthemios of Trales, *architect for Saint Sophia Cathedral.*

535
Stained glass windows
Saint Sophia Cathedral.

550
The Great Arch of Ctesiphon

552
Sailing carriage (first high land speeds)
China.

555
Watermill (outside China)
Belisarius, *general of the Emperor Justinian, while besieged in Rome by the Goths.*

580
Iron-chain suspension bridge
China.

610
Segmental arch bridge
An-chi bridge at Chao-hsien, China. Built by the Sui Emperor Wen-ti *to span the Grand Canal, dug to join the Yang-tze and Yellow River systems.*

640
Ching-Chow Astronomical Observatory
Korea, during the reign of Queen Sondok of Silla.

644
Windmill
Persia, for grinding corn.

672
'Greek Fire'
Kallinikos, *an architect from Heliopolis, Syria, to help the Byzantines destroy the Arab Saracen ships. 30,000 men reported killed by this weapon.*

Byzantines use 'Greek Fire' against Saracens

700
Porcelain (original)
T'ang Dynasty, China.
Ship (stern-post rudder)
China.

770
Block-printed Buddhist charm.

800
Carolingian minuscule script
various scholars under the direction of Alcuin, *Anglo-Saxon abbot of Tours and adviser of the Frankish emperor, Charlemagne.*

808
Bank (money repository)
Jewish merchants in Lombardy.

825
True lock-gates and chambers
China.

862
Slavonic alphabet
Saint Cyril, *alias Constantine the
Philosopher, to convert the Moravians
to Christianity. 42 letters.*

868
Printed book
*The Diamond Sutra, found in
Buddhist caves at Tun-Huang, China.*

879
Gorgonzola cheese
farmers in the Po Valley, *Italy.*

886
Wax candle 24-hour
measurement system
Alfred the Great, *King of the West
Saxons and overlord of England. Six
candles, burning at 3 inches per hour,
were used each full day.*

890
Nailed-on horseshoes
Siberia, Byzantium and Germany

900
Concave, curved iron
mouldboard (plough)
China.

910
Paper money
China.

970
Central Library of Cordoba
Abd-al-Hakam II, *Caliph of
Cordoba, Spain, in his Alcazar (royal
palace). Over 400,000 titles
catalogued in 44 volumes.*

983
Canal locking system
Chiao Wei-Yo, *China.*

990
Giant sextant (radius 55 ft)
Hamid Ibn al-Khidr al-Khujandi,
Arab astronomer.

995
Giant quadrant (22 ft radius)
Abdul-Wafa Albuzjani, *Persian
mathematician.*

1000
Venetian glassmaking system
*first made in the city and then, due to
the danger of fire, transferred to
nearby Murano island.*
Coffee
reported by Avicenna, *Arabian
philosopher and physician.*

1024
Sonnet (poetical form)
Guido d'Arezzo, or Aretino,
*Benedictine monk and musical theorist,
at Pomposa, near Ferrara, Italy. It
was later standardised by* Francesco
Petrarch *(born Arezzo, 1304).*

1026
Musical notation
Guido d'Arezzo.

1049
Printing (clay characters)
Pi Sheng, China.

1090
Silk-flyer (for twisting and
doubling)
China.

1092
Astronomical clock
Su Sung, *after five years' work at the
Imperial Observatory at K'ai Feng. A
cross between the clepsydra and the
spring-driven clock because its time-
keeping ability could be adjusted by*

Frankish warrior, c. 1100

weights. *It struck a gong to indicate the passing hours, by a system of bamboo 'revolving and snapping springs'.*

1100
Rockets and fire-lances
China.

1115
Lodestone-magnetic compass for navigation
China.

1150
Paper-mill
Xativa, Spain.

1155
Oldest printed map of China.

1200
Automaton dancing men and walking peacock
Al-Jazari, *Arabian scholar.*
Ormulum
Ormin, *an Augustinian monk, versifier and spelling reformer, living in the east of England. It was a metrical version of the Gospels and the*

Acts of the Apostles.
Rhythmic musical notation
Franco of Cologne *in his 'Art of Measurable Music'.*

1202
Rosary (prayer beads)
attributed to Domingo de Guzmán, (Saint Dominic).

1220
Practica Geometrica
Leonardo of Pisa, alias Fibonacci.

1221
'Iron fire bombs'
the Kin warriors of China.

1239
Jewish Year 5000.

1241
Hanging, drawing and quartering
first inflicted in England on William Marise, a pirate and nobleman's son.

1245
Mechanical eagle
Villard de Honnecourt. *Turned its head towards the deacon when he read the Gospel.*
Overbalancing wheel (perpetual motion machine)
Villard de Honnecourt, *French architect. One of 325 working drawings in his sketchbook.*

1250
Magnifying glass
Roger Bacon, *philosopher and scientist, working in Oxford, England. He also designed a lighter-than-air machine, an automatic ship and an automatic carriage.*

1253
Alphonsine Tables
King Alphonso X of Castile, *who spent over 400,000 crowns collecting tables composed by Spanish and Arab*

astronomers. *These tables were to be frequently consulted for the next three centuries.*

1259
Bamboo-gun
('fire-spurting lance')
China.

1260
Biblia Pauperum
(Bible for the poor)
Giovanni di Fidanza, *Italian theologian, General of the Franciscans. Later beatified as Saint Bonaventura.*
Gun/cannon
Konstantin Anklitzen (alias 'Berthold Schwarz'). *? Franciscan, Augustinian or Dominican monk;? of Freiburg, Goslar, Ghent, Mainz, Metz, Cologne, Brunswick and/or Prague – not to mention Danish, Greek, Negro and/or Welsh origins. Other dates given for his/her invention 1313, 1320, 1354, 1359 and/or 1380. Details rather hazy!*

1269
Compass
Petrus Peregrinus de Maricourt. *With 360° card graduation and movable sights for taking bearings.*

1276
Giant astronomical gnomon (40 ft)
Kuo Shou-Ching, *Chinese astronomer. Set in the Tower of Chuo Kung to measure the sun's solstitial shadow.*

1280
Belt-driven spinning wheel
Hans Speyer, *German engineer.*

1285
Spectacles
(convex lens for myopia)
Alessandro de Spina, *Dominican monk at the Convent of Santa*

Katerina di Pisa, based on the work of Salvino degli Armati, Florentine physician.

1286
Watermark
(paper identification)
a paper mill in Fabriano, Italy.

1287
Nitric acid (aqua fortis)
Raymond Lully, *Spanish theologian and philosopher.*

1288
Cocchio
(Italian travelling wagon)

1300
Hall-marking
'Gardiens of the Craft', the London Guild of Goldsmiths. *Items of gold and silver struck with a leopard's head, the mark of King Edward I – later a lion's head.*

1305
Acre (standardised)
King Edward I of England.

1310
Blowing-engines for furnaces and forges (crank-handle drive)
China.
Silk machinery (water-powered)
China.

1311
History of the world
Rashid-eddin *of Persia's 'Jami'u't Tawarikh'.*

1313
Indulgences for public sale
Pope Clement V of Avignon.

1326
Iron-bullet-firing metal cannon
Bombards, *as developed by* Rinaldo di Villamagna, *Florence, Italy.*

1330
Caricatures
Buffalmacco Buonamico de
Cristofano, *painter of the early
Florentine School, with a great
reputation as a practical joker. He
drew comic figures and put labels to
their mouths with sentences.*

1335
Automatic striking clock
*installed in the belfry of the Palace
Chapel of the Visconti, Milan, Italy.*
Da Vigevano's Treatise
Guido de Vigevano, *Italian physician
working in France. Contained
drawings of men working the first
known crankshafts inside a road
vehicle, a paddle-wheeled submarine, a
siege tower machine, etc.*

1339
Ship (gun-carrying)
*the 'Christopher', an English cog,
armed with three iron guns and a hand
gun.*

1343
Artillery
the Moors of Algeciras, *Spain. Three
years later King Edward III of
England used four cannons in the
Battle of Crécy.*

1360
**Mechanical clock
(iron movement)**
Henri de Vick *of Württemberg for
King Charles V of France at his royal
palace in Paris.*

1364
Astronomical clock
Giovanni de' Dondi, *astronomer and
physician, for the library of the Duke
of Pavia in Visconti Castle. Showed
the time, the phases of the moon, the
signs of the zodiac and other
astronomical features.*

1367
London's first striking clock.

1381
Padlock
inventor unknown.

1383
Wooden toy cannon
*presented to the child King Charles VI
of France.*

1388
Side-saddles for ladies
introduced by Anne of Bohemia, *wife
of King Richard II of England.*

1392
**Type foundry
(movable bronze characters)**
set up by T'ai Tsung, *King of Korea.*

1410
Wire
Rodolph of Nuremberg, *who devised
the process for drawing out the wire.*

1411
Pasini cipher
Luigi Pasini *of Venice.*

1413
Toy dolls with movable arms
played with by children in Nuremberg

1420
Wagon-fort
Jan Ziska, *leader of the Protestant
Hussites of Bohemia. It was drawn by
teams of horses and armoured with
sheets of iron. Its crew fired crossbows
and guns through loopholes pierced in
the sides. With only 25,000 men and
his wagon-forts, Ziska crushed the
invading force of 200,000 men.*

1421
**Barge with special hoisting gear
(for transportation of building
marble)**
Filippo Brunelleschi, *architect and*

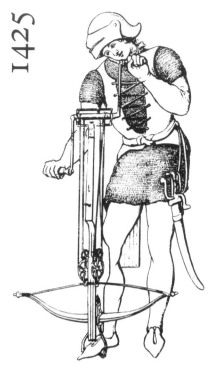

1425

Cocking a cross-bow, *1420*

engineer. *Special privilege granted to him by the Signoria of Florence.*

1425
Giant sextant (180 ft high)
Ulugh-Beg, *Persian prince and astronomer, for his observatory at Samarqand.*

1432
Dome (negative side-thrust)
Filippo Brunelleschi, *architect and engineer, for the Duomo of his native Florence.*

1437
Perspective views (pinhole demonstration)
Leon Battista Alberti, *Italian architect.*

1440
Playing cards (the four suits)
Provence, France. But known as far back as 950 AD *in China, supposedly developed by harem women to stave off boredom.*

1444
Waterless mills
exclusive privilege granted to Antonius Marini de Francia *to construct these by the Venetian Senate.*

1451
Printing press
Johannes Gansfleisch zur laden zum Gutenberg *in Mainz, Germany, with the financial backing of goldsmith Johann Fust and assistance from Peter Schoeffer. Following the printing of an old German poem, Gutenberg threw his energies into the printing of a 1282-page Bible in Latin – an enterprise which bankrupted him out of his workshop.*
Travelling clock
Jean Lieburc, *Paris, with its copper casing by Jean Moulinet.*

1452
Baptistry doors (bronze)
Lorenzo Ghiberti, *sculptor and bronze-caster in Florence.*

1455
Cast-iron pipes
to carry water at the Castle of Dillenburgh, Germany.

1458
Bereguardo Canal
Bertola da Novate, *Italian civil hydraulic engineer. 12 miles long with a complete set of canal locks.*

1464
De Triangulis Omnimodis
Regiomontanus, alias Johannes Müller, *mathematician and astronomer of Nuremberg.*

1470
Jenson typeface
Nicholas Jenson, *French printer and type-founder in Venice.*
Page numbers in a printed book
Werner Rolewinck's *'Sermo ad Populam', published by Hoernen of Cologne.*

1472
Printed music
'Gradual' by unknown Constanz printer.

1474
Automaton eagle and iron fly
Regiomontanus, alias Johannes Müller, *German mathematician and astronomer of Nuremberg.*
Patent Law
The Venetian Senate, *to protect inventors.*

1475
Hourglass set
Nuremberg. Four different sizes to run 15, 30, 45 and 60 minutes.
Lithotomy (high apparatus)
Germain Colot, *French physician, on a criminal in Paris.*

1476
Italic typeface
Francesco Griffo, *typecutter of Venice; used for a series of classics published by Aldus Manutius.*

1477
Printed map
Bologna, Italy.
Printed poster
William Caxton, *London printer – to advertise the thermal cures at Salisbury.*

1478
Set of printed maps
Conrad Sweynheym's *27 maps in Ptolemy's 'Cosmographia', printed by Arnold Buckinck, Rome.*
Spectacle manufacture
Nuremberg.

1481
Wallpaper
Jean Bourdichon, *painter and illuminator, who painted 50 great rolls of paper in blue with the inscription 'Misericordias Domini in aeternum cantabo' and three angels about 3 ft high to hold the rolls in their hands. Bourdichon was paid 24 livres for his work.*

1484
Mine drainage system
Blasius Dalmaticus *from Ragusa. 'Patent rights' were granted him by the Elector Ernst of Saxony.*

1487
Carillon (bells)
First heard in Alost, East Flanders.

1488
Multiple crossbow gun
Leonardo da Vinci, *Italian painter, architect and engineer, in the employ of Lodovico Sforza, 'Il Moro' of Milan. For the previous six years da Vinci, in his thirties, had been keeping detailed*

Leonardo da Vinci

notebooks, *which in special mirror-handwriting show inventions for a longbow shield, a human flying machine, a helicopter, a parachute, a diving bell and suit, articulated chains, giant crossbow, circular armoured vehicle, scythe-wheeled chariots, etc. None of these appear to have been built.*

1489
Plus (+) and minus (−) signs
Johann Widmann's *'Mercantile Arithmetic', published in Leipzig.*

1490
Antimony
Basil Valentine, *alias Thölde, German salt manufacturer and chemist.*

1492
Behaim's terrestrial globe
Martin Behaim, *navigator and geographer, with B Glockenthon, painter – both at Nuremberg.*

1498
Toothbrush
China.

c. **1500**
Bassoon
Afranio, *canon of Ferrara.*

1500
Caesarian section
Jacob Nufer, *sow-gelder of Sigershauffen, Switzerland, on his own wife, using his sow-gelding instruments.*

1501
Hymn-book
Severin of Prague *for the Hussites of Bohemia. Contained 89 hymns in the Czech language.*

1502
Barrel organ
engineers, carpenters and musicians working for Archbishop L. von

Keutschah *of Salzburg, nicknamed 'the 'Salzburg Bull'.*

1510
'The Nuremberg Egg' (round pocket watch)
Peter Henlein, *clockmaker of Nuremberg, after ten years' work, including his invention of a steel mainspring.*

1512
Washing apparatus (for ore)
Sigmund von Maltitz.

1515
Galleas
King Henry VIII of England. *'The Great Galley', 800 tons, with 120 oars manned by 160 oarsmen and 207 guns.*

1518
Fire-engine
Anthony Blatner, *goldsmith for the city of Augsburg.*

1520
Rifle-barrel (spirally-grooved)
August Kotter, *Germany.*

1525
Dental forceps (for extraction)
'Thomaseus'.
Harquebus
Marquis of Pescara, Spain. The first *portable fire-arm: a shotgun having a matchlock operated by a trigger.*

1528
Hodometer
Jean Fernel, *instrument maker, France.*

1529
Bismuth (metal)
'Agricola' alias Georg Bauer, *physician and mineralogist in Joachimstal, Germany.*
Musical bars
'Agricola' in 'Musica Instrumentalis'.

Wandering organists, 1502

Mechanised power –
ball and chain mining pump, 1529

1530
Cristallo glass
Venice.

1531
Diving bell
Guglielmo de Lorena, *Italian*
engineer, in an attempt to raise
Caligula's pleasure galleys from the
bottom of Lake Nemi.

1537
'The New Science'
Niccolo Fontana, *nicknamed*
'Tartaglia', Professor of Mathematics
at Venice University. Demonstrated,
among other things, how cannon balls
follow a curved path rather than
Aristotelian straight lines.
Trucked muzzle-loading gun
carriage
discovered on board the 'Mary Rose,'
warship of King Henry VIII of
England.

1538
Optic nerves
Costanzo Varolio, *anatomist and*
physician of Bologna, Italy.

ANNO·ÆTATIS·
68

Ambrose Paré

van Calcar, *a pupil of the painter Titian.*

1545
Conservatory
Daniel Barbaro *for the Botanical Gardens at Pavia, Italy. A viridarium, heated in winter by a brazier or open hearth.*
Equation of the third degree
Geronimo Cardano, *Professor of Mathematics and Medicine at Padua, Pavia and Bologna.*

1549
Venetian system for making crystal glass
patent protection granted to Jehan de Lame *of Cremona, dealer, residing in Antwerp, to introduce the technique into Brabant.*

1550
Ligature
Ambose Paré, *French military barber surgeon, to replace the blind and brutal use of red-hot cautery.*
Screwdriver
By gunsmiths and armourers, to adjust their gun mechanisms.
Spanner
First designs appeared during this century to deal with the newly developed system of nuts and bolts.

1540
Artificial limbs
Ambrose Paré, *French military barber surgeon, the first to replace amputated limbs with sophisticated mechanical substitutes, using hinges, cog wheels, levers and armatures.*
Fusée (watch-spring device)
Jacob Zech *of Prague.*
Meridionel instrument
'Nicolaus Copernicus', *the Latinised name of* Mikolaj Kopernik, *canon of Frombork, Poland. He used it while evaluating his Heliocentric theory.*
Pistol
Camillo Vettelli, *gunsmith of Pistoia, Italy.*

1541
Zinc
'Paracelsus', alias Theophrastus Bombastus *of Hohenheim.*

1543
Detailed human anatomy book
Andreas Vesalius, *Belgian anatomist, assisted by draughtsman* Jan Stephan

1556
Sealing wax (including shellac).

1557
Enamels
Bernard Palissy, *after sixteen years' research at Saintes, France.*
Equals sign (=)
Robert Record, *Fellow of All Souls College, Oxford, England, in his algebra text 'The Whetstone of Witte'.*

1558
Dolls' house
made by craftsmen for the daughter of the Duke Albrecht of Saxony.

1560
Camera obscura (portable)
Giovanni Battista della Porta,
Neapolitan physicist and philosopher.
Wallpaper (block-printed)
printers *in Delft, Holland and
London, England where there was a
Guild of Paperhangers.*

1561
Dredger
Pieter Breughel *for Brussels
Municipality. Used to excavate the
Rupel-Scheldt canal.*

1563
Wire-manufacturing mill
Nuremberg.

1565
Graphite pencils
Konrad Gesner, *the year after the
discovery of the Borrowdale deposits in
Cumberland, England by German
miners prospecting for copper.*

1567
Silk draw looms
first British use in Norwich.

1568
**Mercator Projection (map
treating the globe as a cylinder
rolled out flat)**
Gerhard Kremer, *Flemish
mathematician, geographer and map-
maker.*

1569
**Screw-cutting and ornamental-
turning lathes**
Jaques Besson, *engineer in the Court
of King Charles IX of France.*

1570
Log line (for navigation).
Modern atlas
Abraham Ortelius *of Antwerp.
'Theatrum Orbis Terrarum' contained
70 maps on 53 plates.*

1571
Theodolite
Leonard Digges, *mathematician,
England.*

1575
**Paper tape-recorder (compass
bearings)**
Thomas Ruckert, *England.*

1576
Magnetic needle dip
Robert Norman, *instrument-maker of
London.*
'Uraniborg' Observatory
Tycho Brahe, *Danish astronomer.
Sited at Hven Island, off the Swedish
coast.*

1578
Prefabricated building
Nonesuch House, *brought from
Holland in sections and re-erected on
London Bridge.*

1581
Pendulum motion
Galileo Galilei, *who timed the
oscillations of the great lamps of Pisa
Cathedral, Italy, swinging from the
roof during an earthquake tremor –
using his pulse beats.*

1582
Calendar reform
Christoph Clavius, *German Jesuit
mathematician, at the request of Pope
Gregory XIII.*
Pumping station (water supply)
*Morris's London Bridge Water
Works.*

1585
Fire-ship (explosion vessel)
*to destroy a bridge of boats at the siege
of Antwerp. Three years later* Lord
Howard of Effingham *used them for
the engagement with the Spanish
Armada.*

Time bombs
Clockwork-activated floating mines used by the Dutch at the afore-mentioned siege of Antwerp.

1586
Giant cannon (bombard)
'The Tsar' of Moscow, *with a 36-inch bore.*

1588
'Reading machine' (vertically revolving bookcase)
Agostino Ramelli, *Italian engineer working in France. One of 195 machines described in a book he published that year in Paris.*

1589
Hosiery-knitting machine
Reverend William Lee *of Woodborough, Nottingham, England. He was refused patronage to develop it, by both Queen Elizabeth I and the London city merchants.*
Stamp mill for hemp (powered by wind, horse, or water)
Elbert de Veer, *burgher of the City of Amsterdam.*
Water closet
Sir John Harington, *godson to Queen Elizabeth I of England and amateur poet. He had been banished from the Court for circulating a risqué story. During exile he built a house at Kelston, Bath and there installed the first flushing lavatory. He next supplied one to his royal godmother at Richmond Palace apparently as a ploy for a literary exercise, for he described his device in a work called 'The Metamorphosis of Ajax – a Cloacinian Satire'. A 'jacks' was a privy. Thus rivalry to the chamber pot was set in motion.*

1590
Compound microscope
Hans Janssen *and his son* Zacharias, *spectacle-makers of Middleburg, Holland. With the use of concave and*
convex lenses it was soon improved by Galileo Galilei.

1592
Armoured ship
Korean Admiral Yi-sun-Sin *for use against the Japanese. With this ship, iron-plated turtle-back, covered with rows of spikes and fitted out with both a ram and archery ports for fire arrows, Yi-sun-Sin virtually wiped out the Japanese fleet in a matter of days.*
Timber-sawing machine (windmill-powered)
Cornelis Corneliszoon *of Uitgeest, Holland.*
Windmills for shammy dressing
Henrick van Zanthen *and* Franz Lambrechtz *from Bois le Duc, residing in Amsterdam.*

Windmill-power

1593
Transportable horse-driven flour mill
Jooris Fries van Embden *and* Merten Jansz *of Dortmund in Westphalia.*

1594
The Back-staff
Captain John Davis, *English navigator and Arctic explorer. Became known as 'Davis's Quadrant'.*

1595
'Flight ship' (swift flat-bottomed yacht)
Pieter J Livorn, *boatbuilder, Hoorn, Netherlands.*
Map atlas
'Gerardus Mercator', alias Gerhard Kremer, *Flemish mathematician, geographer and map-maker.*

1596
'Oorgat' (slot in bridges to allow ships to pass under bridge with mast up)
Hendrick Jacobzoon Staes, *carpenter, Amsterdam.*
Recording map-maker (hodometer-compass)
Christoph Schissler, *instrument-maker of Augsburg.*

1597
Graphometer
Philip Danfrie, *France, chief engraver at the Paris Mint.*

1598
Eternally-running clockwork
Cornelis Jacobzoon Drebbel *of Alkmaar, Netherlands.*

1599
Flageolet (musical instrument)
Juvigny, *instrument-maker of Paris, as a variant of the recorder.*
Silk-knitting machine
Reverend William Lee *of England. Following Lee's obscure death in the Low Countries in 1610, his brother returned to England, and with patronage from a Nottingham merchant, founded the first automatic frame-knitting factory in the North Midlands.*

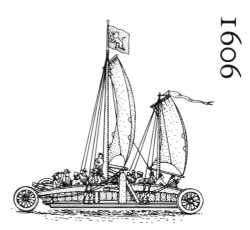

Stevin's sailing chariot

1600
Sailing chariot
Simon Stevin, *mathematician for Stadtholder Prince Maurice of Nassau. The two-masted chariot is credited with a speed of 21 mph on a two-hour journey from Scheveningen to Petten in Holland, with 28 passengers and crew on board.*

1601
Improved printing press
William Blaen *of Amsterdam.*

1602
Measuring compass
Jost Bing *of Hesse.*

1603
Pantograph
Christoph Scheiner, *German Jesuit, for copying, reducing or enlarging plans. An instrument of pivoted levers.*

1605
Multiple-ribbon weaving mill
Willem Dircxz van Sonnevelt *of Leyden, the Netherlands.*

1606
Surveying chain
Edmund Gunter, *English*

mathematician and astronomer of Christ Church, Oxford, England.

1608

Hanging compass
William Barlowe, *English divine and natural philosopher.*

Telescope
Hans Lippershey, *spectacle-maker of Middleburg, Holland. Demonstrated his 'kjiker' (looker) before the Netherlands States General and was subsequently paid 900 florins for the invention.*

1609

Astronomical telescope
Galileo Galilei, *using lenses fashioned by the glassworkers of Murano, Venice. Inspired after hearing of Lippershey's 'looker', Galileo used his instrument to discover the mountains of the moon, and the satellites of Jupiter.*

Newspaper (regularly printed)
simultaneously by Julius Adolph von Söhne, *Keeper of the Royal Press, Saxony, and* Johann Carolus *of Strasbourg. Von Söhne's 'Aviso' lasted until 1624 whilst Carolus's 'Relation' had stopped production two years before.*

Thermostatic incubator (for hatching eggs)
Cornelis Jacobzoon Drebbel, *Dutch physicist working in London.*

c. 1610

Blood circulation
William Harvey MD, *Pavia and Cambridge, physician to St Bartholomew's Hospital, London. He published his findings in 1628.*

1610

Flintlock gun
Marin Le Bourgeois, *France.*

'Ludolph's Number'
Ludolph van Ceulen, *mathematician and Professor of Fortification at Leyden, who discovered it whilst in search of the value of π (pi).*

Saturn observed by telescope
Galileo, *using a magnification of 32 on his largest instrument.*

1611

Coke
Simon Sturtevant, *ironmaker, England.*

Microscope (double convex compound)
Johann Kepler, *Imperial mathematician, Prague, Bohemia; practical instrument built in 1628 by Christopher Scheiner.*

Rainbow (theory of)
Johann Kepler.

Telescope (double convex and darkened glass)
Christoph Scheiner, *German Jesuit at Ingolstadt. Darkened glass enabled him to observe sunspots.*

1612

Circumferentor ('Holland Circle')
Jan Dou, *Amsterdam, Holland.*

1613

Astrolabe
David Fabricius, *astronomer and pastor at Resterhaave and Osteel in East Friesland.*

1614

Logarithms
John Napier, *mathematician, inventor and astrologer of Merchiston, Scotland. He also invented a calculating apparatus nicknamed 'Napier's Bones'. The decimal base for logarithms was invented four years later by Henry Briggs of Oxford, England.*

1615

Oval-turning lathe
Salomon de Caux, *engineer to Henry Prince of Wales, later King Charles I of England.*

Solar motor
Salomon de Caux, *for the*

entertainment of royalty. Used to power a fountain.

Surveying by instrumental triangulation
Willebrord Snell van Roigen, *Professor of Mathematics at the University of Leyden, Holland. He measured an arc of meridian.*

1616

Medical thermometer
Santorio Santorio, *Professor of Theoretical Medicine at Pavia, Italy. It was an adaptation of the air thermometer of his friend, Galileo Galilei. Santorio also invented a pulsimeter.*

1617

Coin-milling engine
Jean Balancier *of France.*
Map manufacture (British towns)
Aaron Rapburne *and* Roger Burges. *British Patent Number 1.*
Oil (to keep armour and arms from rust or canker)
John Casper Wolfen *of King James I of England's Privie Chamber, with Johannes Müller from Germany.*

1618

Ploughing machine
David Ramsay *and* Thomas Wildgoose, *England.*

1620

Double bassoon
Hans Schreiber *of Berlin.*
Merry-go-round
at a fair at Philippopolis, Turkey.
Submarine boat
Cornelis Jacobzoon Drebbel, *Dutch physicist working in London. This was a wooden boat covered with leather and smeared with tallow to render it watertight. It was powered by twelve oarsmen, breathing air from the hollow mast. Demonstrated to King James I in 1624.*

1621

Slide rule (rectilinear)
William Oughtred, *Rector of Albury, Surrey, England. He also invented the trigonometric abbreviations, 'sin' and 'cos'.*

1622

Lacteals
Gaspare Asellio, *Professor of Anatomy and Surgery at Pavia, Italy.*

1623

Drainage or flour mill (driven without wind or manual or animal power)
Lieutenant Laurens Garlyck, *the Netherlands.*
Practical mechanical calculating machine
Professor Wilhelm Schickard *at Tübingen University, Germany.*

1625

Hackney coaches (for hire)
four set up in London by Captain Bailey, *retired sea-captain.*

1626

'Leather' gun
Gustavus Adolphus *of Sweden – complete with internal copper bore.*

1630

Slide rule (circular)
Richard Delamain, *English mathematician, former pupil of William Oughtred.*

1631

Classified advertisements
Théophraste Renaudot, *France, in 'La Gazette'.*
'Fermat's Last Theorem'
Pierre de Fermat, *mathematician and Commissioner of Requests, Toulouse, France.*
Multiplication sign (×)
William Oughtred *of Albury, Surrey, England in his 'Clavis Mathematica'.*

Vernier Auxiliary Scale
Pierre Vernier, *French-born mathematician working for the King of Spain in the Low Countries.*

1632
Automaton lady and cavalier
Philip Hainhofer. *For an art cabinet presented by the town of Augsburg to King Gustavus Adolphus.*
Fire engine ('water-bow')
Thomas Grent.

1634
Coloured-fabric cement
Jerome Lanyer, *textile merchant, England. Called 'Londrindiana'. He also invented 'velvet paper'.*
Combined bow and pike weapon
William Neade, *skilful archer, England.*
Sedan chair
Sir Saunders Duncombe, *who patented and promoted this mode of transport and acquired sole privilege to use, let and hire a number of them for fourteen years.*

1635
Ship salvage engine
Mathewe van Dyck, *Dutch mechanic working in London.*

1636
Blade manufacture (mill-powered)
Benjamin Stone, *swordblade manufacturer, England.*

1638
Micrometer
William Gascoigne, *England (killed at the Battle of Marston Moor in 1644). To measure the dimensions of the sun, the moon and the planets.*

1640
Magic lantern
Athanasius Kirchner, *a German Jesuit.*

Taxi
Nicholas Sauvage, *rue de Saint-Antoine, Paris, France, who had twenty carriages for hire.*

1642
Calculating machine
Blaise Pascal, *nineteen-year-old son of the local president of the Clermont-Ferrand court of the exchequer. To assist his father with his accounts. Patented in 1647, only seventy 'sautoirs' were made, some of them as presents for King Louis XIII of France.*
Mezzotint
Colonel Ludwig von Siegen, *who engraved a portrait of Princess Amelia of Hesse. He disclosed the process to Prince Rupert at Brussels in 1654 and the latter subsequently had craftsmen improve upon it.*

1644
Barometer
Evangelista Torricelli, *Professor to the Florentine Academy, with his colleague* Vincenzo Viviani. *At first known as the 'Torricellian Tube'.*
Monocable ropeway
Adam Wybe, *Dutch engineer, to facilitate the building of the fortifications of Gdansk.*

1647
Bayonet
An unknown armourer at Bayonne, France.
Thoracic duct
Jean Pecquet, *medical student, Montpellier University, France.*
Map of the moon
'Hevelius', alias Johannes Hewelcke *of Danzig, who drew up a catalogue of 1500 stars and observed the planets, the moon and comets from his private observatory.*

Torricellian Tube, later barometer (1644)

Newspaper (daily)
Timotheus Ritzsch *of Leipzig, whose 'Einkommenden Zeitungen' was in production for three months.*
Wheelchair (tricycle)
Johann Haustach *of Nuremberg for the legless Stephen Farfler, who propelled it with hand-cranks.*

1653
Invisible ink recipe
Peter Borel, *although 'sympathetic' or 'secret' inks had been in use since 300 BC.*
Letter-boxes
François Velayer *of Paris, for his 'petite poste'.*

1654
Air vacuum pump
Otto von Guericke, *engineer of Magdeburg. Publicly demonstrated in front of the Imperial Diet of Ratisbon.*

1655
'A Century of the Names and Scantlings of Such Inventions As ... I ... Have Tried and Perfected'
Edward Somerset, *2nd Marquis of Worcester. Written while he was on bail from the Tower of London, it gives a brief account of 100 inventions, including a steam engine and a 14-foot diameter perpetual-motion wheel.*

1657
Pendulum clock
Christiaan Huygens, *Dutch physicist whose timekeeper was built at the Hague by Salomon Coster and installed at Scheveningen. Huygens was the first man to patent the pendulum for clock manufacture.*

1658
Balance spring regulator
Robert Hooke, *engineer and friend of Sir Christopher Wren. Hooke designed a clock mechanism regulated by a balance-spring, whose elasticity caused the balance-wheel to oscillate in equal*

1648
Hydraulic press
Blaise Pascal, *mathematics and physics student, Rouen, France. He also invented the syringe about this time.*
Hydrochloric acid
Johann Rudolph Glauber, *wandering chemist.*

1649
Automaton coach and horses
Camus *for Louis XIV of France when a child. It had a footman, a page and a lady inside. The horses and figures moved naturally and perfectly.*
Frigate (warship)
English naval architects.

1650
Lymphatic glands
simultaneously by Olavs Rudbek *in Uppsala, Sweden,* Thomas Bartholin *in Denmark and* Jolyffe *in England.*

times. A company was formed to manufacture the watch but failed due to mistrust.

Red blood corpuscles
Jan Swammerdam, *Dutch naturalist and pioneer of microscopy in Amsterdam, Holland.*

1659
Cheque (monetary)
Messrs Clayton & Morris, *bankers, law scriveners and estate agents of Cornhill, London. For £400 by Nicholas Vanacher, made payable to Mr Delboe.*

1660
'Berlin' (horse-drawn carriage).
Frictional electrical machine
Otto von Guericke, *engineer of Magdeburg.*
Twin-quill pen
Sir William Petty, *political economist working in Ireland.*

Otto von Guericke's 'Magdeburg Hemispheres'

1661
Methyl alcohol
The Hon. Robert Boyle, *Irish physicist and chemist, working in Oxford, England.*
Postal date-stamp for letters
England. Penny post first set up in London and its suburbs, two years later, by Robert Murray, *upholsterer.*
Bank notes
Issued by the Bank of Stockholm, Sweden.

1662
'Boyle's Law'
The Hon. Robert Boyle.
Buses (eight-seaters)
Blaise Pascal, *French philosopher and scientist, with his friend and chief financial backer, the* Duc de Roannez. *First scheduled service ran between Porte de Saint-Antoine and Porte de Luxembourg, Paris. Called 'carrosses à cinq sous' (five-cent coaches).*
Catamaran yacht
Sir William Petty, *founder member of the Royal Society of England. The following year, crewed by 30 men, 'Experiment' won the first open yacht race.*
Spirit level (air-bubble system)
Jean de Mechisedech Thévenot, *French researcher and traveller.*

1664
Hygrometer
Francesco Folli, *Italy.*
Lens-grinding lathe
Giuseppe Campani, *engineer of Rome.*
Theatre curtains
Ichimura-za Theatre, *Edo, Japan.*

1665
Blue sugar paper
Charles Hildeyard, *England.*
Syphon wheel barometer
Robert Hooke, *Curator of the Royal Society, Oxford University. Further developed in 1670 by* Sir Samuel Morland, *'master mechanic' to King Charles II.*

1666
'Fluxions'
(principles of integral calculus)
Isaac Newton, *24-year-old mathematics graduate of Trinity College, Cambridge, England.*

**Open-plan city
(piazza-radiating streets)**
Sir Christopher Wren, *architect and
Surveyor-General, following the
Great Fire of London. Never
constructed due to vested interests.*

**1667
Blood transfusion
(animal-to-human)**
Jean -Baptiste Denys, *Professor of
Philosophy and Mathematics at
Montpellier University, France, and
personal physician to King Louis
XIV. A lamb's blood was given to a
fifteen-year-old boy, who recovered.*
Wind gauge
Christian Förner *of Germany, to
measure the air pressure inside the
bellows of a church organ.*

**1668
Cast-iron pipeline**
*for supplying water to Versailles. It
stretched for 5 miles between the
reservoirs of Picardy and
Montbauron.*
Reflecting telescope
Isaac Newton, *mathematician, while
Fellow of Trinity College, Cambridge,
England.*

**1669
Phosphorus**
Hennig Brand, *alchemist, Germany.*

**1670
Megaphone**
Sir Samuel Morland, *'master
mechanic' to King Charles II of
England. Manufactured by Simon
Beal, Suffolk Street, London.*
Roberval Balance
Gilles Personne de Roberval,
mathematician and engineer, France.
Wheel-cutting machine
Robert Hooke, *Curator of
Experiments to the Royal Society,
London. To replace the laborious
process of cutting and filing the wheels
and pinions by hand.*

**1671
Spun silk**
Edmund Blood, *England – he
developed a machine for carding and
spinning silk.*

**1672
Hose-pipe (fire-fighting)**
Nicholas *and* Jan van der Heijden *of
Amsterdam, Holland. They devised a
suction hose to draw water out of the
canals and into a funnel, a gravity
system to feed the pump and a pressure-
hose to direct the water at the fire.*
Weather barometer
Otto von Guericke, *engineer-
showman, who mounted his 34 ft-high
brass tube containing water with a
closed glass section on the side of his
Magdeburg house so that passers-by
could see the little figure of a man
floating high on the water in good
weather, and sinking when the weather
was stormy.*

**1673
Artificial horse
(equestrian teaching machine)**
simultaneously by John Wells *and*
Raphael Folyart, *both of London.*

**1674
Musical time-signature**
Giovanni Maria Bononcini, *composer
and violinist in the Court Orchestra of
Modena, Italy.*
Tourniquet
Morel, *a French surgeon, to aid
amputation.*

**1675
Crystal-clear flint glass**
George Ravenscroft, *England,
commissioned by the Glass Seller's
Company to make better glass than the
Venetians. In 1676 Ravenscroft's
raven head seal appeared on the
Company's glassware.*
**The foot-rule
(measuring instrument)**
unknown inventor. A wooden 2-ft rod

calibrated in inches and subdivided into eighths of an inch.
Spiral balance spring
Christiaan Huygens. *Incorporated in a timepiece by Isaac Thuret of Paris.*
Water filtration system
William Woolcott, *England.*

1676
Repeating clock (rack-and-snail striking mechanism)
Reverend Edmund Barlow *residing at Parkhall, Lancashire, England.*
Universal joint
Robert Hooke, *Curator of Experiments to the Royal Society, London. To manipulate his helioscope or safe, sun-observing machine.*

1677
'Glassis' (very hard floor plaster)
Kenrich Edisbury, *of Hammersmith, Middlesex, England.*

1678
Mechanical shuttle change loom
M. de Gennes, *France.*

1679
Amputation (flap method)
Lowdham *of Exeter, England.*

1680
'Anchor' escapement
William Clement *of London, who made the first practical long pendulum clock with the new escapement. But Robert Hooke claimed that with the assistance of Thomas Tompion, he had invented the anchor pallet in 1666, the year of the Great Fire of London.*
Lithotomy (lateral operation)
Franco.
Pressure cooker
Denis Papin, *French assistant to the Hon. Robert Boyle in London. Papin's 'steam digester' was a container with a tightly fitting lid, which increased the pressure inside and considerably raised the boiling point of water. The safety*

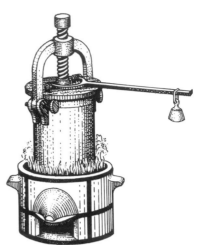

Denis Papin's 'steam digester', the original pressure-cooker

valve on the lid was also invented by Papin, who claimed that his 'bone digester' could make the oldest and hardest cow-beef 'as tender and savoury as young and choice meat'.

1681
Combined grenade discharger and musket
John Tinker, *England.*

1682
Brine-pumping machine
William Marbury *of Cheshire, England.*
Halley's Comet
Edmund Halley, *astronomer and mathematician at Oxford, England. He subsequently decided that it must be identical with comets previously seen in 1607 and 1531. He accurately predicted its return for 1758. It has been seen since in 1835, 1910 and 1986.*

1683
'Animalcules' (little animals)
Antonie van Leeuwenhoek, *a secretive former cloth warehouse clerk of Delft, Holland. He observed these, using meticulously ground, high-powered lenses (× 2–300*

A   B

C

Van Leeuwenhoek's 'animalcules'

multiplications) on saliva, teeth-scrapings, cow dung, etc. Unwittingly he had discovered bacteria.

Croissant (crescent-shaped bun)
Kulyezisk, *Polish baker in Vienna.*

Ironware-manufacturing machinery (mill-powered)
William Palin *and* William Loggins *of London.*

Sawing engine (for use by the disabled)
John Booth, *merchant, England.*

1684
Geometric-block communications
Robert Hooke. *Boards of different shapes were hung up in a large, square frame divided into four compartments. The characters could be varied in 10,000 ways and observed through telescopes. The same character might be seen in Paris the minute after it was shown in London. But the President of the Royal Society made the criticism that its use would be hindered by fogs and mists.*

Gravitational theory
Isaac Newton, *Lucasian Professor of Mathematics at Cambridge University, England. His book 'Principia' was published in 1687, but did not include the initial inspiration: an apple falling.*

Thimble
Nicholas van Benschoten, *Amsterdam.*

1685
Iron pens.

1687
Statistics
Sir William Petty, *economist, England.*

1688
Cast plate glass
Bernard Perrot, *glazier of Orléans, France. Following casting and rolling, the plates had to be laboriously ground and polished by hand.*

Champagne
Dom Pierre Pérignon, *monk and cellarer at the Benedictine Abbey of Hautvilliers, Champagne, France.*

Continuous-pattern wallpaper
Jean-Michel Papillon, *interior designer, France.*

Experimental rockets (50–100 lb)
Friedrich von Geissler, *Berlin.*

Mural arc
John Flamsteed, *English clergyman and first Astronomer Royal, with his assistant Abraham Sharp for use at the Greenwich Observatory. Since 1676, using sextants, quadrants, clocks and telescopes, Flamsteed had been measuring the motions of the heavens and locating the positions of over 2000 stars by means of tens of thousands of observations. It was not until 1725, six years after his death, that Flamsteed's epic catalogue was published.*

1690
Clarinet
Johann Christoph Denner, *instrument-maker of Nuremberg. Originally called the 'chalumeau'.*

Hydrapsis (water-shield)
Johannes Christoph Wagenseil *of Nuremberg 'as a machine by the help of which a person may walk upon the water without fear of sinking'. King Christian V of Denmark tried the hydrapsis and went more than a mile with it on the open sea.*

Light (finite velocity)
Ole Rømer, *Royal Mathematician and Professor of Astronomy at Copenhagen, Denmark.*

Denis Papin

Steam engine (experimental)
Denis Papin, *French Huguenot physicist working in exile at Marburg University. Dogged by bad luck, Papin ultimately died in poverty and melancholy in the slums of London.*

1691
Rollers to replace carriage wheels
Kendrick Edisbury, *of Hammersmith, Middlesex, England.*

1692
'Fives' (ball game)
Thomas Samborne, *England.*
Languedoc Canal
Pierre Paul Riquet, *France. 150 miles long, with over 100 locks and a 515 ft long tunnel, it took 26 years to complete.*

1694
Pollen (in plants)
Rudolph Jakob Cammerarius, *German physician and botanist.*

1695
Bank notes
Issued by the Bank of England.
Epsom salts
Nehemiah Grew, *vegetable physiologist, England.*

1697
Pantaleon
Pantaleon Hebenstreit, *tutor at Maseburg, Germany, as a drum with tuned strings.*

1698
Inflated girdle
Francis Cryns, *who demonstrated his life-preserver at Bristol and Portsmouth and in front of King William III at Windsor. Priced at a guinea and a half.*
Steam-powered mining pump
Captain Thomas Savery, *English military officer. It was used with partial success in pumping water out of Cornish mines, and also into buildings.*

1699
Fire pump (portable)
Dumaurier Duperrier *of Paris. By 1722 there were 30 such engines in Paris.*

1701
Machine seed drill
Jethro Tull, *Oxford graduate and barrister, while running his father's Howberry Farm, near Wallingford, Oxfordshire, England. It sowed seeds in precise rows. The farm labourers went on strike in protest, fearing the efficiency of the machine.*

1702
Borax/Boron
Guillaume Homberg, *alchemist in the laboratory of Philip II, Duke of Orleans, Paris, France.*
Tidal pump
George Sorocold *for the London Bridge Water Company, River Thames, England.*

1703
Binary arithmetic
Gottfried Wilhelm Leibnitz *of Leipzig.*

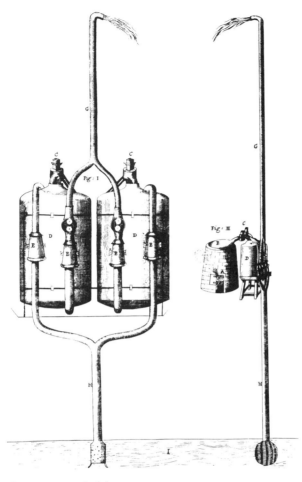

Savery's steam-powered mining pump,
1698

1704
Jewelled watch
Niccolo Facio, *Italy*.
Prussian blue pigment
Diesbach, *Berlin colour
manufacturer, and his assistant,*
Dippel. *They managed to keep their
manufacturing method a secret until
1724. At first it was used as an artist's
pigment and later on in the century as a
dye.*

1709
Anemometer
Wolfius, *to measure the strength and
velocity of the wind.*
Coke smelting (iron)
Abraham Derby *at Coalbrookdale,
Shropshire, England. He had recently
come from Bristol to take over an
existing charcoal blast furnace set up in
1638.*
Hammerklavier
Bartolommeo Cristofori, *harpsichord
builder in Florence.* Gottfried

Silbermann, *a German organ builder, further developed Cristofori's idea.*

1711
Tuning fork
John Shore, *Sergeant Trumpeter to the Court of Queen Anne of England and lutenist at the Chapel Royal.*

1712
Steam engine piston
Thomas Newcomen, *Dartmouth ironmonger and Baptist preacher, with his friend* John Calley, *a plumber and glazier. Their first practical pumping station was at Tipton, Staffordshire to pump water from the Earl of Dudley's Coneygre Colliery.*
Swell pedal (for church organs)
Abraham Jordan *of London. First box built at St Magnus the Martyr, London Bridge.*

1714
Fahrenheit scale
Gabriel Daniel Fahrenheit, *modest German physicist residing in Holland. He made important improvements in the construction of thermometers and introduced the thermometric scale known by his name by choosing zero as the lowest temperature obtainable with a mixture of common salt and ice.*
Writing machine
Henry Mill, *England. Machine and patent lost.*

1715
Dead beat escapement (clock device)
George Graham, *engineer, London.*
Diving suit (practical)
Andrew Becker. *Demonstrated in the River Thames, England.*

1716
True porcelain
Johann Friedrich Böttger, *German alchemist in the Royal Laboratory of Friedrich August II, Elector of Saxony. In 1711 Böttger came across* kaolin and felspar *being sold as part of the wig industry. From these he made true 'hard paste' porcelain. To maintain secrecy, Friedrich the Strong compelled the alchemist to work in private, under an armed guard. By 1716 Meissen porcelain had been placed on the market.*

1717
Diving bell
Sir Edmond Halley, *astronomer and mathematician.*
Perpetual motion drum
'Offyreus', *alias* Johann Ernst Elias Bessler. *'Tested' over two months in a carefully locked room in the castle of Charles, Landgrave of Hesse-Kassel.*

1718
Machine-gun (flintlock)
James Puckle, *London lawyer, for use on board ships. At a public demonstration in 1722, a Puckle gun discharged 63 bullets out of its eleven pre-loaded chambers in 7 minutes. Puckle's gun, manufactured at White Cron Alley factory, was found to be too heavy to be practical. It was designed to shoot round bullets against Christians and square bullets against Turks.*

1719
Full-colour printing process
Jakob Christof Le Blon *of Frankfurt, Germany.*

1720
Optical harpsichord
Louis Bertrand Castel. *Coloured tapes were fitted to each finger key so that, as light passed through these during a recital, a coloured light show was also enjoyed. Called the 'clavecin oculaire'.*
Pedal harp
unknown instrument-maker *in Germany.*
Types cast
William Caslon, *London gun engraver*

and toolmaker. *He cut his first 'founts' in 1720 and the first specimen of printed work appeared in 1734.*

1725
Stereotyping
William Ged, *Scottish printer and former goldsmith, Edinburgh. Developed a prototype built by Van der Mey of Leyden 200 years before.*

1726
Compensated mercury pendulum
George Graham, *watchmaker, London.*

1728
'Temple spectacles' (with rigid side pieces)
Edward Scarlett.

1729
Aberration of light
Reverend James Bradley, *Savilian Professor of Astronomy, Oxford University, England.*
Tensile-testing machine
Pieter van Musschenbroek, *Physics Professor at Utrecht, Holland.*

1730
Rotherham plough
Joseph Foljambe *of Holland. It was adapted from a Dutch plough with a wooden mouldboard.*
Thermometer (spirit)
René Antoine Ferchault de Réamur, *scientist and Director of the Paris Academy of Sciences – after whom the 80-degree scale is named.*

1731
Octant
John Hadley, *mathematical instrument maker of East Barnet, London. Known as 'Hadley's Quadrant'.*
Pyrometer
Pieter van Musschenbroek, *Physics Professor at Utrecht, Holland.*

1732
Copper-zinc alloy
Christopher Pinchbeck, *London clockmaker.*
Pitot Tube
Henri Pitot, *French physicist.*
Threshing machine
Michael Menzies *at Edinburgh. It was driven by a water wheel.*

1733
Arsenic
Georg Brandt, *Professor of chemistry, Uppsala University, Sweden.*
Child's perambulator
William Kent *for William Cavendish, 3rd Duke of Devonshire, England.*
Composite glass lens (crown-flint)
Chester Moor Hall, *Essex, England.*
Flying shuttle
John Kay, *woollen-mill owner's son and clockmaker at Colchester, England. Kay's invention was plagiarised by the Yorkshire clothiers who formed the 'Shuttle Club' to cover each other's costs should they be prosecuted. In a succession of (successful) lawsuits, Kay bankrupted himself and was pursued by angry weavers from Leeds and so to France, where he died in debt.*

1734
Fire-extinguisher bombs
M Fuchs *of Germany. Water-filled glass balls were thrown on the fire.*
Metal-plate rolling machinery (water-powered)
Kristofer Polhem, *Swedish scholar, engineer and industrialist.*

1735
Marine chronometer
John Harrison, *a self-educated Yorkshire carpenter, as an entry for a £20,000 prize offered by the British Admiralty's Board of Longitude for a clock which would have an error of less than 2 minutes on a voyage to the West Indies and back.*

Platinum
Don Antonio de Ulloa de la Torre
Giral, *mariner, Spain.*

1737

Contour lines
Philippe Buache, *French cartographer, on a chart of the English Channel.*

Didot Points
François-Ambrose Didot, *French printer working in Paris, to standardise type sizes.*

Linnaean system (botany)
'Linnaeus', alias Carl Linné, *Swedish botanist of Uppsala, resident in Holland.*

1738

Automaton duck (including digestive system)
Jacques Vaucanson *of Paris.*

1739

Jewish Year 5500.
Pile-driving machine
Charles Valoue, *clockmaker, London.*

1740

Curare drug
Charles Marie de Lacondamine, *French geographer, during a visit to Peru and the Amazon. It was originally used by the Peruvian Indians on their arrow tips.*

Dividing engine
Henry Hindley *of York, England, for cutting the teeth of clock wheels.*

Microscope (pocket-sized)
Benjamin Martin, *instrument maker, Chichester, Sussex, England.*

1741

Fizzy mineral water
Dr William Brownrigg, *chemist, of Whitehaven, Cumbria, England.*

Ventilators
Reverend Stephen Hales, *perpetual curate of Teddington, Middlesex, England. He also invented machines for distilling sea-water, preserving meat, etc.*

1742

Ballistic pendulum
Benjamin Robins, *English mathematician. This was the first accurate means of testing the force of gunpowder.*

Celsius scale
Anders Celsius, *Swedish astronomer at Uppsala – to express temperatures on a Delisle thermometer as degrees (°C) on the International Practical Temperature Scale. Celsius's scale is believed to have been invented by his compatriot, Carl Linné.*

Crucible steel technique
Benjamin Huntsman, *a clockmaker from Doncaster, England.*

Indoor swimming pool
'The Bagnio', *Lemon Street, Goodman's Fields, London.*

1743

Carding machine (wire-toothed revolving cylinders)
David Bourne, *Leominster, Herefordshire, England.*

Cart weighbridge (compound lever)
John Wyatt, *carpenter, Staffordshire, England.*

'Chaise volante' (elevator)
installed in the private apartments of King Louis XV of France in the Petite Cour du Roi at Versailles, to give the king access to his mistress, Madame de Châteauroux, on the floor above. It was worked by weights and pulleys.

Sheffield plate
Thomas Boulsover *of Sheffield, England. It was originally used for the manufacture of buttons.*

Whist (card game)
Edmund Hoyle, *England.*

1745

Windmill (automatic fantail)
Edmund Lee, *England. To keep the sails square to the wind.*

1746
Leyden jar (prototype electrical condenser)
Simultaneously by Pieter van Musschenbroek, *Dutch scientist, of Leyden and* Dean E G von Kleist *of Cammin, Pomerania.*

1747
Boxing gloves
Jack Broughton, *London pugilist.*
Diet deficiency
James Lind, *British naval surgeon on board* HMS *Salisbury, who discovered that by making sailors suffering from scurvy eat a diet of oranges and lemons, there was at once a return to good health. Lind went out of his way to bring this cure to the attention of the British Admiralty.*
Multiple-mirror solar furnace
Count Louis Leclerc Buffon. *Using between 100 and 360 small flat mirrors, individually pointed to send sunlight to a common focus, Buffon was able to ignite woodpiles over 60 metres away. The same year, using a burning glass 112 cm in diameter, Jacques Cassim of the Paris Observatory was able to melt silver to such a fluid state that the drippings formed hairlike strands when chilled in water.*

1748
Sea quadrant
B Cole *in the Orrery, Fleet Street, London, England.*

1750
Papier-mâché
John Baskerville, *printer of Birmingham, England, who also designed 55 typefaces.*
Prony brake (dynamometer)
Gaspard de Prony, *France.*
Wove paper
John Baskerville, *using woven wire sieves to produce the 'laid' effect.*

1751
Nickel
Axel Fredrik Cronstedt, *metallurgist, Swedish Academy of Sciences.*
Planing machine (for iron)
N Focq, *France.*

1752
Geometrical pen
G Suardi, *Italy.*
Lightning conductor
Benjamin Franklin, *for his own house in Market Street, Philadelphia, Pennsylvania. His steel-pointed, 7 ft iron rods were also mounted on the State House and the Academy Building during the same year. Their erection followed experiments with moistened silk kites and iron keys to assess electric discharge from thunderstorms.*

Benjamin Franklin

Parabolic reflector
William Hutchinson, *marine consultant, England.*

1753
Writing machine
Professor Friedrich von Knauss, *Director of the Physical and Mathematical Laboratory, Vienna.*

1754
Lightning conductor
'Procopius Divis', *Moravian monk, independently of Benjamin Franklin's experiments.*

1755
Bridge (iron-girdered)
M Garvin, *French engineer. Of a 25 metre span across the River Rhone, only one of the three arches was of forged iron, the remaining two being of wood.*
Kaolin
William Cookworthy, *druggist and Quaker at St Austell, Plymouth, England.*

1756
Carbon dioxide
Joseph Black, *chemist, Glasgow University, Scotland.*

1757
Sextant
Captain John Campbell, *as a navigational improvement on John Hadley's quadrant of 1731.*

1758
Achromatic lens combination
John Dollond, *Huguenot silk weaver turned London optician. The soda-lime type of lens was used for crown-glass and the potash-lead type for flint glass.*
Achromatic refractor telescope
John Dollond.
Aneroid barometer (prototype)
Zaiker, *for use at sea.*
Revolving stage
Namiki Shozo *for the Kado-za doll theatre in Osaka, Japan.*
Ribbed stocking-frame machine
Jedediah Strutt, *former Blackwell farmer, on the suggestion of his brother-in-law.*

1759
Lighthouse (Portland stone and hydraulic cement construction)
John Smeaton, *civil engineer and former instrument maker, England. The third 'Eddystone' lighthouse.*

1760
Cacodyl
Louis Claude Cadet de Gassicourt. *Described as 'one of the worst smells in chemistry'.*
Cast-iron cog wheels (for mills)
Carron Iron Works, *Glasgow, Scotland.*
Parallelogram window blinds
Dr Gowin Knight, *scientist, London.*
Roller-skates
Joseph Merlin *of Huy, Belgium, a violinist with an ill-fated sense of showmanship.*
Screw-manufacturing machine
Job *and* William Wyatt, *England.*
'The Young Writer' (automaton child doll)
Pierre Jaquet-Droz, *Swiss watchmaker. It could write a letter of 50 words.*

1761
Clavecin électrique (electric harpsichord)
Jean Baptiste Delaborde *of Paris. It was unsuccessfully powered by Leyden jars.*
Marine chronometer (pocket-watch size)
John Harrison, *Yorkshire carpenter. During a voyage to Jamaica, it lost 5 seconds during 81 days at sea. But it was only through the personal intervention of King George III that the British Admiralty paid Harrison the prize money they had originally offered in 1713.*
Percussion (medical diagnostic technique)
Joseph Leopold Avenbrugger, *physician, Spanish Hospital, Vienna, Austria.*
Scissors and shears (mass-produced in cast steel)
Robert Hinchliffe *of Sheffield, England.*
Vaulting horse
Lund University, Sweden.

1762
Brewing thermometer
Michael Combrune, *France.*
Fire-extinguisher
Dr Godfrey *of London. Sal ammoniac-filled containers were burst by gunpowder.*
Latent heat
Joseph Black, *Professor of chemistry, Glasgow University, Scotland.*
Rubber solvent
François Fresnau *at Marennes, France.*
Sandwich
John Montagu, *4th Earl of Sandwich, during a 24-hour gaming session in London.*

1763
Cork lifejacket
Dr J Wilkinson. *It was accepted by the British Admiralty for use in the Fleet.*
Jigsaw puzzle
John Spilsbury *of Drury Lane, London, 'Engraver and Map Dissector in wood', to facilitate the teaching of geography.*
Printed cheque
Hoare's Bank, *London.*
Queen's Ware pottery
Josiah Wedgwood, *English potter. He bought from several inventors a process which drastically reduced the time and cost of producing decorated creamware. He also invented Jasperware.*

1764
Bolting machine
John *and* Thomas Morris, *hosiers of Nottingham, England, were financed by* John *and* William Betts.
Road-making system
Pierre Trésaguet *of the Corps des Ponts et Chaussées. He developed a base of flat stones set vertically in a rough voussoir-arch shape – later adopted by Thomas Telford in Great Britain.*

Spinning jenny
James Hargreaves, *illiterate weaver and carpenter of Standhill, near Blackburn, England. The number of spindles on his first machine was eight, reproducing the actions of eight hand-spinners. Later it rose to twenty or 30 and eventually 120. Whether the word 'jenny' refers to his little daughter or his wife, or is a nickname for the engine, is unresolved.*

1765
Hard porcelain (French system)
Louis-Félicité de Brancas, Comte de Lauraguais, *using kaolin or china-clay from Alençon in France. Three years later,* Pierre Jean Macquer *identified kaolin at St Yrieix-la-Perche.*
Steam engine condenser
James Watt, *Glasgow instrument maker, originally commissioned to repair a small Newcomen engine. Watt made his engine more fuel-efficient by producing a separate condensation chamber. Together with a Birmingham manufacturer called Matthew Boulton, 'Jimmy' Watt changed the face of the industrial world, their Soho factory turning out hundreds of steam engines.*

1766
Fire escape
David Marie, *watchmaker, Westminster, London.*
Glass harmonica
Benjamin Franklin, *American ambassador and scientist, living in London. A treadle-operated 'vérillon' with self-tuned glasses. Following the addition of a keyboard in 1785, both Mozart and Beethoven wrote for it.*
Hydrogen (gas)
Henry Cavendish, *35-year-old natural philosopher, living at Clapham Common, South London. His discovery of what he called 'inflammable air' led to early balloon experiments – as hydrogen is 10.8 times lighter than common air.*

Lightning conductor
placed on top of St Mark's, Venice.
Mariner's direct-reading compass
Dr Gowin Knight. *Better steel for the needle for increased magnetic strength, better support for the compass bowl and a gimbal system were the features of the Knight compass as adopted by both the British Royal Navy and Germany.*
Mesmerism
Friedrich Anton Mesmer *and* Professor Hehl, *a Jesuit priest and astronomer in Vienna, from their interest in animal magnetism.*
Watercolour pigment cakes
William *and* Thomas Reeves *of London. They were rectangular in shape, and by 1781 there was a choice of 40 colours.*

1768
Areometer
Antoine Baumé, *master apothecary, Paris. For measuring specific gravity of fluids.*
Dividing engine
Jesse Ramsden, *instrument maker, Haymarket, London.*

1769
Automaton chess player
Baron Wolfgang von Kempelen. *His automaton was dressed as a Turk and the Baron had people believe that it was all done with levers, gears, drums and cylinders. In fact it was done by a confederate 'out of sight' using a system of balls and magnets. Emperor Joseph II of Austria, Empress Catherine of Russia and even Napoleon were convinced it was an automaton.*
Lifejacket
L'Abbé de La Chapelle, *French priest.*
Model boat (steam-powered)
William Henry, *gunsmith of Lancaster, Pennsylvannia. It was tested on the Conestoga Creek.*
Tartaric acid
Carl Wilhelm Scheele, *chemist in Stockholm, Sweden.*
Venetian blinds
Edward Bevan, *England.*
Water frame
Richard Arkwright, *former wig-maker and hair-dyer of Bolton, England, with* John Kay, *Warrington clockmaker. While their first spinning frame mill at Nottingham (1771) was horse-powered, a second plant at Cromford, Derbyshire was powered by the River Derwent – hence the term, water frame.*
Water wheel shaft (cast iron)
John Smeaton *civil engineer, Austhorpe, England.*

1770
Electric battery (135 Leyden jars)
John Cuthbertson, *English instrument maker in Amsterdam, Holland.*
Eraser (india-rubber)
Mr Nairne, *mathematical instrument maker, the Royal Exchange, London.*
False teeth (mineral paste)
A Duchâteau, *apothecary, France.*
Nail violin
Johannes Wilde, *St Petersburg, Russia. Metal nails instead of strings.*
Phrenology
Dr Franz Joseph Gall, *German-born medical student in Vienna. He began with his schoolfellows, then studied the heads of criminals and others, delivering his first lecture in Vienna in 1796. In 1802 the Austrian government prohibited his teaching. He was joined by* Dr Johann Caspar Spurzheim *and their researches led to an increased study of the brain.*
Ramsden's lathe
Jesse Ramsden, *Halifax-born instrument maker, Haymarket, London.*
Steamcar
Nicholas Joseph Cugnot, *French artillery officer, whose 'voiture en petit' built for the National Arsenal could easily carry four people at 2½*

mph though it had to stop every 15 minutes for the boiler to be re-filled by hand.

Sulphur dioxide
Joseph Priestley, *chemist and Presbyterian minister, Mill Hill Chapel, Leeds, England.*

1771
Fluorine
Carl Wilhelm Scheele, *chemist, Uppsala, Sweden.*
Picric acid
Woulfe.

1772
Automaton musician
Henri-Louis Jaquet-Droz *and his associate* Leschot. *The lady played five tunes on the organ, with a curtsy between each tune.*
Coloured inks patent.
Weighing machine with dial
John Clais, *London.*
'Empyreal air'
Carl Wilhelm Scheele, *chemist, Uppsala, Sweden.*
Nitrogen ('noxious air')
Daniel Rutherford, *Scottish physician and botanist.*
Printing press (cast-iron)
Wilhelm Haas, *typefounder at Basle.*
Varnish formulary
Jean Felix Watin, *France.*
Windmill-spring sail
Andrew Meikle, *millwright, Houston Mill, Dunbar, Scotland. His hinged shutters replaced sailcloths.*

1773
'The Designer' (automaton draughtsman)
Henri-Louis Jaquet-Droz, *Swiss automaton-maker, and his associate* Leschot.

1774
Ammonia
Joseph Priestley, *chemist and Presbyterian minister at Mill Hill Chapel, near Leeds, England.*

Barium
Carl Wilhelm Scheele, *chemist, Uppsala, Sweden.*
Chlorine gas
Scheele.
'Dephlogisticated air'
Joseph Priestley.
Manganese
Scheele.
Tea fountain
John Wadham, *England.*

1775
Borer and polisher (steam-engine cylinders)
John Wilkinson, *Bersham Furnace, Wrexham, England. Based on Smeaton's prototype of 1765.*
Chain-driven machine
Crane, *England. Used to manufacture ladderproof stockings.*
Digitalis (as drug)
William Withering, *botanist and chief physician of Birmingham General Hospital, England.*
Electrophorus
Count Alessandro Volta, *Italian physicist and Professor of Natural Philosophy at Pavia. For obtaining friction electricity.*
Pressure anemometer
Dr James Lind, *Scottish physician, Windsor, England. For measuring extreme velocity of the wind at 93 mph.*
Valve water closet (S-shape pipe)
Alexander Cummings, *watch and clockmaker of Bond Street, London. Modified three years later, this WC sold 6000 units during the next two decades.*

1776
Automatic lace-making machine
Leture, *French engineer.*
Breech-loading rifle
Colonel Patrick Ferguson, *Scotland.*
Submersible
David Bushnell, *Saybrook, Connecticut,* USA. *The 'Turtle' was*

submerged by taking water into its tanks and raised by pumping it out again. Hand-cranked by Sergeant Ezra Lee, 'Turtle' attempted the first submarine attack on Lord Howe's flagship 'Eagle'. Unable to fix an explosive charge on 'Eagle's' bottom, 'Turtle' had to make a quick getaway.

Threshing machine
Andrew Meikle, *millwright, Houston Mill, Dunbar, Scotland.*

Torsion balance
Charles Augustin de Coulomb, *French scientist working in Martinique.*

1777

Circular saw
Samuel Miller, *England. It did not come into regular use until the application of steam power, after which it sold in numbers around the sawmills of Great Britain and the Continent.*

Harecastle Tunnel
James Brindley, *to carry the Trent and Mersey Canal under Harecastle Hill near Stoke-on-Trent, England. Just wide enough for a narrowboat, it was 2926 yd long and took eleven years to construct.*

Iron boat
12 ft pleasure craft on the River Foss, Yorkshire, England.

'Oxygine' (Greek, acid maker)
Antoine Laurent de Lavoisier,

Lavoisier's apparatus for analysing 'Atmospheric Air'

French chemist and director of the government powder mills.

1778

Copying machine
James Watt *at the Soho Works, Birmingham, England. 150 machines were sold before the end of the year, and it was patented in 1780.*

Molybdenum
Carl Wilhelm Scheele, *chemist and apothecary, Köping, Sweden.*

Mortise lock (many-tumbler)
Robert Barron, *locksmith of Shoreditch, London.*

1779

Carronade (naval gun)
General Robert Melville, *cast by the Carron Company of Falkirk, Scotland.*

Cast-iron bridge
'Ironbridge' over the River Severn at Coalbrookdale in Shropshire, England.

Glycerine
Carl Wilhelm Scheele, *chemist and apothecary, Köping, Sweden.*

Rotative steam engine
Matthew Wasborough *of Bristol, England. It was used to power machinery at James Pickard's button manufactory in Birmingham.*

Spinning mule
Samuel Crompton, *26-year-old farmer's son from Firwood, near Bolton, Lancashire, England, who was only granted government compensation for his invention – some £5000 – when aged 59.*

1780

Argand lamp (circular wick)
Aimé Argand, *Swiss physician, chemist and balloonist. A patent for this lamp was granted in 1784 but its validity was disputed and in 1787 Argand died in poverty at Versoix, near Geneva.*

Artificial insemination
Lazzaro Spallanzani, *Italian priest*

Carronade, 1779

and biologist working in Pavia.
Lighthouse (glass reflectors).
Pyrometer
Josiah Wedgwood, *pottery manufacturer of Hanley, England.*
Two-horse swing plough
James Small, *Scotland.*
War rockets
Hyder Ali, *ruler of Mysore, India, against the British at Guntar.*

1781
Compound steam engine
Jonathan Carter Hornblower, *Cornish mining engineer, Radstock Colliery, Bath, England.*
Eidophusikon
Philip James de Loutherbourg, *French-born stage- and scenery-designer for David Garrick, the actor. Natural phenomena represented by moving pictures. Exhibition at Lisle Street, Leicester Square, London.*
Hair hygrometer
Horace Bénédict de Saussure, *physicist and geologist, Professor of Philosophy, Geneva, Switzerland.*
Sun-and-planet gearing system (steam engines)
William Murdoch, *engineer-assistant to James Watt, Soho Works, Birmingham, England.*

Uranus (planet)
Frederick William Herschel, *German-British astronomer, former oboist in the Hanoverian Guards band and former organist and music teacher in fashionable Bath, England. It was first called 'Georgium Sidus' after King George III, next 'Herschel' and finally Uranus.*

1782
Tellurium
Franz Joseph Müller, *chemist, Director of the State Mines of Transylvania, although given its name by Martin Heinrich Klaproth.*

1783
Balloon
Joseph-Michel *and* Jacques-Etienne Montgolfier, *paper-makers of Annonay, near Lyons, France, who gave the first public demonstration of their 'hot-air' balloon (of 110 ft circumference) in their town square. It was made of cloth with paper and fastened together with buttons. Following the aerial flight of a cock, a sheep and a duck suspended in a basket, the first men-carrying balloon, the 'Montgolfière', ascended from the Bois de Boulogne and drifted some 5 miles across Paris.*

1783

Balloon (hydrogen-filled and valve-controlled)
Jacques Alexandre Charles, *Parisian physicist, and the* Robert brothers. *Its first flight was 27 miles.*

Centrifugal governor for windmills
Thomas Mead, *carpenter of Sandwich, Kent, England.*

Cotton printing (cylindrical roller process)
Thomas Bell, *Scotland.*

Paddleboat (steam-engined)
Marquis Claude de Jouffroy d'Abbans *and* Comte Charles de Follenay. *Equipped with an engine built by Frères Jean of Lyons, France, 'Pyroscaphe steamed several hundred yards against the current of the River Saône at Lyons.*

Pianoforte (sustaining and damper pedals)
John Broadwood, *London cabinet-maker.*

Wolfram (tungsten)
Don Fausto d'Elhuyar, *Spanish chemist, and his brother,* Juan José.

1784

Bifocal lenses
Benjamin Franklin, *while* US *minister in Paris.*

Gas-lit room
Professor Jean Pierre Minckeler *in the lecture-room of the University of Louvain, France.*

Mail coach
John Palmer, *theatre manager of Bath and Bristol, England.*

Model helicopter (twin rotors)
Launoy, *French academician and naturalist, and his mechanic Bienvenue. Demonstrated to the Academy of Sciences.*

Puddling furnace (iron)
Henry Cort, *ironmaster of Funtley, Hampshire, England, and also a navy agent in London. He also invented a prototype rolling mill.*

Rope-making machine
Richard March, *hosier, London.*

Saccharometer in brewing
John Richardson *of Hull, England.*

Safety combination lock
Joseph Bramah, *lame Yorkshire mechanic. Bramah was so confident about his burglar-proof barrel-shaped lock, with its 494 million combinations, that he offered a prize of 200 guineas to the first person who could pick it. The prize remained unclaimed for 67 years until a* US *locksmith, Alfred C Hobbs, picked the lock after one month's work.*

Shrapnel shell
Lieutenant Henry Shrapnel, *Royal Artillery, England. His shell, filled with tiny lead balls, was designed to burst in mid-air and disperse metal fragments into the bodies of the enemy. It was adopted by the British Artillery in 1803.*

Steam winding colliery engine
a Boulton & Watt engine was adapted at the Walker Colliery on Tyneside, England.

Windmill-powered carriage
Bishop W Wilkins MA. *Hand-cranked to turn a four-bladed 'windmill' puller propeller.*

1785

Beer-pump handle
Joseph Bramah, *London.*

Grist mill factory (fully automatic)
Oliver Evans *of Philadelphia,* USA.

Human artificial insemination
M Thouret, *doyen of the Medical Faculty at Paris University. An intravaginal injection of sperm on his sterile wife with a tin syringe resulted in the birth of a healthy baby.*

Methane and ethylene
Comte Claude Louis Berthollet, *France.*

1786

Breech-loading musket
Henry Nock *of Ludgate Hill, London.*

Barbeu du Bourg's 'lightning conductor umbrella'

Lightning-conductor umbrella
Barbeu du Bourg, *having recently translated the writings of Benjamin Franklin into his native French.*
Match (phosphoric or philosophical bottle technique).
Raised letters for the blind
Valentin Haüy, *founder of the National Institute for Blind Youth, Paris, using embossed Roman letters.*
Steamboat (endless paddlechain)
John Fitch, *gunsmith of Windsor, Connecticut,* USA. *The Philadelphia-built hull was propelled by twelve vertical paddles, steam-operated by Fitch's own engine. Bankrupted in an attempt to interest the French*

government in his invention, Fitch worked his passage back to the USA *and later poisoned himself.*

1787
Eau de Javel
Comte Claude Louis Berthollet, *French chemist, for bleaching clothes.*
Flax-spinning machinery
Kendrew, *an optician, and* Porthouse, *a clockmaker, both of Darlington, England.*
Iron ship
John Wilkinson's *70 ft barge, 'Trial'.*
Power loom
Dr Edmund Cartwright, *Rector of Goodby, near Marwood, Leicestershire, England. The prototype was built by both the local carpenter and smithy, then powered by two strong men. Attempts to employ steam-engine versions at Doncaster and Manchester (400 power looms) met with fierce opposition and arson.*
Roller-bearings
John Garnett, *merchant of Redland, Gloucestershire, England.*
Steamboat (waterjet-propelled)
James Rumsey, US *engineer. Travelled upstream on the River Potomac.*
Theodolite
Jesse Ramsden, *Fellow of the Royal Society, and instrument-maker for the Trigonometrical Survey of England and Wales.*

1788
False teeth (wax mould and plaster model)
Dubois de Chemant, *Paris dentist.*
Panorama
Robert Barker, *Irish portrait painter. Paintings around the wall of a circular building. His first work was a view of Edinburgh.*
Phaeton (carriage).
Steamboat (multi-hulled)
Patrick Miller, *inventor, with* William Symington, *engine-builder, and* James Taylor. *It successfully*

1789

steamed on Loch Dalswinton on Miller's Dumfries estate, Scotland.

1789

Anémocorde
J J Schnell, *who took the then popular Aeolian harp of ancient origin and instead of allowing the natural wind or breezes to produce music from its ten strings, fitted a bellows and a keyboard. The Polish Aeolopantaleon of 1825 and Niccolo Isouard's 'piano éolien' of 1837 were modified versions of this.*

Bourbon (drink)
Eliza Craig, *Kentucky, Massachusetts, USA.*

Mural circle (5 ft)
Jesse Ramsden, *English instrument-maker, for Palermo Observatory, Italy.*

Roller reefing sail (windmills)
Stephen Hooper, *England, who also invented a horizontal windmill.*

Steam-engined power loom
John Austin, *Glasgow, Scotland. A variation on Cartwright's loom of two years before.*

Uranium (metal)
Martin Heinrich Klaproth, *Professor of Chemistry at Berlin University, Germany.*

Zirconium
Klaproth.

1790

Boot-sewing machine
Thomas Saint, *London cabinet-maker.*

Crane-neck phaeton (horse-drawn carriage)
'Highflyer', *built in London.*

Galvanic electricity
Professor Luigi Galvani *of Bologna, Italy. His wife was skinning frogs with a scalpel and was horrified to notice that when the scalpel touched the frog's salinated thighs resting on a zinc platter, the thighs twitched. From this incident, Galvani developed his theory of 'animal electricity'.*

Ice yacht
Oliver Booth, USA.

Machine for spinning and roving cotton
William Pollard, *Philadelphia. The earliest US patent.*

Tower clock (windmill-powered)
Benjamin Hanks, *New York, USA.*

Two-wheeled wooden hobby horse
Le Comte de Sivrac, *Paris. Called 'Célérifère' (quick walker).*

Wool-combing machine
Revd Edmund Cartwright, *England.*

Wristwatch
Henri-Louis Jaquet-Droz *and* Leschot, *Swiss watchmakers of Geneva.*

Zinc white
Bernard Courteois, *French chemist.*

1791

Contour lines (land map)
Jean-Louis Dupont Triel, *French surveyor.*

Gas-turbine (design)
John Barber *of Nuneaton, England. Although fully patented, it was never built.*

Guillotine
Joseph Ignace Guillotin, *French physician, for efficiently inflicting capital punishment by decapitation during the French Revolution, thus avoiding the cruel feudal use of the wheel, the stake and the gibbet. At first the machine was called the 'Louisette', but was later nicknamed 'Madame La Guillotine'.*

Ophicleide
Louis Alexandre Frichot, *French instrument-maker working in London. It was a keyed bassoon-cum-bugle and was later replaced by the tuba.*

Titanium in ilmenite
William Gregor, *chemist, mineralogist and Rector of Diptford, Devonshire, England.*

1792
Ambulance
Baron Dominique Jean Larrey, *personal surgeon to Napoleon. Larrey's Ambulance Corps, with twelve fully manned spring-mounted ambulances per division, was first used in Napoleon's Italian campaign of 1796–7.*
'Cordelier' (rope-spinning machine)
Reverend Edmund Cartwright, *later improved by* Captain Huddart.
Cottage (lit by coal gas)
William Murdoch, *Scottish-born area manager for Boulton & Watt steam-engines in Cornwall, achieved this at his Redruth home with help from the local physician,* Dr Boaze.
Graphite-clay pencil
Jacques-Nicolas Conté *of France, and independently by* Josef Hardtmuth *of Austria.*
Optical telegraph
Claude Chappe, *French mechanic, who perfected his T-type moving-arm machine with the help of the noted clock and watchmaker,* Abraham Louis Breguet. *Following its demonstration to the authorities, the first Chappe Telegraph Line was set up between Paris and Lille in 1794.*
Water-chemical fire extinguisher
von Ahen *and* Nils Moshein, *Sweden.*

1793
Astigmatism
Thomas Young, *21-year-old medical student at St Bartholomew's Hospital, London. For this discovery he was made a Fellow of the Royal Society.*
Daltonism (colour-blindness)
John Dalton, *teacher of mathematics and science in New College, Manchester, England.*
Public zoo
Le Jardin des Plantes, Paris.
Revolving stage (full-size)
Nakamura-za theatre, Edo, Japan.

Strontium
Thomas Charles Hope, *Professor of Chemistry, Glasgow, Scotland.*
Veneer planer, shaver, etc
Samuel Bentham. *His brother, Jeremy Bentham, the famous writer on political economy, was in charge of several industrial prisons, whose inmates were incapable of learning manual trades. Samuel therefore developed woodworking machines which the prisoners could handle. These included a veneer planer for plywood panels, the first jig-saw, circular saw tables, moulders, shapers and lathes.*

1794
Ball bearings
Philip Vaughan, *who also designed an axle using ball bearings.*
Cam-operated weaving loom
William Bell *of Dumbarton, Scotland.*
Cotton gin
Eli Whitney, *Massachusetts University law graduate, with* Phineas Miller, *at Mulberry Grove cotton plantation, Georgia, USA. By 1800, US cotton output had risen from 140,000 lb to 35 million lb per year, thanks to Whitney's gin.*

Whitney's prototype cotton gin

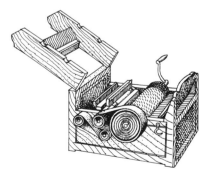

Machinery powered by inflammable vapour
Robert Street, *London varnish-maker.*

Maximum and minimum thermometer
James Six, *mathematical instrument-maker, Maidstone, Kent, England.*
Safety curtain (theatre)
Drury Lane, London. Nicknamed 'The Iron'.
Yttrium
Professor Johann Gadolin, *Finnish chemist, in a mine at Ytterby, Sweden. The metal was first obtained by* Friedrich Wöller, *German chemist, in 1828.*

1795
Hydraulic press
Joseph Bramah, *England.*

1796
Carbon disulphide
Lampadius.
Lithography
Alois Sennefelder, *unsuccessful Bavarian actor. Although he originally used porous limestone, later on zinc plates became more popular.*
Smallpox vaccination
Dr Edward Jenner *at his native Berkeley, Gloucestershire, England, after twenty years' research. The first successful inoculation was given to eight-year-old James Phipps.*

1797
Alcohol engine
Dr Edmund Cartwright, *who combined a steam engine with a 'Scotch' fast-going still.*
Chromium
Louis Nicolas Vauquelin, *chemistry professor, School of Mines, Normandy, France.*
High beaver hat
John Hetherington, *haberdasher of London, who was arrested for creating a disturbance when Londoners rioted at the sight of such an innovation.*
Hydrotherapy (revival)
Dr James Currie, *Scottish physician practising in Liverpool, England.*

Orange marmalade (commercial production)
James Keiller & Son, *Dundee, Scotland. Based on Janet Keiller's recipe.*
Parachute
André Jacques Garnerin, *former balloon inspector in the French Army, who made his first descent from a balloon at a height of 2230 ft above Paris.*

1798
Charles's Law
Jacques Alexandre César Charles, *Professor of Physics, Conservatoire des Arts et Métiers, Paris.*
Gaslit factory
Boulton & Watt's Soho factory, Birmingham, England. Based on developments by their employee, William Murdoch.
Magnetic variometer
Henry Cavendish, *wealthy bachelor grandson of the 2nd Duke of Devonshire, also chemist and physicist. It was invented to record variations of the earth's magnetic field between two points.*
Mass production
Eli Whitney, *inventor, New Haven, Connecticut, USA. In order to make 10,000 muskets for the US Army in fifteen months, he subdivided the work into a great number of small jobs so that all the component parts of the musket could be made to such an accuracy as to be interchangeable. To eliminate the element of human imprecision, Whitney had the various milling, boring, grinding and polishing machines specially adapted.*
Optical glass manufacture
Peter Louis Guinard, *bell founder, Switzerland.*
Slide rest lathe
Henry Maudslay, *English toolmaker, Oxford Street, London.*
Transparent soap
Andrew Pears, *Cornish-born hairdresser working in Soho, London.*

1799

Clavicylindre
Ernst F F Chladni, *physicist of Wittenberg. Tuned glass rods were pressed against a rotating cylinder by means of a keyboard and thereby set in vibration.*

Enamel kitchen utensils
Samuel S Hickling, *Birmingham, England.*

Engine driven by condensed wind
George Medhurst, *England.*

Metric system
French Academy of Sciences, *after nine years' research. Committee members, Délambre and Méchain, measured an arc of the meridian between Dunkirk and Barcelona and from subsequent calculations they arrived at the metre in 1795.*

Paper-making machine
Nicholas-Louis Robert, *clerk at Pierre-Francois Didot's Essone Paper Works, France. Revolutionary France did not favour further development, so Didot took the invention to England.*

Pedometer
Ralph Gout.

Thermolampe (gas-fuelled)
Philippe Lebon *of Bruchay, near Joinville, France.*

1800

Cavalry sword (curved blade)
Colonel Jean Gaspard le Marchant, *France.*

Clock movement manufacturing machine (water-powered)
Eli Terry, *Plymouth, Connecticut, USA.*

Continuous current pile (prototype electric battery)
Alessandro Volta, *Professor of Natural Philosophy at Pavia, Italy.*

Infra-red region (light spectrum)
Sir William Herschel, *German-British astronomer.*

Iron-frame piano
John Isaac Hawkins, *English*

instrument-maker residing in Philadelphia, *USA.*

Mechanically-produced pasta
Naples, Italy.

Paper (vegetable fibre)
Matthew Koops. *He developed a process for pulping straw, nettles, wood and even waste papers (recycled) to replace the system using old cotton linen rags. Koops's paper was, however, off-white hence not of the best quality.*

Screw-cutting lathe
Henry Maudslay, *English toolmaker, Oxford Street London.*

Screw-down tap (stop-cock)
Thomas Gryll, *Cornish-born ironmonger working in Birmingham, England.*

Seppings blocks (for suspending vessels in dock)
Robert Seppings, *master shipwright assistant, Plymouth Dockyard, Devon, England.*

Submarine (cigar-shaped)
Robert Fulton, *USA. The 'Fulton' had a pointed hull covered with copper plates and oxygen to support four men and two small candles for 3 hours. Propulsion was by hand-cranking and trials were made off New York.*

Tram-roads
Benjamin Outram, *civil engineer, constructed an iron tram-road from Croydon to Wandsworth, Surrey, England, which was completed in 1801.*

1801

Asteroids
Guiseppe Piazzi, *Italian astronomer, Director of the Palermo Observatory in Sicily. During the compilation of a star catalogue, he discovered Ceres, lying between Mars and Jupiter. This was the first of many new asteroids to be discovered.*

Engineering micrometer
Henry Maudslay, *UK. It was nicknamed 'The Lord Chancellor', for from it there was no appeal.*

Niobium
Charles Hatchett, *chemist*, UK. *It was originally called 'columbium'.*

Pulley-block-making machinery
Marc Isambard Brunel, *designer, and* Henry Maudslay, *toolmaker, to produce at least 100,000 pulley blocks for the Royal Navy during the Napoleonic crisis. By 1808 the plant was in operation in Portsmouth and the output of 130,000 blocks per year required ten unskilled men to do the work of 110 skilled craftsmen. A 30 hp steam engine powered the sequence of machinery. The government was saved £18,000 per year for a capital outlay of £54,000.*

Pumping station (steam-powered)
the Fairmount Water Works, *to supply Philadelphia,* USA, *with water.*

Semaphore
Dépillon, *former French artillery officer, with a three or four-arm machine. Dépillon's system furnished 301 signals, 105 more than the Chappe telegraph. By 1803 the Semaphore had been installed along the whole of the French coastline for ship-to-shore communication.*

Sugar-beet extraction factory
Franz Karl Achard, *Berlin chemist of Huguenot descent. Sited at Cunern, near Steinau in Silesia, the factory was sponsored by Frederick William III of Prussia.*

Wave theory of light
Thomas Young, *whilst working at Emmanuel College, Cambridge,* UK.

1802
Commercial steamboat
William Symington, *Scottish engineer for Lord Dundas of Kerse, to see whether steam could replace towpath-horse-towing. Tried on the Forth and Clyde canal, the 'Charlotte Dundas' was able to tow two 70-ton barges for six hours against a headwind.*

Perpetual log (marine)
Edward Massey *the younger, watchmaker of Hanley, Stoke-on-Trent, England.*

Tantalum
Anders Gustaf Ekeberg, *Swedish chemist and mineralogist. He chose the name because of the tantalising work involved in finding something to react with it.*

Ultra-violet rays
Johann Wilhelm Ritter, *scientist, Gotha, Germany, and separately,* William Hyde Wollaston, *philosophical writer, London.*

Wood-planing machine
Joseph Bramah, *London.*

1803
Atomic theory
John Dalton, *Quaker chemist of Manchester,* UK, *also a keen meteorologist.*

Cerium
Martin Heinrich Klaproth, *chemistry teacher, Berlin Artillery School, and named after Ceres, the asteroid – also discovered simultaneously by* Jöns Jakob Berzelius, *chemist, Stockholm School of Surgery, and* William Hisinger, *mining chemist, Sweden.*

Cinchonine
Andrew Duncan, *the younger, physician and professor at Edinburgh University,* UK.

Cotton-dressing machine
Thomas Johnson *of Bredbury, an ingenious but dissipated young weaver employed by Radcliffe & Ross of Stockport,* UK. *Johnson's machine, called the 'Dandy loom', enabled the warp to be dressed before it was put on the loom. Johnson received a bonus of £50 from the patentees, his employers, for his ingenuity.*

Fourdrinier paper-making machine
Bryan Donkin, *Bermondsey engineer, commissioned by John Gamble and financed by London stationers, Henry and Sealy Fourdrinier. The first*

machine, at Two Water Mills,
Frogmore, Hertfordshire, UK,
produced 600 lb of paper in 24 hours.
Iridium
Descotils *and* Smithson Tennant,
chemist, Cambridge, UK.
Spray gun
Dr Alan de Vilbiss, *medical*
practitioner of Toledo, Ohio, USA,
while searching for the ideal method of
applying medication to oral and nasal
passages without the use of a swab.
Three-furrow plough
Reverend Edmund Cartwright,
whilst in charge of the Duke of
Bedford's experimental farm, UK.

1804
Bomb-proof round defence
tower (martello)
Brigadier-General William Twiss,
commanding engineer of the Southern
District of the UK. *It was based on a*
model of the defence tower at Mortella
Point, Corsica, brought back in 1794
by Admiral Jervis of HMS *'Victory'.*
Over 70 towers were built, but they
were never used for defence.
Catamaran fire ship
Sir Sydney Smith, *in an unsuccessful*
attempt to destroy the Boulogne flotilla
established by Napoleon to invade the
UK.
Double-cylinder expansion
steam engine
Arthur Woolf, *mining engineer,* UK.
Elliptic carriage spring
Obadiah Elliott, *coachbuilder,*
Lambeth, London.
Embroidery machine
John Duncan, *weaver, Glasgow,* UK.
Fishnet-making machine
Joseph Marie Charles Jacquard,
Lyons-born mechanic. Having seen the
machine, Emperor Napoleon asked,
'Are you the man who can do what God
Almighty cannot – tie a knot in a taut
string?' 'I can do not what God cannot,
but what God has taught me to do,'
replied Jacquard.

Flinders bar
Captain Matthew Flinders, UK,
following his observation regarding
compass deviation during a voyage to
Australia on the iron-hulled HMS
'Investigator'. Flinders's solution was
a vertical bar of soft iron.
Gay-Lussac's Law
Louis-Joseph Gay-Lussac, *assistant*
to Count Claude Louis Berthollet,
Arcueil, France.
Osmium
Smithson Tennant, *chemist,*
Cambridge, UK.
Palladium
Dr William Hyde Wollaston,
chemist, UK.
Quinquet lamp
Aimé Argand, *Swiss physician and*
chemist working in the UK.
Steam locomotive (railway)
Richard Trevithick, *Cornish-born*
steam-road-carriage engineer for the
Penydaren Iron Works, South Wales,
UK. *Pulling five wagons, 10 tons of iron*
and 70 men, it travelled at 5 mph on
the 10-mile track.

Trevithick's locomotive

1805
Amphibious vehicle
Oliver Evans, *whose 'Orukter*
Amphibolos' (amphibious digger) was
designed and built for the commercial
purpose of dredging the Philadelphia
Docks. A single cylinder steam engine
with a 20-inch boiler was geared to
drive either a wooden-wheeled wagon
or an eight-bladed stern paddlewheel.

Chimney-sweeping machine
Smart.
Circular-blade mower for corn or grass
Thomas James Plucknett, *of Deptford, England.*
Gas-lit cotton mill
Phillips & Lee *of Manchester*, UK.
1000 separate burners facilitated night-shift work.
Mechanical silk loom
Joseph Marie Jacquard, *silk-weaver of Lyons, France. Based on weaving machines of* B Bouchon, H Falcon *and* Jacques Vaucanson *and also through study of Vaucanson's automata toys at the Paris Conservatorium. By 1834, some 30,000 Jacquard looms were at work in Lyons.*

Jacquard's punch-card loom

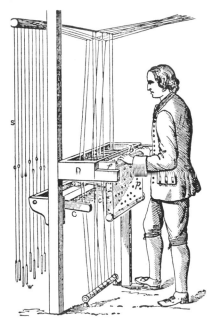

Morphine
Friedrich Wilhelm Adam Sertürner, *apothecary, Hameln, Germany.*
Rhodium
Dr William Hyde Wollaston, *chemist, London.*

War rocket (gunpowder in sheet-iron casings)
Sir William Congreve, *son of the Comptroller of the Woolwich Arsenal, London. In 1807, during the siege of Copenhagen, 40,000 such rockets were launched.*

1806
Bank-note-numbering machine
Joseph Bramah, *so enabling the Bank of England to dismiss 100 clerks.*
Beaufort wind scale
Rear-Admiral Sir Francis Beaufort, *hydrographer, as a set of numbers to indicate the strength of the wind from a Calm 0 to a Hurricane Force 12, 'that which no canvas could withstand'. It was accepted by the British Admiralty in 1838 and by the International Meteorological Committee in 1874.*
Carbon paper
Ralph Wedgwood *of London.*
Cast-iron printing press (rounded cheek frames)
Charles Mohun, *3rd Earl of Stanhope*, UK, *manufactured by Robert Walker.*
Coffee pot (built-in metal sieve)
Benjamin Thompson, Count Rumford, *Anglo-American administrator and scientist.*
Reflecting goniometer
Dr William Hyde Wollaston, *chemist, London.*

1807
Camera lucida
Dr William Hyde Wollaston, *chemist, London.*
Gas-lit street
Golden-lane, London, UK.
Patent for a gas-driven automobile
Isaac de Rivez.
Percussion-cap (arms)
Reverend Alexander John Forsyth, *former minister of Belhevie, Aberdeenshire*, UK. *He was pensioned by the British government when he refused to sell the secret to Napoleon.*

Potassium
Humphrey Davy, *Professor of
Chemistry, Royal Institution, London.*
**Sensory versus motor nerves
(brain)**
Charles Bell, *Scottish anatomist and
surgeon.*
Sodium
Humphrey Davy, *Professor of
Chemistry, Royal Institution, London.*
Steamboat (long-distance)
Robert Fulton, *former miniaturist
and landscape painter,* US. *His 150 ft
paddlewheeler, 'Clermont', travelled
150 miles on the Hudson River between
New York and Albany at an average 5
mph.*
Windmill patent sail
William Cubitt, *miller, millwright
and cabinet-maker of Norfolk,* UK. *It
combined hinged shutters with remote
control by chains from the ground. The
patent sail and fantail were adopted in
Denmark, Germany and the
Netherlands.*

1808
All-iron plough
Robert Ransome, *engineer, of
Ipswich,* UK.
Band-saw
William Newberry, UK. *It did not
come into regular use until 1850 when
better methods of steel-making and
tempering removed the danger of the
snapping of the band.*
Barium
Humphrey Davy, *Professor of
Chemistry, Royal Institution, London.*
Bermuda sail (yacht)
Harvey, *a local resident of the Islands
of Bermuda, who rigged his schooner
with high, lateen-like sails, instead of
the usual gaff rig.*
Boron
Joseph Louis Gay-Lussac, *Professor
of Chemistry, École Polytechnique,
Paris, and his associate* Louis Jacques
Thénard.
Calcium
Humphrey Davy, *Professor of*

Chemistry, Royal Institution, London.
Circular railway
Richard Trevithick, UK *engineer near
Euston Road, London. His locomotive
'Catch-Me-Who-Can' achieved 12
mph.*
**Cotton-lace/bobbin-net
machinery**
John Heathcoat. *He set up a factory in
Nottingham,* UK, *which was destroyed
by the Luddites, forcing him to move
business to Tiverton, Devon.
Heathcoat also innovated ribbon- and
net-making machinery.*
Electric-arc lamp
Humphrey Davy, *Professor of
Chemistry, Royal Institution, London.*
Magnesium
Humphrey Davy, *Professor of
Chemistry, Royal Institution, London.*
Steamboat (offshore)
John Stevens, US. *'Phoenix' steamed
from New York out into the Atlantic
around Cape May and up the
Delaware Bay to Philadelphia.*
Typewriter (for the blind)
Pellegrino Turri *of Castelnuovo,
Italy, for his blind friend, Contessa
Carolina Fantoni da Fivizzano. It
was a 27-charactered machine.*

1809
Heat-bottled food
Nicholas Appert, *chef and
confectioner, Massey, France,
following fifteen years' research. He
was awarded 12,000 francs by the
French Manufacturers' Bureau,
provided he publish his process in
detail. His pamphlet was translated
and published into German, English
and Swedish and was widely adopted.
The House of Appert was formed in
1812.*
Lamarckism
Jean Baptiste Pierre Antoine de
Monet, *Chevalier de Lamarck,
naturalist and evolutionist, France.*
**Paper-making machine
(cylindrical)**
John Dickinson *of Kings Langley,*

Hertfordshire, UK. *This type was developed for the manufacture of higher-grade papers.*

1810

Ammonia-soda reaction
Augustin Jean Fresnel, *French chemist.*

Electrolytic telegraph
S T von Sömmering, *Bavarian Academy of Sciences, Munich, at the command of Napoleon's ally, Margrave Leopold of Bavaria. This slow and complicated system was soon replaced by the electric telegraph.*

Sömmering's electro-chemical telegraph

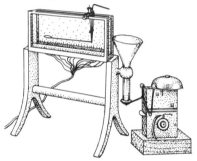

Hemp and flax-spinning machine
Philippe de Girard.

Homeopathy
Dr Samuel Hahnemann of Leipzig, *who propounded his theories in the 'Organon of Medicine' and other works.*

Hydro-aeronaut (navigator's lifebuoy)
Thomas Cleghorn, UK.

Metronome
Dietrich Nikolaus Winkel, *German-born master organ-builder living in Amsterdam. Winkel demonstrated the instrument to Johann Nepomuk Maelzel, who patented it as his own invention and set up a factory to produce it as the Maelzel Metronome, despite an investigation in Paris which upheld Winkel as the true inventor.*

Maelzel's friend Beethoven was the first composer to include metronome markings on his scores in 1816.

Mowing machine
Peter Gaillard, *Lancaster, Pennsylvania,* USA.

Panomonico (or panharmonium)
Johann Nepomuk Maelzel *of Vienna, after seven years' development. 300 separate musical instruments were built into a one-man console. It was sold for 100,000 francs to the Archduke Charles of Austria for the exclusive purpose of annoying his friends. Beethoven wrote his 'Wellington's Victory at Vitoria' (1823) to be specifically played on this instrument.*

1811

Animal bone charcoal as a filter for sugar
Louis Figuier, *French chemist at the Paris School of Pharmacy.*

Avogadro's Law
Amedeo di Quaregna e di Ceretto, Count Avogadro, *scientist, Turin, Italy.*

Iodine
Bernard Courteois, *chemist and saltpetre manufacturer near Paris, France.*

Iron dome
Over the Paris Corn Market, France.

London Mint coin-milling machinery.

1812

Canned food
Bryan Donkin, *partner in John Hall's Dartford Iron Company,* UK. *He used iron cans coated with tin. The* UK's *first cannery was set up in Blue Anchor Road, Bermondsey. Among the first users of canned food – particularly preserved meat – were naval ships on longterm exploration voyages.*

Locomotive (rack and pinion gear)
John Blenkinsop, *Middleton Collieries, near Leeds,* UK.

Rotary printing press (steam-powered)
Friedrich König, *German engineer working in London, with chief mechanic,* André Friedrich Bauer, *and their financial backer,* Thomas Bensley. *The first commerical model was acquired by John Walker II for 'The Times' newspaper, London, in 1814. Its output was 1400 printed sheets per hour on one side only.*

Steamship (regular commercial service)
Henry Bell's *30-ton 'Comet' on the Clyde between Glasgow and Helensburgh,* UK.

Sugar-refining vacuum pan
the Hon. Edward Charles Howard.

1813
Air beds and water beds (india-rubber cloth)
John Clark, *grocer, Bridgewater, Somerset, England.*

Armoured warship (steam-powered)
Robert Fulton's *'Demologos'. It was 140 ft long with thirty 32-pounder cannon on board.*

Cartridge (practical)
Samuel Pauly, *Paris, for a .59 breech loader.*

Gas meter
Samuel Clegg, *chief engineer to the Chartered Gas Light and Coke Company in London, formed the previous year by an energetic German, Friedrich Albert Winzer, who had anglicised his name to Winsor.*

Liquid-filled compass
Francis Crow, *Faversham,* UK.

Teak-built ship
HMS *'Cornwallis', launched from the Bombay dockyard, India.*

Toxicology
Mathieu Joseph Bonaventure Orfila, *French chemist.*

1814
Planimeter
J M Hermann. *The first instrument was constructed to his designs in 1817.*

Plough (replaceable cast-iron parts)
John Jethro Wood, *Poplar Ridge, New York,* USA.

Railway gauge (4 ft 8½ in)
George Stephenson, *colliery engine-wright at Killingworth,* UK *for his locomotive 'My Lord', which attained a speed of 6 mph.*

1815
Alkalimeter
Andrew Ure, *Professor of Chemistry, Anderson's College, Glasgow,* UK.

'The Dennett Gig' (horse-drawn carriage)
Bennett, *coachbuilder of Finsbury, London. It was named after Miss Dennett, a popular dancer.*

Jaunting car (Irish)
Carlo Bianconi, *Milan, Italy. To drive between Clonmel and Caher, Tipperary, Ireland, each day, charging 2d a mile. By 1837 Bianconi*

Davy's miner's safety lamp (see overleaf)

had established 67 such conveyances, drawn by a total of 900 horses.

Miner's safety lamp
Independently by Sir Humphrey Davy, *chemist for the Society for Preventing Accidents in Coal Mines, and by* George Stephenson, *colliery engine-wright. The former became known as 'The Miner's Friend', while the latter was nicknamed 'The Geordy'.*

Pruning shears (sécateurs)
Bertrand, *Marquis of Moleville, France.*

1816

British Admiralty semaphore telegraph
Admiral Sir Home Riggs Popham. *Although he produced the first edition of his Telegraphic Signals in 1800, the official trial between London and Chatham, with semaphores mounted on ships, took place almost twenty years later. Once proved, lines were established from Whitehall to Portsmouth, to Plymouth, to Chatham and to Dover. Popham received £2000 for his invention.*

Circular saw (tempered steel teeth)
Auguste Brunet *and* Jean-Baptiste Cochot, *France.*

Compressed-air fire extinguisher
Captain George William Manby *of London.*

Copper percussion cap (arms)
Joshua Shaw, *English portrait painter.*

Electric telegraph
Francis Ronalds *in his garden at Hammersmith, London. Although he published a description of it in 1823, nobody showed any interest.*

'Heliographie' (sun-drawing)
Joseph Nicéphore Niépce, *amateur scientist of Chalon-sur-Saône, France. He adapted the conventional 'camera obscura' to fix his images on a zinc or pewter plate, made light-sensitive by bitumen of Judea, a kind of asphalt normally used in engraving and here dissolved in oil of lavender. Niépce's exposure time was eight hours on a summer's day and his invention was the forerunner of photography.*

Kaleidoscope
Dr David Brewster, *Scottish physicist and Edinburgh magazine editor.*

Oxyhydrogen blowpipe
Robert Hare, *Professor of Chemistry, medical faculty, Pennsylvania University,* USA.

Print roller
manufactured following a suggestion of William Nicholson, *editor of the 'Philosophical Journal',* UK.

Psychrometer
Joseph Louis Gay-Lussac, *French chemist and physicist, for measuring atmospheric humidity.*

Stethoscope (wood-turned cylinder)
René Theophile Hyacinthe Laënnec, *pupil of Napoleon's personal physician, Corvisart, and chief physician to the Hôpital Necker, France.*

Stirling-cycle engine
Reverend Robert Stirling, *Dumbarton, Scotland, as a hot-air external combustion engine, with a regenerator that prevented dissipation of heat between cycles.*

1817

Apollonicon
Benjamin Flight *and* Joseph Robson, *organ-builders of Leicester Square, London. The most elaborate barrel-organ yet built, performing with either three barrels or six performers with one console each.*

Dental plate
Anthony A Plantson, US.

Hobby-horse (steerable front wheel)
Baron Karl von Drais de Sauerbrun, *Germany. It was called either 'draisina' or 'velocifère'.*

Lithium
John August Arfwedson, *chemist and metallurgist, Sweden.*
Parkinson's Disease
James Parkinson, *surgeon and palaeontologist, UK.*
Pin-manufacturing machine
Seth Hunt, *Syracuse, New York, USA. Not in operation until 1824 when Samuel Wright took out a UK patent.*
Treadmill (prison)
William Cubitt *of Ipswich. The first such punishment machine was erected at Brixton prison, London.*

1818
Blood transfusion
Dr Thomas Blundell *of Guy's Hospital, London, who was unable to make his system work successfully through insufficient knowledge of the nature of blood.*
Cadmium
Friedrich Strohmeyer, *Inspector General of the Apothecaries, Hanover, Germany.*
Detector lock
Jeremiah Chubb, *ironmonger in Portsea, the dockland area of Portsmouth, UK. Despite being given any tool necessary, blank keys, duplicate lock, as well as the offer of a free pardon from the government and £100 from Mr Chubb, a lockpicker on board the Portsmouth prison ship was unable to pick the detector lock after three months' work.*
Encke's Comet
Johann Franz Encke, *astronomer, Seeberg Observatory, Germany, mathematically predicted the orbit of a faint comet and successfully predicted its return.*
Geothermal energy experiment
F de Larderel, *French engineer in Tuscany, Italy. Larderel Power Station went into action in 1904.*
Hydrogen peroxide
Baron Louis-Jacques Thénard, *Professor of Chemistry, University of Paris.*

Lens-polishing machine
Josef von Fraunhofer, *physicist and toolmaker, Utzschneider Optical Institute, near Munich, Germany.*
Profile lathe
Thomas Blanchard *of Middlebury, Connecticut, USA. The machine did the work of thirteen operators and allowed a reduction in wood-working prices.*
Selenium
Johan Jakob Berzelius, *chemist and Secretary to the Swedish Academy of Science.*
Strychnine
Pierre-Joseph Pelletier *and* Joseph Bienaimé Caventou *at the School of Chemistry, Paris.*

1819
Alternation of generations
Adelbert von Chamisso, *French-born biologist and lyric poet, in Berlin.*
Blocks of chocolate (factory produced)
François-Louis Cailler *at Vevey, Switzerland.*
Dental amalgam
Charles Bell.
Dioptric system
Augustin Jean Fresnel, *French physicist, chief of the Department of Public Works, Paris. This was an arrangement of lenses for refracting light in lighthouses. Fresnel also invented 'Fresnel's rhomb', a special prism for producing circularly polarised light.*
Macadamising
John Loudon Macadam, *a tacksman of the Kaims colliery, Muirkirk, Ayrshire, UK – as an improved method of road-making. In his early experiments with firmly embedded layers of granite chips, Macadam was working alongside his second cousin, Lord Dundonald.*
Magnetic field
Hans Christian Oersted, *Danish physicist, while giving an evening lecture on the galvanic battery at Kiel University. A wire from the battery*

accidentally fell on a mariner's compass, deflecting its needle from Polar North. Thus Oersted found the link between electricity and magnetism.

Naphthalene
John Kidd, *Aldrichian Professor of Chemistry at Oxford*, UK.

Open diving dress
Augustus Siebe, *Saxon engineer working in Soho, London.*

Siren
Baron Charles Cagniard de la Tour, *Parisian physicist, for measuring sound frequency.*

Transatlantic ship (sail with auxiliary steam)
Francis Fickett, *boatbuilder, and* Stephen Vail, *engine-wright. The 'Savannah' took 29½ days from Savannah, Georgia,* USA *to Liverpool,* UK.

Trigonometrical Ordnance Survey map of the UK
Printers as directed by Colonel Mudge, *following commencement of the survey back in 1783.*

1820

Arithmometer
C X Thomas *of Colmar, Alsace. Sold in large numbers 40 years later to assurance companies.*

Corn cultivator
Henry Burden, *of Troy, New York,* USA.

Diphtheria
Pierre Fidèle Bretonneau, *physician and bacteriologist, Tours, France.*

Gas-lit city
London; Paris at the same time; Dublin by 1825; Sydney, Australia by 1841.

Hygrometer (ether)
John Frederic Daniell, *chemistry professor at King's College, London.*

Medical stitching wire
Pierre-François Percy, *French surgeon.*

Quinine
Pierre Joseph Pelletier *and* Joseph Bienaimé Caventou *at the School of Chemistry, Paris.*

Soda-water-making apparatus
Charles Cameron, UK.

Tilbury gig (carriage)
Thomas Tilbury, *coachbuilder of Marylebone Road, London for the Hon. Fitzroy Stanhope.*

1821

Anchor (stockless patent)
Richard Francis Hawkins, *Master Mariner, Plumstead, Kent,* UK.

Caffeine
Pierre Joseph Pelletier *at the School of Chemistry, Paris.*

Dark lines (solar spectrum)
Josef von Fraunhofer, *Director of the Physical Cabinet, Munich, Germany.*

Eidograh
William Wallace, *Professor of Mathematics, Edinburgh University,* UK.

Enchanted lyre
Charles Wheatstone, *musical instrument-maker, London. It conveyed the sounds of a musical box from a cellar to upper rooms by means of a deal rod.*

Heliotrope
Carl Friedrich Gauss, *German mathematician, astronomer and physicist, to conduct a trigonometrical survey of Hanover.*

Mouth organ
Friedrich L Buschmann *of Berlin, as a free-reed musical instrument he called the 'Mundäoline'.*

Multiplying galvanometer
Johann Christian Poggendorff, *German physicist and chemist, aged 25.*

1822

Accordion
Friedrich L Buschmann *of Berlin, as a bellow-vibrated piano keyboard. He called it the 'Handëoline'.*

Aquatic tripod
William Kent *of Lincolnshire,* UK. *It was used to sit on and paddle while duckshooting.*

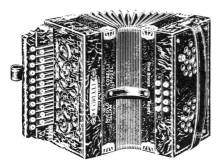

Buschmann's 'Handëoline' or accordion

Bauxite
Pierre Berthier, *French mineralogist,
near Baux, Provence, France.*
Diorama
Bouton *and* Louis Jacques Mande
Daguerre, *scene painter in Paris.*
Thermocouple
Thomas Johann Seebeck, *physicist,
Berlin Academy. He found that when
one pair of dissimilar wires with
intertwined ends is placed in a hot
surround, and another pair in a cold
surround, an electric current flows.*
**'Volontas' (horse-drawn
carriage)**
Miln Parker, US *for Cuba and Mexico.*

1823
Bowed zither
Johann Petzmayer, *Munich,
Bavaria.*
Döbereiner lamp
Johann Wolfgang Döbereiner,
German chemist in Jena.
Double-escapement piano
Sébastien Érard *of Paris.*
Hackney cabriolets
David Davies, *coachbuilder of Albany
Street, London.*
Meteorological kite
Reverend George Fisher *at Igloolik,
while on Captain William Parry's
second Arctic exploration voyage. He
attached a Six thermometer to a paper
kite.*
Omnibuses
Stanislaus Bawdry, *Proprietor of the*

*hot water baths in a suburb of Nantes,
France.*
Robinson's Patent Barley Water
Matthias Archibald Robinson, *grocer
of St George the Martyr, London.*
Roller-skates (five wheeled)
Robert Tyers, *London fruiterer.*
Silicon
John Jakob Berzelius, *Swedish
chemist.*
Talking doll
Johann Nepomuk Maelzel. *She said
'Maman' and 'Papa' and cost 10
francs.*
Waterproof rubber
Charles Macintosh, *manufacturing
chemist, Glasgow, UK, using the
rubber-dissolving naphtha by-product
from the recently established Glasgow
Gasworks. He went into partnership
with H H Birley to build a factory for
the manufacture of waterproofed capes
and coats.*

1824
Artificial cement
Joseph Aspdin, UK. *Called 'Portland
cement', it was further perfected in the
1840s by Aspdin's son William, and by
Isaac Thompson, London builder.*
Electromagnet
William Sturgeon, *lecturer in
Science, Royal Military Academy,
London. He found that any piece of
soft iron could be turned into a
temporary magnet by putting it in the
centre of a coil of insulated wire and
making electric current flow through
the coil (see overleaf).*
Galvanometer
André-Marie Ampère *at the
National Institute of Paris, to measure
the flow of electricity with a free-
moving needle. The amp is named after
him.*
High-speed self-acting mule
Richard Roberts, *inventor,
Manchester, UK.*
Magnetic pull
François Jean Dominique Arago,
Parisian astronomer, whose non-

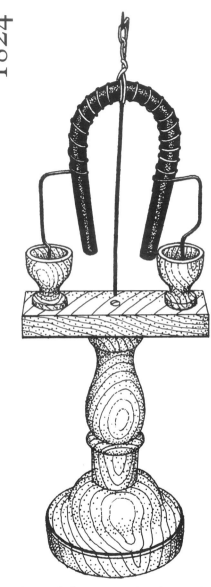

Sturgeon's electro-magnet (previous page)

magnetic disc could bring a vibrating magnetic needle to rest and could then, mysteriously, pull the needle round with it when rotated – but without contact.

Plaited candle wick
Jean Jacques Cambacères, *Paris, France.*
Steam carriage (four-wheel-drive)
Timothy Burstall *and* James Hill.
Steam gun
Angier March Perkins, UK. *However, this proved ineffectual.*
Veneer-cutting machine
Alexander Craig, UK.
Wine-pressing machine
Ignazio Lomeni, *Italy.*
Wrought-iron I-shaped beam
Thomas Tredgold, *engineer,* UK. *It was not used practically until 1841 by Robert Stephenson for railway bridge construction.*

1825
Actinometer
Sir John Herschel, UK *astronomer, to measure the heating power of solar rays.*
Binocular telescope (production model)
J P Lemière, *Paris.*
Concertina
Professor Charles Wheatstone, *London.*
Double-action pedal-harp
Sébastien Érard, *piano-maker, Paris.*
Frigate (iron-masted)
HMS *'Phaeton'.*
Regular locomotive route
George Stephenson, *consulting engineer for the Stockton and Darlington mineral railway, near Newcastle,* UK. *His 'steam blast' device gave both his locomotives a practical investment.*
Rubber driving belts (industrial)
Thomas Hancock, UK.
Thaumatrope (optical toy)
Dr John Ayrton Paris, UK.

1826
Aniline (distilled indigo)
Otto Unverdorben, *chemist, Potsdam, Germany.*

Biela's Comet
Wilhelm von Biela, *Austrian army officer and amateur astronomer.*
Bromine
Antoine-Jérôme Balard, *chemistry demonstrator, School of Pharmacy, Montpellier, France.*
Drummond Light
Lieutenant Thomas Drummond, *following a design by* Goldsworthy Gurney. *In 1820 Drummond had joined the Ordnance Survey project and used his 'limelight' and his improved heliostat for the work in hand. At night the Drummond Light could be seen at a distance of 112 miles.*
Friction match
John Walker, *chemist of Stockton-on-Tees, County Durham,* UK. *Originally made of cardboard coated with potash and antimony, packed by the 100 in tin tubes, and ignited using glass paper, they were nicknamed 'Lucifers'.*
Gas stove
James Sharp, *assistant manager of the Northampton Gas Co.,* UK. *He installed it in the kitchen of his Northampton home.*
Illuminated clock dial
St Bride's Church, *Fleet Street, London; lit by twelve gas-burners.*
Kite-carriage
Colonel Viney *and* George Pocock, UK. *Using one or more kites they made a journey from Bristol to London. The following year, on the Reading to Windsor road, the galloping horses pulling the Duke of Gloucester's carriage were unable to keep pace with the kite-carriage.*
Tea in packets
John Horniman *of Ryde, Isle of Wight,* UK.
Velocity of sound (in water)
Jacques Charles François Sturm, *Swiss mathematician, by means of a bell submerged in Lake Geneva.*

1827
Aluminium
Friedrich Wöhler, *chemistry teacher,*
Berlin, Germany.
Artificial synthesis of the urea molecule
Friedrich Wöhler, *as above.*
Astigmatic lens
George Biddell Airy, *Lucasian Professor of Mathematics, Cambridge,* UK. *It was manufactured the same year by Messrs Fuller of Ipswich.*
Electrical resistance
George Simon Ohm, *while professor at the Jesuit College in Cologne. On publication of his discovery, Ohm was so disappointed at its poor reception that he resigned his post. His name now represents the unit of electrical resistance.*
Floorboard-planing machine
Malcolm Muir *with* William Thomson, UK.
Mechanically-produced pasta factory
Buitoni *of Sansepolcro, Italy.*
Microphone
Professor Charles Wheatstone, *London.*
Microscope (dioptric-achromatic)
Giovanni Battista Amici, *Director of the Observatory, Royal Museum, Florence, Italy.*
Shorthand-writing machine
Gonod, *librarian at Clermont-Ferrand, France.*
Trifocal lenses
John Isaac Hawkins.
Water turbine
Bénôit Fourneyron, *student of Professor Claude Burdine, St Etienne School of Mines, France. By 1833 Fourneyron had developed a 50 hp turbine spinning at an unheard-of 2300 rpm.*

1828
Artificial ultramarine blue (pigment)
simultaneously by Guimet *and* Leopold Gmelin, *Professor of Chemistry, Heidelberg, Germany.*

1828

Beryllium
Friedrich Wöhler, *chemistry teacher,
Berlin, Germany.*

Chairs on small carriage springs
Samuel Pratt, UK.

Cocoa
Coenrad van Houten *of Amsterdam,
by extracting excess cocoa butter from
the crushed cacao bean.*

Corrugated iron
Richard Walker, *builder, of
Rotherhithe, London.*

Differential gear
Onésiphore Pecqueur, *France, for
steam-carriages.*

Hot-blast iron smelting
James Beaumont Neilson, *foreman
and manager of Glasgow Gasworks,
and* Charles Macintosh.

Reaping machine
Patrick Bell, *divinity student and son
of a farmer in Angus,* UK, *together with
the local blacksmith. By 1832 only ten
Bell reapers had between them cut 400
acres of corn in Scotland.*

Ring-spinning frame
J Thorpe, UK.

Stethoscope (with earpiece)
Pierre Adolphe Poirry, *France.*

Thorium
Johan Jakob Berzelius, *chemist and
Secretary of the Swedish Academy of
Science.*

1829

Aeolina
Professor Charles Wheatstone, *as a
free-reed hexagon-shaped musical
instrument.*

Electro-magnetic motor
Joseph Henry, *Professor of
Mathematics and Natural Science at
the Albany Academy, New York,* USA.
*Henry used a powerful short coil
magnet for his innovation.*

Filtration plant for water
James Simpson, *Chelsea Water
Works, London.*

The Graham cracker (biscuit)
Sylvester Graham, US *food doyen.*

**'The most improved locomotive
engine'**
George Stephenson's *Newcastle-built
'Rocket', beating both the 'Novelty'
and 'Sans Pareil' locomotives at
Rainhill, near Liverpool,* UK, *by
averaging 15 mph for the total of 60
miles, hitting a top speed of nearly 30
mph on the last lap.*

**Steam-powered fire engine
(horse-drawn)**
John Braithwaite *and* John Ericsson
*of London. Although it was successful
at the Argyll Rooms, the English
Opera House and Barclay's brewery,
orders for the engine were not
forthcoming in the* UK, *although
Braithwaite sold an improved version
to the King of Prussia for use in Berlin.*

Steam-jet-blast coach
Sir Goldsworthy Gurney *of London.
Two years later, Charles Dance ran a
steam-coach service between
Gloucester and Cheltenham (four
round trips daily). Before it was
suspended, his fleet had travelled a
total 4000 miles and carried 3000
passengers.*

**Swinging-sector writing
machine**
William Austin Burt, *Michigan,* USA.
Called 'the Typographer'.

'True water cure'
Vincent Priessnitz, *Grafenberg,
Silesia.*

1830

Bi-metal thermostat
Andrew Ure, *Professor of Chemistry,
Anderson's College, Glasgow,* UK.

Elastic web
Messrs Rattier *and* Guibal *at their
waterproof cloth factory in Saint-
Denis, Paris, assisted by machinery
and workmen supplied by Thomas
Hancock* UK *two years earlier.*

Lawn mower
Edwin Beard Budding, *textile
engineer, Gloucester,* UK, *in
partnership with* John Ferrabee.
'Country gentlemen may find in using

Budding's patent grass-cutting machine

my machine themselves an amusing, useful and healthy exercise.'
Net-weaving machine (variable pitch mesh)
Alexander Buchan, *Scottish engineer.*
Paraffin
Karl, Baron von Reichenbach, *German natural philosopher and industrialist of Blansko, Moravia.*
Phosphorous matches
Charles Sauria, *student at Dole College in the Jura Mountains, France. They were manufactured commercially by Frederick Kammerer in Germany, and known as 'Congreves'.*
Pig-iron (boiling)
Joseph Hall *of Tipton, Staffordshire,* UK.
Sewing machine
Barthélemy Thimmonier, *poor tailor of St Etienne, France, chiefly for the purpose of making army clothing. Although he took out patents in both the* UK *and* USA *in 1848, Thimmonier was unable to market his invention and died in poverty in 1857.*

Tubular boiler
James Napier, *marine engineer, Glasgow,* UK.
Vanadium
Nils Gabriel Sefström, *Swedish physician and chemistry teacher, School of Mines, Stockholm, Sweden.*

1831
Alizarine red
Robiquet *and* Colin, *as a crystalline body, the colouring principle of madder. It was synthesised by* Graebe *and* Liebermann *in 1869.*
'Bogie' truck (railroads)
John Jervis, *locomotive builder of New York,* USA.
Electric bell
Joseph Henry, *mathematics teacher at Albany, New York,* USA.
Electro-magnetic balance
Antoine César Becquerel, *Paris.*
Electro-magnetic induction
Michael Faraday FRS, *Fullerian Professor of Chemistry at the Royal Institution, London. He also discovered the principle of the*

transformer.

Magnetic North Pole
Commander James Clark Ross,
during Sir John Ross's second voyage.

Reaping machine
Cyrus McCormick, *22-year-old son
of Robert McCormick, farmer and
inventor of Virginia*, USA. *The
McCormicks sold their first reaper as
late as 1840 and by 1844 50 had been
sold. Soon after, they moved to
Chicago and set up a factory which by
1871 was turning out 10,000
harvesters a year.*

Steam-powered omnibus
Walter Hancock, UK. *Called 'The
Infant', it ran between London and
Stratford.*

Thermo-multiplier
Macedonio Melloni *and* Leopoldo
Nobili, *Italian physicists.*

Waistcoat-pocket pedometer
William Payne, *watch and
clockmaker, London.*

1832
Alcohol still
Aeneas Coffey, *distiller, the Dock
Distillery, Dublin.*

**Artist's watercolour cake
(glycerine-based)**
William Winsor *and* Henry Charles
Newton, *artists' colourmen of
Rathbone Place, London.*

Collapsible silk opera hat
Antoine Gibus, *Paris hatter.*

Hydraulic-powered factory
E Egberts *and* Timothy Bail, US.

Hydrostatic bed
Neil Arnott, *Scottish physician to the
Spanish Embassy, London.*

Kindergarten
Frederick Wilhelm August Froebel,
*former science teacher at Blankenburg,
Switzerland.*

Magneto-electric machine
Hippolyte Pixii, *Paris, and in 1833 by*
Joseph Paxton, *Cambridge,* UK.

**Motion-picture machine
(primitive)**
simultaneously by Dr Joseph Antoine

Ferdinand Plateau *(Phenakistoscope)
in Ghent, Belgium, and* Simon Ritter
von Stampfer *(Stroboscopic disc) in
Vienna, Austria.*

**Multiple-effect sugar
evaporation**
Rillieux.

**Passenger-tramway (horse-
drawn)**
New York to Harlem, USA.

**'Wheel of Life/Daedalum'
(optical toy)**
W H Horn *of Bristol,* UK. *Patented in
the* USA *in 1867 by* Milton Bradley.

1833
Book dust-jacket
Messrs Longman *of London for their
annual 'The Keepsake'.*

Creosote
Karl Baron von Reichenbach,
*German natural philosopher and
industrialist of Blansko, Moravia.*

Cycloidal paddle-wheel
Joshua Field, *civil engineer, Lambeth,
London.*

Diastase
Anselm Payen, *Professor of Industrial
Chemistry in Paris, and* J F Persoz,
*Professor of the School of Pharmacy at
Strasbourg.*

Differential calculating machine
Charles Babbage, *Lucasian Professor
of Mathematics at Cambridge,
working in London. Babbage began
work in 1821 on a machine that would
prevent errors occurring in
mathematical and astronomical tables.
During the next decade the government
subsidised him to the tune of £15,000
and would have continued, but in 1834
Babbage had yet another argument
with his skilled instrument-maker, who
resigned. The portion completed was
placed in the library of King's College,
London.*

Electric telegraph key
Carl Friedrich Gauss, *Professor of
Mathematics, and* Wilhelm Eduard
Weber, *Professor of Physics,
Göttingen University, Germany.*

Hansom's patent safety cab, 1834

Fluorescence in chlorophyll
Sir David Brewster, *Scottish chemist.*
Glacial fossil fishes
Jean Louis Rodolphe Agassiz, *Swiss naturalist.*
Kryptographic pen
Xavier Progrin *of Marseilles, France.*
Lock-stitch sewing machine
Walter Hunt, US.
Nervous reflex
Marshall Hall, *physician and physiologist, London.*
Nitro-cellulose
Théophile-Jules Pelouze *and* Henri Braconnot, *French chemists.*
Oakey glasspaper
John Oakey *of North London. A piano-maker's son, some 30 years before he had experimented with gluing powdered glass on to old ledger sheets. For his trademark, Oakey chose the profile of the then popular Duke of Wellington.*

1834
Adhesive postage stamp
James Chalmers, *bookseller and newspaper publisher, Dundee,* UK.
Chinese white (watercolour pigment)
Messrs Winsor & Newton, *artists' colourmen of Rathbone Place, London.*
Continuous electric light
(galvanic cells)
James Bowman Lindsay, *self-taught scientist, Dundee,* UK.
Free-standing kitchen range
Philo Penfield Stewart, *missionary and teacher in Ohio,* USA. *It was called the 'Oberlin stove'.*
Hansom cab (carriage)
Joseph Aloysius Hansom, *inventor and architect, as his Patent Safety Cab. Substantial improvements were made by* John Chapman *in 1836.*
Machine-made tiles
Joseph *and* Xavier Gilardoni *at their factory in Altkirch, Alsace, France.*
Nocturne (musical)
John Field, *Irish composer.*
Pantascopic spectacles
George Richard Elkington, *optician and gilt toy-maker of Birmingham,* UK. *They were half-eye spectacles with joint angles.*
Powered roller-conveyor
Mr Grant *of the British Navy's Deptford Victualling Department, for fully mechanised 'hardtack' biscuit-baking.*
Threshing/fanning machine
John A *and* Hiram Abial Pitts, US.
Tonometer
H Schreiber, *Crefeld, for tuning musical instruments. It incorporated 52 tuning forks.*

Valve gear (marine steam
engine)
Samuel Seaward, *London.*

1835

1835
Automatic revolving cylinder gun
Samuel Colt, *former merchant seaman and chemistry lecturer of Hartford, Connecticut,* USA. *In 1846, for the Mexican War, the* US *government ordered 1000 Colt 'revolvers' for army use.*

Colt's revolver

Tonic sol-fa
Sarah Glover, *musical educationalist of Norwich,* UK: 'doh, re, mi, fa, so, la ti, doh'.
Tunnelling shield (cast-iron)
Marc Isambard Brunel *for the completion of the abandoned Rotherhithe-Wapping tunnel under the River Thames,* UK. *It was completed by 1843.*
Wallpaper (cylinder-printed)
Bumstead, UK. *The machine was hand-cranked and could print 200 rolls of wallpaper per day. Introduced to France by Isidore Leroy.*
Whit classification (standardised nuts and bolts)
Joseph Whitworth, *engineer, Manchester,* UK.

1836
Acetylene
Edmund Davy, *Professor of Chemistry, Dublin, Ireland.*
Bullet (conoidal cup rifle)
Captain Claude Etienne Minié *and* Captain Delvigne *of Vincennes, France. The first long bullet to replace the round metal ball.*

Colour printing
George Baxter. *In some of his illustrations to his 'Pictorial Album', he employed twenty different blocks.*
Combine harvester
H Hoare *and* J Hascall *of Michigan,* USA. *Although it initially failed because of climatic conditions, later it was taken to California and used successfully in 1854*
Daniell Cell
John Frederick Daniell, *first Professor of Chemistry at the newly founded King's College, London. Daniell's non-polarising design gave a more constant flow of electric current.*
Filing/shaping machine
James Nasmyth, *Scottish engineer, at his Bridgewater foundry, Patricroft, near Manchester,* UK.
Horse-box (purpose-built)
Messrs Herring, *coachbuilders of Long Acre, London, as commissioned by Lord George Bentinck, race-horse owner.*
Long-haired-silk-spinning machine
Gibson *and* Campbell *of Glasgow,* UK.
Standard test for arsenic (forensic analysis)
James Marsh, *chemist, Royal Military Academy, Woolwich,* UK.
Steam ram (for naval warfare)
James Nasmyth, *Scottish engineer.*
Steam shovel
William Smith Otis, *to dig the railbed of the Western Railroad, Massachusetts,* USA. *Soon after, it was used on railway projects in Maryland and Canada.*
Stroboscope
Joseph Antoine Ferdinand Plateau, *Professor of Physics at Ghent.*
Yeast growth during fermentation
Baron Charles Cagniard de la Tour, *Parisian physicist.*

1837
Axial turbine
Fontaine-Baron, *France.*

Bolt-action rifle ('needle gun')
Johann von Dreyse *of Sommerada, Prussia.*
Braille
Louis Braille, *blind professor at the Institute for Blind Youth, Paris. The 43-symbol system was based on artillery captain Charles Barbier's dot-dash punching machine for battle-field communication by night.*
Closed diving dress
Augustus Siebe, *Saxon engineer working in London.*
Daguerreotype (photographic process)
Louis Jacques Mandé Daguerre, *former Parisian scene-painter, with Claude de St V M F Niépce. Exposure time was reduced to 20–30 minutes by using mercury vapour and fixing with a solution of common salt.*
Electric motor
Thomas Davenport *of Rutland, Vermont, USA, for drilling holes and turning hardboard. Three years later, this motor powered the first electric printing press.*
Electric telegraph
William Fothergill Cooke *and* Professor Charles Wheatstone *of King's College, London. In 1846 Cooke established the Electric Telegraph Company and after six years some 4000 miles of telegraph wire had been installed.*
Galvanised iron
William Henry Crauford, *London.*
Propeller (practical marine)
simultaneously by Francis Pettit Smith, *Middlesex farmer, with the three-masted schooner 'Archimedes',* and Jon Ericsson, *ex-Swedish Army officer with the 10-knot 'Francis B Ogden' and the 13-knot 'Robert F Stockton' – both in the UK.*
Shorthand system (phonetic)
Isaac Pitman, *aged 24, UK.*
Steam tram
New York and Harlem Railway, USA.
Steel plough
John Deere, *blacksmith and plough*

maker of Illinois, USA. The entire plough was steel except its braces, beams and handles.
Worcester sauce
John Lea *and* William Perrins, *pharmacists of Worcester, UK, to a recipe of* Sir Marcus Sandys.

1838
Brougham (carriage)
Named after Henry, 1st Baron Brougham and Vaux, *UK politician and barrister, and allegedly built to his own design, either by Robinson & Cook of Mount Street, London or Sharp & Bland of South Audley Street, London.*
Children's hosiery factory (steam-powered)
John Button, *UK emigrant at Germantown, USA.*
Cipher-writing machine
Antoine Dujardin, *Lille, France. It was called the 'Tachygraphe'.*
Equirotal phaeton (horse-drawn carriage)
W Bridges Adams, *sponsored by the Duke of Wellington.*
Fluorescence in fluorspar
David Brewster, *Scottish physicist.*
Morse Code
Samuel F B Morse, *art professor, and his student* Alfred Vail *at New York City University, USA. It was not until 1843 that the Morse Bill for the first telegraph line from Washington to Baltimore was passed in Congress by a very narrow margin (see overleaf).*
Paddleboat (electric)
Professor Moritz Hermann Jacobi, *German physicist and engineer at St Petersburg, Russia. It attained 2 mph along the River Neva.*
Photographic prints (on silver chloride)
William Henry Fox Talbot *of Lacock Abbey, Wiltshire, UK. Because the Talbotype or Calotype involved a positive-negative process allowing for more than one print, it had soon ousted the direct positive, unique-print*

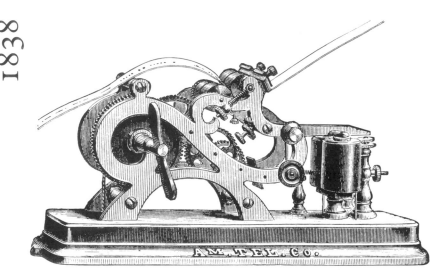

Morse telegraph recorder

Daguerreotype. *A similar system was evolved simultaneously by* Hippolyte Bayard, *civil servant, France.*

Plant cells
Matthias Jakob Schleiden, *Professor of Botany, University of Jena, Germany.*

Secondary electric currents
Joseph Henry, US.

Steam hammer
James Nasmyth, *Scottish engineer, for forging an enormous wrought-iron paddleshaft that was never used. In 1842 he found his steam hammer at work in Le Creusot, France. It had been adapted from his own scheme-book. Nasmyth patented his invention and it was adopted by the Admiralty in 1843.*

Stereoscope (3-D pictures)
Charles Wheatstone, *Professor of Experimental Physics, King's College, London.*

1839
Animal cells
Theodor Schwann, *German physicist and Professor of Anatomy at the Roman Catholic University, Louvain, France.*

Artificial fertiliser
John Bennet Lawes, *for the cultivation of turnips at Rothamsted, Harpenden, UK.*

Babbitt Metal
Isaac Babbitt, *inventor, Boston, Massachusetts, USA.*

Brick-making machine
Messrs Cooke & Cunningham, *whose machine could make 18,000 bricks in ten hours.*

Chenille weaving process
James Templeton, UK.

Electrotype
Professor Moritz-Hermann Jacobi, *physicist and engineer at St Petersburg, Russia.*

Gas cooker
At Leamington Spa, UK, *where a dinner was cooked for 100 guests.*

Harrow (iron)
William George Armstrong, *inventor and solicitor, Newcastle, UK.*

'Jet-propelled' ship
J *and* M V Rutheven. *First tests were held on the Scottish Union Canal, near Edinburgh.*

Lanthanum
Carl Gustav Mosander, *Professor of Chemistry and Mineralogy,*

Stockholm, Sweden.
Manganese steel (crude)
Thomas Heath.
Odontology
Professor Richard Owen, *Professor of Comparative Anatomy at the College of Surgeons, London.*
Pedal bicycle (prototype)
Kirkpatrick MacMillan, *blacksmith, of Courthill, Dumfriesshire, Scotland. A one-off, its furthest distance travelled being 40 miles.*
'Protoplasm' (the term or word)
Jan Evangelista Purkinje, *Professor of Physiology, Prague, Czechoslovakia.*
Silurian system
Roderick Impey Murchinson, *Scottish geologist.*
Tenor-bass trombone
C F Sattler, *Leipzig.*

1840
Astro-photography
Dr John William Draper, *New York,* USA. *He took a primitive daguerreotype of the moon. Three years later he took a daguerreotype of the solar spectrum.*
Atmospheric railway (pneumatic)
Samuel Clegg the Younger *and the* Samuda brothers, Joseph and Jacob. *A line of rail was laid across Wormwood Scrubs, London, between Shepherd's Bush and the Great Western Railroad, to test the efficiency of atmospheric tubes, air pump and the speed of carriages.*
Box-girder construction
Andrew Thomson, *for a road bridge across the Pollock and Govan Railway, Glasgow,* UK.
Chronoscope
Professor Charles Wheatstone – *to measure small intervals of time.*
Clarence carriage (horse-drawn)
Named after the Duke of Clarence.
Collapsible metal tube (squeeze-and-fold)
John Rand, US. *The following year,*

Messrs Winsor & Newton, *artists' colourmen of London, adapted the process for their pigments.*
Electroplating (silver-gold etc.)
John Wright. *Commercialised by the* Elkington *cousins,* George Richard and Henry, *cutlers of Birmingham, replacing Sheffield plate.*
Ozone
Christian Friedrich Schönbein, *Professor of Chemistry at Basle, whilst experimenting with William Grove's battery.*
Photographic lens (f/3.4)
Joseph Petzval, *Hungarian optician working for Voigtländer in Vienna.*
Pimm's Number One (drink)
James Pimm, *London restaurateur.*
Prefabricated glass and iron construction
Joseph Paxton, *gardener-architect, and* Decimus Burton. *They built the Royal Conservatory at the Duke of Devonshire's Chatsworth Estate,* UK. *By 1850 Paxton had developed his system for the Crystal Palace, London, using 300,000 panes of glass, erected in only eight months.*
Pre-paid postage at low and uniform rate
Rowland Hill, UK. *Called the 'Penny Post'.*
Silvered glass process (mirrors)
Justus von Liebig, *Professor of Chemistry at Giessen.*
Tonic water
Schweppes Co. *of England. Sugar and quinine were added to the soda water recipe developed half a century before by* Jakob Schweppe, *German amateur chemist working in London.*
Type-composing machine
James Young, UK, *and* Adrien Delcambre *of Lille, France. Called the 'Pianotype', it was used in 1841 to set a weekly magazine, 'The Phalanx', published in London.*
Xylophone (modern orchestral)
Zinc-coated ('galvanized') iron sheet.

1841

1841
Clod-breaking roller
William Croskill, UK.
Didymium
Carl Gustaf Mosander, *chemist, Stockholm.*
Disposable book
Christian Bernhard Tauchnitz *of Leipzig with the 'Collection of English Authors' series.*
Street-cleaning machine
Joseph Whitworth, *mechanical engineer. It was introduced by the Road and Street Cleaning Co. at Manchester, UK.*
Vulcanised rubber
Charles Goodyear, *former hardware merchant, Philadelphia, USA, after a decade of struggle, poverty and ridicule.*

1842
Ether anaesthesia (surgery)
Crawford Williamson Long *of Jefferson, Georgia, USA, physician, who first used it while operating on the neck tumour of Mr James Venable, but did not reveal his discovery until 1849.*
Friction match-making machine
Reuben Partridge.
Galvanic battery (carbon electrodes)
Robert Wilhelm Bunsen, *German physicist and Professor of Chemistry at Marburg.*
Hyposulphite of soda (photographic fixative)
John F W Herschel, UK *astronomer.*
Steam-powered planing machine
Joseph Whitworth, *engineer, in Manchester, UK.*
Submarine cable
Samuel Morse, US. *It was laid in New York harbour between the Battery and Governor's Island.*
Superphosphate (artificial fertiliser)
Sir John Bennet Lawes *at Rothamsted, UK. There were further*

experiments with J G Gilbert throughout the 1840s.
Torpedo (static mine)
Samuel Colt, US *inventor. It was exploded underwater in New York harbour by means of a galvanic battery and later blew up a 500-ton ship on the River Potomac, with Colt detonating the circuit from 5 miles' distance.*
Wagonette
Mr Lovell, *coachbuilder, of Amersham, UK, for Lord Curzon.*

1843
Aneroid barometer (spring-type)
Lucien Vidi, *France.*
Calculating machine
Pehr Georg Scheutz and his son Edward, *engineers, of Stockholm, following a study of the Babbage machine. Their machine was purchased by J F Rathbone, US merchant, to be presented to Dudley Observatory, Albany, New York.*
Christmas card
John Calcott Horsley RA, *on the suggestion of Henry Cole. 1000 were printed by Messrs Jobbins of Warwick Court, Holborn, London, and individually hand-coloured.*
Cigarettes (commercially-produced)
Manufacture Français des Tabacs, France.
Derringer (pocket pistol)
Henry Derringer, *small-arms manufacturer, Philadelphia, USA.*
Erbium
Carl Gustaf Mosander, *chemist and mineralogist, Stockholm.*
Gutta-percha
made known in Europe by Dr Montgomery *and Dr de Almeida, who brought it back from Malaya some 200 years after* John Tradescant, *English naturalist and gardener, had brought small specimens to England.*
Iron steamship (screw-propelled)
Isambard Kingdom Brunel. *His* SS

'Great Britain' made the crossing from Liverpool to New York in $14\frac{1}{2}$ days.

Joule
James Prescott Joule, *brewer and self-taught physicist, Manchester,* UK: *an energy unit to describe his first law of thermodynamics.*

Pantelegraph
Alexander Bain, *London.*

Photographic enlarger
Alexander Wolcott *and* John Johnson, US.

Quaternions (mathematics)
Sir William Rowan Hamilton, *mathematician and President of the Royal Irish Academy.*

Ridley-Bull wheat stripper
simultaneously and independently by John Ridley *and* John Wrathall Bull, *Southern Australia.*

Terbium
Carl Gustaf Mosander, *chemist and mineralogist, Stockholm.*

Textile mill (water-turbine-powered)
Uriah A Boyden, *New England,* USA.

Wood pulp (mechanically ground)
Friedrich G Keller, *weaver, of Haynich, Saxony. Keller sold his secret process to Heinrich Voelter for a pittance and the latter became rich by perfecting the process commercially.*

1844

Adhesive envelopes
to replace sealing wax.

Adjustable pitch propeller
Bennet Woodcroft, *consulting engineer and patent agent.*

'Aerial wheels' (pneumatic and bolted on)
Robert William Thompson, *civil engineer, of London. Of india-rubber or gutta-percha coated on to canvas and filled with air, they were hand-made, hence expensive and not easily replaceable.*

Boat-cloak
Lieutenant Peter Alexander Halkett, Royal Navy, *as a garment which could be turned into an inflatable boat. Halkett's many demonstrations of the boat-cloak culminated in the Bay of Biscay where he successfully paddled with a convertible umbrella-sail from* HMS *'Caledonia' to* HMS *'Vincent' of the Experimental Squadron.*

Cigarette-manufacturing machine
Le Maire, *France.*

Crèches
established in Paris, to receive the young children of working mothers, they came to London in 1863.

Napier's coin-weighing machine
for use in the Bank of England.

Nitrous oxide anaesthetic
Horace Wells, *dentist, and* Gardner Quincy Colton, *chemist, at Hartford, Connecticut,* USA.

Paper-cutting machine
Guillaume Massicot, *France.*

Ruthenium
Karl Karlovitch Klaus, *chemistry lecturer, Kazan University, Russia.*

Saxophone
Antoine Joseph Sax, *instrument-maker of Belgium. In 1857 Sax was appointed saxophone instructor at the Paris Conservatoire (see overleaf).*

Spinning rocket (angled exhaust outlets)
William Hale, UK.

1845

Ball-valve (reliable)
Edward Chrimes *of Rotherham, Yorkshire,* UK.

False teeth (individual porcelain/steel springs)
Claudius Ash, US.

Giant telescope
William Parsons, *3rd Earl of Rosse, astronomer, at Birr Castle, Parsonstown, Ireland. It had a 72-inch reflector and was 54 ft long, and he used it to discover the spiral nebulae.*

Horizontal turret lathe
Stephen Fitch, *Middlefield, Connecticut,* USA.

1845

Saxophone, 1844

Hydraulic crane
William George Armstrong, UK *inventor and solicitor.*

Mustard mill (steam-powered)
Jeremiah *and* James Colman, *Stoke Mill, Norwich,* UK.

Number-stamping machine
William Shaw, UK.

Panoramic camera
F von Martens, *France. It was particularly useful for city scenes and school photographs.*

Rotary printing press
Richard M Hoe *of New York,* USA. *It was first used in 1847 on the Philadelphia Public Ledger.*

Rubber bands (vulcanised rubber)
Stephen Perry *of Messrs Perry &*

Co., *rubber manufacturers, London.*
Steam pile-driver
James Nasmyth, *Scottish engineer.*
Typewriter ribbon feed system
Alexander Bain, *London.*

1846
Adams patent omnibus.
Computing telegraph
J E Fuller *of New York,* USA, *combining Palmer's circular slide rule with his own time telegraph.*
Cup anemometer
Thomas R Robinson, *Irish astronomer.*
Ether vapour (anaesthetic)
Dr William Thomas Green Morton *of Massachusetts General Hospital,* USA. *Having tried it on his pet spaniel, then on his friends, then on himself, then on a patient at the hospital, Morton patented his discovery as 'letheon'.*
Gun cotton
Professor Christian Friedrich Schönbein *of Basel, German chemist, and* Rudolph Christian Böttger – *working independently of each other.*
Hexameter machine (automated poetry producer)
exhibited at the Egyptian Hall, Piccadilly, London. 86 wheels of all shapes and sizes, visible through a glass window, churned out 1440 hexameter verses per day. It was a complete hoax.
Neptune (planet)
Johann Gottfied Galle *and* Heinrich Ludwig d'Arrest *at the Berlin Observatory.*
Parchment paper (vegetable based)
Figuier *and* Poumarède *of Paris.*
Protoplasm (full investigation)
Hugo von Möhl, *Professor of Botany at Tübingen, Germany.*
Sponge rubber
Charles Hancock, *London.*
Vertical casting of iron pipes
D Y Stewart, *Scottish engineer.*

1847

Carbolic acid
Auguste Laurent, *France, by distilling pit coal.*

Chloroform anaesthetic
Jacob Bell *of London and* Dr James Young Simpson, *Scottish obstetrician, Head of the Midwifery Department of Edinburgh University. Although at first opposed on medical and theological grounds, Simpson's method was finally given official approval in 1853 after he had administered it to Queen Victoria during the birth of Prince Leopold.*

Double-decker bus
Adams & Co. *of Bow, London, for the Economic Conveyance Co. with accommodation for fourteen upper-deck passengers, sitting back-to-back on the roof.*

Ether inhaler
Attlee & Co, UK.

'Fish joint' (for uniting the ends of rails)
William Bridges Adams, *London engineer.*

Flute (present-day type)
Theobald Boehm, *distinguished Munich flautist, and his friend, the physicist* K F E Schafhäutl.

Iris (asteroid)
John Russell Hind, *astronomer, Royal Observatory, Greenwich, London. The first of eleven such asteroids he discovered.*

Kymograph
Karl Friedrich Ludwig, *physiologist, Marburg, Germany. It traced pulse beats on a graph.*

Nitro-glycerine
Ascanio Sobrero, *chemistry student, Italy.*

Railroad bridge truss
Squire Whipple, US *engineer.*

Rocket plane (experimental)
Ernst Werner von Siemens, *Berlin.*

Saccharimeter
Soleil, *Parisian optician, for determining the amount of sugar in solutions – using rotary polarised light.*

Steel gun
Alfred Krupp *at his cast steel factory at Essen, Rhenish Prussia.*

Taxi-meter cab
Patent Mile Index *fitted to a London cab.*

.22 rifle
Nicholas Flobert, *Parisian arms manufacturer.*

Vulcanised toy balloons
J G Ingram, *London.*

1848

Alpaca as umbrella-covering material
William Sangster, UK.

Chewing gum (spruce-based)
John Curtis *on a Franklin stove in the kitchen of his home at Bangor, Maine,* USA. *He called his product 'State of Maine Pure Spruce Gum'.*

Concrete boats
Joseph Louis Lambot, *French horticulturalist, used the method for rowing skiffs. He also made flowerpots in the same way – plastering sand-cement mortar over a framework of iron bars and mesh.*

Corliss valve (steam engine)
George H Corliss, *self-taught engineer from Rhode Island, New York,* USA.

Department store
Alexander Turney Stewart, *Irish ex-schoolmaster on lower Broadway, New York,* USA. *Called the 'Marble Dry Goods Palace'.*

Folding propeller (marine)
John Seaward, *civil engineer, Millwall,* UK.

Milling machine
F W Howe *for the Robbins & Laurence Co. of Vermont,* USA.

Telekouphonon
Francis Whishaw, *civil engineer of Hampstead, London. A speaking telegraph consisting of piping of gutta-percha, caoutchouc, glass or earthenware with a terminal mouthpiece of ivory, bone, wood or metal, it was used for long-distance*

communication in dockyards and large establishments.

1849

Amines
Karl Adolf Würtz, *Professor of Organic Chemistry, Mineral Chemistry and Toxicology at the Paris School of Medicine.*

Artificial leather-cloth
Messrs J R *and* C P Crockett *of Newark,* USA. *Unbleached cotton was coated with a mixture of boiled linseed oil and turpentine, then coloured.*

Benzole (or benzene)
Charles B Mansfield, *student of A W von Hofman at the London College of Chemistry. Distilling it from coal tar, Mansfield was the first to produce benzene on a large scale. Tragically he died six years later from burns received during a further experiment on it.*

Bloomers (women's trousers)
Mrs Amelia Bloomer, *(née Jenks) of New York,* USA.

Bowler hat
Thomas *and* William Bowler, *felt-hat makers of Southwark Bridge Road, London, as commissioned by Lock & Co. of St James's, Piccadilly for their customer, William Coke of Holkham, Norfolk.*

Building (cast-iron construction)
James Bogardus *for his foundry, New York City,* USA.

Fire annihilator
T Phillips, *to put out a fire using steam and carbonic acid.*

Methyl gas
Simultaneously by Edward Frankland, *organic chemist, Royal College of Engineers,* UK, *and* Hermann Kolbe, *Germany.*

Milk (condensed or evaporated)
Gail Borden, *near New York,* USA. *A factory went into operation in 1851 and successful manufacture was established by 1856. Cecil Borden invented the meat biscuit in 1850.*

Polar clock
Professor Charles Wheatstone – *an optical apparatus whereby the hour of the day was found by the polarisation of light.*

Safety pin
Walter Hunt, US, *and* Charles Rowley, *Birmingham,* UK.

Shrapnel shell (with time-fuse and diaphragm)
Captain Edward Boxer, *British Royal Artillery, at the Royal Laboratory, Woolwich,* UK.

Telephone (experimental)
Antonio Meucci, *Italian engineer working in Havana, Cuba.*

Velocity of light
Armand-Hippolyte-Louis Fizeau, *French physicist.*

1850

Binaural stethoscope
G P Caniman, *New York,* USA.

Derricks
A D Bishop, *of New York,* USA.

Foucault's Pendulum
Jean Bernard Léon Foucault, *French physicist, to prove that the earth rotates.*

Grain binder
John E Heath, US. *The first in which twine was used.*

Jeans (trousers)
Levi Strauss, *Bavarian immigrant to San Francisco, as hard-wearing work trousers for Gold Rush miners.*

Machine-made watch (interchangeable parts)
Aaron Dennison *and* Edward Howard *of Roxbury, Massachusetts,* USA. *It was later called the 'Waltham'.*

Mustard powder (in tins)
J & J Colman, *Stoke Mill, Norwich,* UK.

Myographion
Hermann von Helmholtz, *Professor of Physiology at Königsberg, Germany: a machine for determining the velocity of the nervous current.*

Photographic slides
Frederick Langenheim,

Philadelphia, USA.
Refrigerator
simultaneously by James Harrison,
newspaper publisher, at Rodney Point,
Victoria, Australia, and Alexander
Catlin Twining at Cleveland, Ohio,
USA.
Stearine candles
Messrs Price, Battersea, London.
This was achieved by the steam-
distilled saponification of fats.
Submarine lamp
Siebe & Gorman, London.
Teledynamic transmitter
M Hirn. He assembled water-wheels,
endless wires and pulleys for using the
power of waterfalls at a distance.
Thirteen-month calendar
Auguste Comte, French philosopher
and socialist, in his 'System of Positive
Politics'. Each month was dedicated
successively to Moses, Homer,
Aristotle, Archimedes, Caesar, Paul,
Charlemagne, Dante, Gutenberg,
Shakespeare, Descartes, Frederic and
Bichat; an eminent person was
commemorated every day.
Tubular suspension bridge
Robert Stephenson and William
Fairbairn, engineers – 1 mile south of
the Menai Suspension Bridge, and
called 'The Britannia'.
Vortex turbine
James Thomson, Scottish engineer.

1851
Carpet loom
Erastus Brigham Bigelow, inventor,
after six years' work at Lowell,
Massachusetts, USA.
Carriage odometer
William Grayson, for measuring the
distance covered by carriages.
Collodion process
Frederick Scott Archer, portrait
sculptor, UK, had learnt calotype
photography in order to have portraits
of his sitters as studies. His improved
Ambrotype soon ousted the calotype.
Colour-printing machine
G C Leighton, using aquatinted plates

and electro-typed silver and copper
surfaces to obtain purity of colour.
Compound ureas
Karl Adolf Würtz, chemist, École de
Médecine, Paris.
Cross-channel telegraph cable
Professor Charles Wheatstone,
taking up Faraday's suggestion to
insulate with gutta-percha, and
assisted by Joseph Brett. It was laid by
the steamship 'Goliath'.
**Dairy ice-cream (mass-
produced)**
Jacob Fussell, dairyman of
Baltimore, Massachusetts, USA.
Doppler's Principle
Christian Doppler, Professor of
Physics at Vienna, Austria.
High-speed flash photography
Henry Fox Talbot of Laycock Abbey,
UK. 1/100,000-second exposure,
electric spark photo of a page of 'The
Times', attached to a spinning
cylinder.
Ophthalmoscope
Herman von Helmholtz, physician
and physicist, Germany. Initially the
machine directed a beam of light into
the eye by means of a mirror in which
there was a tiny lens aperture through
which an observer could study the
interior of the human eye.

Helmhotz's opthalmoscope

Printing machine
Thomas Nelson, *Edinburgh publisher, whereby paper was printed using curved printing plates affixed to revolving cylinders.*

Quadricycle (man-pedalled)
Willard Sawyer *of Deal, Kent,* UK.

Stereoscopic still camera (lenticular)
Louis Jules Duboscq, *France, as proposed by* Sir David Brewster, UK. *Binocular version was introduced two years later by* J B Dancer *of Manchester,* UK *and* A Quinet *of Paris.*

Vulcanite/ebonite (hard rubber)
Nelson Goodyear, US, *and* Charles Macintosh, UK.

Vulcanite dentures
in France.

Wrought iron manufacture (fuel-less pneumatic process)
William Kelly, *ironmaster, Kentucky,* USA.

1852

Airship (steam-powered)
Henri Giffard, *aeronautical engineer, Paris. The 144-ft dirigible, powered by a 3 hp steam engine, voyaged 17 miles at 6 mph.*

Compressed air caisson
Isambard Kingdom Brunel, *to assist construction of a bridge at Chepstow, Monmouthshire,* UK. *Fourteen cast-iron cylinders filled with concrete or brickwork, used to lay foundations under water.*

Cork lifebelt (multi-section)
Captain John Ross Ward RN *of the National Life Boat Institution. 800 units were supplied to life boat crews and remained standard issue from 1856 to 1904.*

Eau Grison (fungicide)
M Grison, *French gardener.*

External screw top (for bottles)
François Joseph Beltzung, *engineer, Paris, France.*

Fluorescence
Professor George Gabriel Stokes, *mathematician and physicist, Fellow of Pembroke College, Cambridge.*

Glider (man-carrying)
Sir George Cayley *of Brompton Hall, near Scarborough, Yorkshire,* UK. *The fixed-wing glider, piloted by Sir George's coachman, became airborne after being towed by manpower down a hill against a slight breeze. After touchdown, the terrified coachman is reported to have given in his notice, shouting 'Please sir – I was hired to drive, not to fly!'*

Goods elevator
Elisha Graves Otis, *master mechanic of a New York bedstead company.*

Lutetia (asteroid)
Hermann Goldschmidt, *German astronomer in Paris. This was the first of fourteen asteroids he discovered by poking small telescopes through his attic window.*

Microfilm
John Benjamin Dancer *of Manchester,* UK, *using collodion film.*

'Odic force'
Karl Baron von Reichenbach, *who discovered a new force intermediate between electricity, magnetism, heat and light and recognisable only by the nerves of sensitive persons.*

Rake (mechanical)
Samuel Johnston, US.

Rotary hook (sewing device)
Allen B Wilson, US, *financed by Nathaniel Wheeler.*

Ship (caloric-engined)
Jon Ericsson, *Swedish inventor working in the* USA.

Thesaurus of English words and phrases
Peter Mark Roget, *scholar and physician, London.*

Thetis (asteroid)
R Luther, *astronomer; the first of twenty such asteroids he discovered.*

Water ejector pump
James Thomson, *Scottish engineer.*

1853

Chuck lathe
E K Root *at the Colt Armory,* USA.

Cigarette facory (steam-powered)
Don Luis Susini *in Havana, Cuba.*
Capacity rose to 2,583,000 per month.
Cutting-out machines
Frederick Osbourn, *and manufactured by Messrs Hyams for cutting out clothes.*
Dark-field microscope condenser
Francis H Wenham, UK.
Duplex telegraphy
Dr Wilhelm Julius Gintl, *Austria.*
Gas engine (internal combustion)
Eugenio Barsanti *and* Felice Matteucci *of Florence, Italy. The first working unit was installed at the Maria Antonia Railway Station, Florence, three years later.*
Gaslit cystoscope (medical).
Hypodermic syringe (silver construction)
Charles Gabriel Pravaz, *French surgeon of Lyons, France.*
Hypodermic syringe
Alexander Wood, *physician and lecturer, Edinburgh University,* UK, *to administer morphia.*
Multiplex telegraphy
Moses Gerrish Farmer, *of Salem, Massachusetts,* USA.
Panopticon of science and art
T H Lewis, *architect, in Leicester Square, London – for lectures and musical performances. It included a very large electrical machine and battery. The speculation failed and the business was sold off in 1857.*
Railway carriage (corridor approach)
Messrs Eaton & Gilbert *of Troy, New York for the Hudson River Railroad.*
Rocking chair (revolving with adjustable seat angle)
Peter Ten Eyck, *New York,* USA.
Saltaire (model village)
Sir Titus Salt, UK *alpaca wool manufacturer and benefactor. He built the village for his employees, around his factories beside the Aire River.*
Steam-powered fire engine (self-propelled)
A B Latta *of Cincinnati,* USA.
Trotman's anchor
John Trotman *of Dursley, Gloucestershire,* UK.

1854
Air filters and respirators (charcoal-based)
Dr John Stenhouse, *chemical lecturer, St Bartholomew's Hospital, London. First used at the Mansion House, London.*
Aluminium (chemically produced)
Henri Sainte Claire Deville, *France, based on the pioneer work of Davy, Oersted and Wöhler. The first aluminium factory went into operation at Glacière in 1856.*
Annular-sail wind pump
Daniel Halliday.
Boolean algebra
George Boole, *Professor of Mathematics, Queen's College, Cork, Ireland.*
Cartes de visite (photographic)
Disdéri *of Paris, who succeeded in taking eight photos on one negative. They were popularised in 1857 when the Duke of Parma had his portrait, as taken by Ferrier in Nice, France, placed on his visiting cards.*
Chemical wood pulp
Watt *and* Burgess, *who used caustic soda to digest wood pulp directly under pressure to make paper.*
Compound marine engine
John Elder & Co., *marine engineers, Clyde,* UK – *for SS 'Brandon'.*
Cornflour
John Polson, *later of Brown & Polson Ltd,* UK.
Electric loom
Cavaliere Gaëtano Bonelli *of Turin – employing magnets and electro-magnets in weaving, attempting to supersede the costly Jacquard system of cards.*

Electro-magnetic manufacture
Breguet, *France – of mathematical instruments.*

Four-motion feed (sewing device)
Allen B Wilson, US.

Giant train shed
New Street Station, *Birmingham,* UK.
The span of its iron roof was 211 ft.

Letter-printing telegraph
David Edward Hughes, *London-born Kentucky music professor.*

Palaeozoic fossils
Reverend Adam Sedgwick, UK *geologist.*

Paraffin lamp
John H *and* George W Austen, *New York,* USA.

Perforated postage stamps
UK.

Planimeter (constant length tracing arm)
Amsler. *12,000 such planimeters had been manufactured by 1884.*

Psychic pressometer
Dr Robert Hare, *Professor of Chemistry at Pennsylvania University,* USA, *to expose fraudulent 'table-tapping' mediums. Several mediums 'forced' the machine to register the equivalent of 18–40 lb, ultimately collapsing the apparatus. When a convinced Hare attempted to present his evidence to the American Association for the Advancement of Science, he was ridiculed as senile.*

Telephone (experimental)
Charles Bourseul, *French telegraphy engineer.*

1855

Armoured vehicle (steam-powered)
Cowan, UK.

Bichromate battery
Heinrich Ruhmkorff, *German physicist.*

Bunsen burner
Robert Wilhelm Eberhard von Bunsen, *Professor of Chemistry at Heidelberg. In that year Heidelberg came to be lit by gas. Bunsen researched a controllable-flame burner to use in his laboratory, the former refectory of an ancient monastery.*

Cased-in water turbine
James Bicheno Francis, UK *émigré engineer to the* USA.

Chronograph (stopwatch)
Edward Daniel Johnson, *watch manufacturer of London.*

Cupola (turret) ship
Captain Cowper Coles, *British naval officer. Several such ships were constructed by E J Reed, including 'Sovereign', 'Monarch' (armour-clad) and 'Captain', which sank in a storm in the Bay of Biscay with the loss of 472 lives, including the inventor.*

Dry-cleaning (clothes)
Jean Baptiste Jolly, *French dyeworks owner in Paris, France. His maid accidentally upset a paraffin lamp on a tablecloth and Jolly noted that the area soaked became cleaner. Operating from his dyeworks, he offered the new process as 'dry-cleaning' as opposed to the soap and water process.*

Fire-fighting vessel (steam-powered)
James Braidwood, *Superintendent of the London Fire Engine Establishment, based on an idea by* John Braithwaite, *engineer,* UK.

Glider (sea-going)
Jean Marie Le Bris, *French sea captain, who made a glide of 200 metres near Douarnenez in a boat-shaped machine, patterned after the albatross.*

Locomotive speedometer
A W Forde.

Pedal-operated machine kettledrum
M Gautrot, *Parisian instrument-maker.*

Portable prospector (hand, horse or steam power)
Thomas Oultram *and* Jacob Braché, *engineers, Melbourne, Victoria, Australia. Australian Patent Number 2.*

Powdered milk (industrial manufacture)
Grünwald, Germany.
Safety match
Johan Edvard Lundström of Jönköping, Sweden, using red phosphorous for the striking surface as suggested ten years before by Gustave Pasch of Sweden and Professor R Bottger of Frankfurt-on-Main, Germany.
'Scaphandre' (diving apparatus)
Joseph Martin Cabirol of Narbonne, France.
Smith & Wesson revolver
Daniel Baird Wesson at Springfield, Massachusetts, USA.
Spinal anaesthesia
J L Corning, US, who used a solution of cocaine to inject into nerve endings.
Timber-bending machine
T Blanchard of Boston, Massachusetts.
'Very Powerful Electric Machine' (Leyden-jar-powered)
Dr Henry Minchin Noad at the Panopticon, Leicester Square, London.
Z-crank steam engine
Morton & Hunt, Glasgow, UK.

1856
Aerated bread
Dr John Dauglish, Edinburgh. Marketed by the ABC (Aerated Bread Co.) at their London tea-rooms in 1861.
Aniline dye
William Henry Perkin, aged eighteen, while studying at the London College of Chemistry under the German chemist, A W von Hofman. The following year, with capital from his father and brother, Perkin set up a chemical works to turn out 'aniline purple', later called 'mauve'.
Bessemer converter
Henry Bessemer, UK engineer, who set up his own steelworks at Sheffield to promote the process.

Clockwork toy train
George Brown, US clockmaker.
Equatorial sextant
William Austin Burt, surveyor, Mount Vernon, Michigan, USA.
Fuchsine dye
Wladyslaw Natanson, Poland.
Glycogen
Claude Bernard, French physiologist.
Glycol
Charles-Adolphe Würtz, Dean of the École de Médecine, Paris.
Parkesine
Alexander Parkes, chemist of Birmingham, UK – the prototype celluloid. The Parkesine Co. Ltd was formed in 1866, but lasted only two years.
Patent safety railway points and signals (interlocking)
John Saxby, railway engineer of Brighton, Sussex, UK. In 1862 he went into partnership with John Stinson Farmer to manufacture from a factory in Kilburn, London.
Practical gas fire
Pettit and Smith, UK.
Sarrusphone
Sarrus, French military-bandmaster, together with P L Gautrot, Parisian instrument-maker – 'a brass oboe'.

1857
Compressed-air drilling (tunnels)
for the Mount Cenis Tunnel, Switzerland.
Concrete mixer
for construction of a bridge over the River Tisza at Szeged, Hungary.
Domestic sewing machine (production model)
Isaac Merritt Singer of Pittsdown, New York, USA. It was based on his treadle-operated lock-stitch machine of 1851.
Lamplit street (electric power)
Messrs Lacassagne et Thiers in the rue Impériale, Lyons, France.
Laryngoscope
Johann Nepomuk Czermak, Polish

physiologist working in Vienna.

Passenger elevator
Otis Steam Elevator Co. *in E V Haughwout & Co., a five-storey china shop in Broadway, New York,* USA. *It raised six people at 40 ft per minute.*

Phonautograph
Léon Scott de Martinville, *French painter – to analyse sound waves.*

Photoheliograph
simultaneously by Sir John Herschel *at Kew Laboratory, London, and* H van der Weyde, US *artist.*

Silicon nitride
independently by Henri Etienne Sainte-Claire Deville, *French chemist, and* Friedrich Wöhler, *German chemist.*

Steam-powered franking machine.

Vibration microscope
Jules Antoine Lissajous, *Professor of Physics at the College St Louis, Paris. This showed visually the 'Lissajous figures'.*

Wood pulp (alkaline process)
Houghton *at Cone Mills, Lydney, Gloucester,* UK.

1858

Burglar alarm
Edwin T Holmes, US. *Installed in Boston, Massachusetts.*

'Cell-from-cell' theory
Rudolf Virchow, *Professor of Pathological Anatomy at Berlin.*

Donati's Comet
Dr Giovanni Battista Donati *of Florence, Italy.*

Gas-lit railway carriages
Eastern Railway Co., *France.*

Harvester
Charles Wesley Marsh, *and his brother* William, *Illinois,* USA.

Lighthouse (magneto-electric)
Professor Holmes. *South Foreland lighthouse, Dover,* UK.

Mechanised washtub and dolly
Hamilton Smith, *Pittsburg,* USA.

Möbius strip
Professor August Möbius, *whilst*

investigating polyhedra. Not published until 1865.

Pencil with attached eraser
Hyman Lipman *of Philadelphia,* USA.

Pistolgraph (still camera)
Thomas Skaife. *Its 'snap-shooting' inventor was nearly arrested for aiming it at Queen Victoria.*

Quadrivalent nature of carbon
Archibald C Couper, *Scottish chemist.*

Seismometer
Robert Mallet, *civil engineer, Dublin, Ireland.*

Shoe-sole-sewing machine
Lyman R Blake, *Abington, Massachusetts,* USA. *It was perfected soon after with financial assistance from Gordon McKay.*

Steam plough
John Fowler, *agricultural engineer,* and Jeremiah Head, UK.

True atomic and molecular weights
Stanislao Cannizzaro, *Professor of Chemistry in Genoa, Italy.*

1859

Accumulator/storage battery
Gaston Planté, *France, using lead-acid. Based on* Johann Wilhelm Ritter's *discovery of the secondary cell, as early as 1803.*

Aerial screw propeller
J B Lassie, *whose model for an aerial ship was submitted to the Paris Academy of Sciences, although the full-scale craft was never built.*

Aniline yellow dye
Peter Greiss, *a German working in a brewery in Burton-on-Trent,* UK.

Binoculars (central focus dial)
A A Boulanger, *France.*

Breech-loading gun (converted)
Jacob Snider, *wine merchant, Philadelphia,* USA. *His invention was finally adopted by the government in 1866 and an order placed for 100,000 breechloaders.*

Cathode rays
Julius Plücker, *mathematical*

1858

Steam organ

physicist at Bonn, Germany.

Horizontal-arch reservoir dam
Zola, *French engineer, at Aix-en-Provence, based on recommendations by F Delocre.*

Merchant ship (bessemer steel construction)

Miniature camera
T Morris *of Birmingham,* UK.

Oil well
'Colonel' Edwin L Drake *at Titusville, Pennsylvania,* USA, *whilst working for George Bissel, whose small company had been marketing lamp oil made from crude oil obtained from surface seepages. Success at Titusville was achieved when the hole partly filled with oil at a depth of 69½ ft.*

'On the origin of the species'
Charles Robert Darwin, UK *naturalist, following 25 years' research.*

Pneumatic dispatch
for conveying letters and parcels through tubes by means of atmospheric pressure and a vacuum.

Primitive fluorescent lamp
Edmond Becquerel, *France, who placed fluorescent materials inside a Geissler discharge tube.*

Pullman cars (railway)
George Mortimer Pullman, *Chicago, Illinois,* USA. *He made his first sleeping cars using two old day coaches divided into ten compartments. In 1864 he patented the folding upper berth. In 1867 the Pullman Palace Car Company was formed; in 1868 dining cars were introduced.*

Spectroscope
Gustav Robert Kirchoff *and* Robert Wilhelm Bunsen, *German physicists working in Heidelberg.*

Steam organ (calliope)
Arthur Dennis, US. *Powered by a 30 hp engine, developing a steam pressure of 125 lb per square inch, kalliopes were installed at St Louis and New Orleans.*

Steam road-roller
Louis Lemoine, *France.*

Wide-angle lens (camera)
Thomas Sutton, UK *editor of 'Photographic Notes'*.

1860
Caesium
Robert Wilhelm Bunsen *and* Gustav Robert Kirchoff, *using spectrum analysis*.
Debusscope (pattern-making device for calico-printers)
France.
Linoleum
Frederick Walton, *who built his first factory at Staines, Middlesex*, UK. *It was patented by a rubber manufacturer to replace Kamptulicon, an oil-rubber product.*
Press-stud fastener
John Newnham, UK.
Solar boiler (for distillation)
August Mouchot, *France*.
Synthetic ruby
Edmond Frémy, *French chemist*.
Winchester repeating rifle
B Tyler Henry, *chief designer for Oliver Fisher Winchester*, US *gun and ammunition manufacturer*.

1861
Enamel-mosaics
Dr Antonio Salviati, *while reviving the glass industry of Murano, Venice*.
Esparto pulp
Thomas Routledge *of Eynsham Paper Mills, Oxford*, UK, *from esparto grass grown in the Mediterranean lands. The innovation proved so successful that in 1862, 1000 tons of esparto grass were imported into the UK.*
Harpoon log
Thomas Walker, UK.
Hydrofoil boat
Thomas Moy, UK – *on the Surrey Canal between Rotherhithe and Camberwell*.
Kinematoscope (stereoscopic photos)
Coleman Sellers *of Philadelphia*, USA.

Mechanical road transport (army use)
R E Crompton, *whilst at Harrow School*, UK. *His 'Blue Belle' was shipped out to India in 1867, where Crompton was serving with the Rifle Brigade.*
Motor speech centre of the brain
Pierre Paul Broca, *surgeon to the Paris hospitals*.
Photography (underwater)
William Thompson *of Weymouth*, UK. *His initial experiment was made 5 years previously, by lowering a watertight box from a boat in Weymouth Bay.*
Pneumatic drill
Germain Sommelier, *chief engineer on the Mont Cenis railway tunnel project*.
Postcard
John P Charlton, *Philadelphia*, USA.
Rubidium
Robert Wilhelm Bunsen *and* Gustav Robert Kirchoff, *Germany, using spectrum analysis*.
Telephone (partially articulate)
Philipp Reis, *physics teacher at a private school near Frankfurt-am-Main, Germany, for the enlightenment of his pupils. Although Reis lectured on and demonstrated his machine publicly, he was unable to realise its full potential, dying aged only 40 after a long illness which eventually robbed him of his voice.*
Thallium
William Crookes, *chemistry lecturer and editor of 'Chemical News', London*.
Transatlantic oil cargo.
Universal milling machine
Joseph R Brown, US, *to manufacture musketry parts for the* US *Civil War*.

1862
Auriscope
John Brunton.
Bread-making machine
Ebenezer Stevens *of Cheapside, London*.

Ironclad ship
Jon Ericsson's *'Monitor'*, *complete
with a gun turret, as invented by* T R
Timby, US, *some twenty years before.
That spring, 'Monitor' engaged with
the Confederate ironclad 'Merrimac'
at close quarters. Each hit the other
with shells over twenty times, and they
withdrew in stalemate.*
Machine-gun
Richard Jordan Gatling *of North
Carolina*, USA. *His manually rotated
ten barrels fired bullets, gravity-fed
from a drum at the rate of 350 rounds
per minute.*

Cartridges for Gatling's machine gun

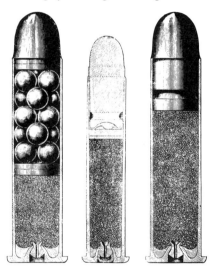

Milling machine
Joseph Brown *of Brown & Sharpe
Co., Providence, Rhode Island*, USA.
Phonoscope
Dr Rudolph Koenig *of Paris, to test
the quality of musical strings.*
'Reciprocal electrophorus'
Cromwell Fleetwood Varley,
electrical engineer, UK.
Rubber stamps
John Leighton FSA, *but patented two
years later in the name of Alfred
Leighton. They were made of
vulcanised india-rubber.*

**Steel-hulled merchant ship
(transatlantic)**
*paddle-steamer, 'Banshee', 325 tons,
built in Liverpool as a blockade-runner
for the* US *Civil War.*
Washing machine
*100 garments were washed in a few
minutes by steam and dried by a
centrifugal machine at a New York
hotel.*

1863
Anti-cyclone (meteorology)
Francis Galton, *scientist and explorer*,
UK.
**Dental drill (clockwork-
powered)**
George Fellows Harrington,
*subsequently manufactured at his own
factory in Lind Street, Ryde, Isle of
Wight*, UK.
Harpoon gun (practical)
Sven Foyn *of Tönsberg, Norway. It
was fitted aboard the whaler 'Spes et
Fides' in 1864.*
Indium
Ferdinand Reich, *German physicist,
and* Hieronymous Theodor Richter,
German chemist.
Medical photography
Dr Henry Goode Wright, *London.*
Motor carriage (coal gas)
Etienne Lenoir, *France, who fitted
one of his 1½ hp gas engines to a
carriage and used to take 1½ hours to
drive 6 miles from his home to his
works.*
Multi-plate clutch (locomotive)
Thomas Aldridge Weston,
mechanical engineer, Birmingham, UK.
Pepper's ghost
John Henry Pepper, *analytical
chemist at the Royal Polytechnic
Institution of London. Pepper adapted
his invention from a design by* Henry
Dircks, UK *civil engineer, who used
mirrors and lenses to 'project' a ghost.*
Phonograph
Fenby. *A machine to be attached to
pianos, etc. by which any music played
was written down on blank paper, since*

1863

it ruled and printed the notes simultaneously.

Regenerative smelting oven (Siemens-Martin process)
initiated in 1856 by William *and* Friedrich Siemens, *brothers, for glass and steel manufacture, and perfected by* Emile *and* Pierre Martin, *father and son, in Sireuil, France.*

Roller-skates (four-wheeled)
J L Plimpton, *New York,* USA.

Sodium carbonate production process
Ernest Solvay, *Belgian chemist.*

Sphygmograph
Jules-Etienne Marey *of Paris, for investigating disease by showing the state of the pulse.*

Submarine (explosively charged)
Horace L Hunley's *'David', propelled by an eight-man, hand-cranking crew. It was destroyed while blowing up a Federal vessel in the* US *Civil War.*

TNT (explosive)
J Wilbrand, *Germany.*

Underground steam railway
a broad gauge line from Victoria Street to the Great Western Railway at Paddington, London. The carriages and stations were gas-lit. The exhaust from the steam engines was returned to the tanks by big pipes from the cylinders, based on the idea of George Pearson, *solicitor.*

1864

Anthropoglossus
an alleged automaton talking machine was exhibited at St James's Hall, London, but proved to be a complete hoax.

Driving chains
James Slater, *owner of a small textile-machine factory at Salford,* UK. *In 1880,* Hans Renold, *a Swiss who had acquired Slater's factory, patented the bush-roller chain to give a much greater load-bearing surface.*

Electromagnetic wave transmission
Dr Mahlon Loomis, *who succeeded in transmitting and receiving electromagnetic waves without wires over 14 miles at Loudon County, Virginia,* USA.

Narrowboat (steam-powered)
Grand Union Canal Company's 'Dart', UK.

Nitro-glycerine explosive
Alfred Nobel, *Swedish inventor and manufacturer in St Petersburg.*

Open hearth furnace (pig and scrap melting)
Emile *and* Pierre Martin, *father and son, in Sireuil, France.*

Pasteurisation
Louis Pasteur, *Dean of the Faculty of Sciences, Lille, France, the centre of the brewing industry.*

Louis Pasteur

Photography (using magnesium flash)
Alfred Brothers *of Manchester,* UK.

Smokeless powder
J F E Schultz, *Germany – very effective for hand grenades and shotguns.*

Solar-powered steam engine
August Mouchot, *French engineer.*
This was the prototype for a series he
built during the next fifteen years.

1865

Aérophore (compressed-air diving apparatus)
Benoît Rouquayrol, *mining engineer,*
and Auguste Denayrouze, *French naval officer.*

Automatic turret lathe
Christopher M Spencer, US.

Canoe (European pleasure)
John MacGregor, *who persuaded*
Searle's of Lambeth, London, to build
him his first Rob Roy canoe, influenced
by his seeing the canoes of North
America and the Kamschatka.

Cyclic structure of benzene
August Friedrich Kekulé, *German*
Professor of Chemistry at Ghent.

'Duplex burner' (oil lamp)
James *and* Joseph Hinks,
Birmingham, UK.

Electric arc welding
Wilde.

Ferryboat (air-cushioned)
running between New York and New
Jersey, USA. *While the vessels went*
with less power applied to the paddles,
the power required to provide
compressed air along channels in the
keel was just about equal to what was
saved in driving the paddles – so the
service was discontinued.

Genetics
Gregor Johann Mendel, *Austrian*
biologist and Augustinian Abbot of
Brun, who combined his love of
mathematics with his knowledge of
plants.

Motorboat (gas-engined)
Jean Joseph Etienne Lenoir, *France.*

Oil pipeline (commercial)
Samuel van Syckel, *pioneer refiner –*
about 6 miles, from a flush oil field at
Pithole to a railroad loading point at
Miller Farm, Pennsylvania, USA.

Pedal bicycle (front-wheel-drive

production model)
Pierre Michaux, *Parisian coach and*
perambulator manufacturer, and his
son, Ernest. *Over 400 velocipedes sold*
that year.

Reinforced concrete
W B Wilkinson *of Newcastle-upon-*
Tyne, UK *who built a house in which*
the concrete beams were reinforced
with wire colliery rope.

Spinnaker yachting sail
William Gordon, *sailmaker of*
Southampton, UK, *who first used a*
prototype in his yacht 'Niobe' in a
match race off Cowes, Isle of Wight.

Steam road-roller
Thomas Aveling, *former farmer and*
agricultural mechanic from Kent, UK.
For two years Aveling's machine was
attacked and banned. In 1867 the City
of Liverpool ordered the first steam-
roller.

Tinplated food cans.

Uninflammable gunpowder
Mr Gale, *a blind gentleman from*
Plymouth, Devon, UK, *by combining*
gunpowder with finely powdered glass
which could be readily separated by a
sieve when the powder was to be used.
Gale's Protected Gunpowder
Company was formed in 1865 and
wound up in 1867.

Web-offset printing machine
William Bullock, *Pittsburg,*
Pennsylvania, USA, *who lost his life*
soon after in an accident with one of his
presses.

Western ('10 gallon') hat
John B Stetson, US.

Yale cylinder lock
Linus Yale Jr, US – *enabling the mass-*
production of locks to become a reality.
Sadly Yale died, aged only 47, from
heart failure in 1868, just before his
Lock Manufactory went into
operation (see overleaf).

1866

Aerial cablecar
Ritter, *so that people could inspect*
turbine installations on the Rhine,

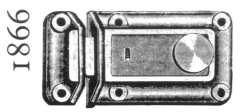

Yale Pattern night lock, 1865

Germany. It had a track of 101 metres, four steel-carrying cables and conveyed two people in one cabin. Its hauling rope was hand-cranked.

Atlantic telegraph cable
Cyrus West Field, US, *with* Samuel Canning *and* Daniel Gooch, UK. *Over 3000 miles of cable were laid by the 'Great Eastern' from Valentia, Ireland to Heart's Content, Newfoundland. It was said that the engineer of the cable passed signals through it by means of a battery formed in a lady's thimble.*

Chassepot rifle
Alphonse Chassepot, *France. By 1867, 10,000 chassepots had been issued to French troops.*

Cigar-shaped steamship
William Louis *and* Thomas Winans, *formerly of Baltimore, USA, but resident in London; built by Hepworth and launched on the Thames, UK.*

Depressor nerve
Karl Friedrich Wilhelm Ludwig, *Professor of Physiology, Leipzig, with* Élie de Cyon.

Electro-magnetic road vehicle
E Poitevin.

Fire-extinguisher (carbon dioxide gas)
François Carlier, *French doctor – by mixing sodium bicarbonate with sulphuric acid.*

Ironclad gunboat (waterjet-propelled)
HMS *'Waterwitch'*, British Admiralty *– propelled by the hydraulic reaction arrangement introduced by J Ruthven.*

Key-opening canned food
J Ousterhoudt, *New York, USA.*

Self-acting dynamo
Werner von Siemens, *Germany.*

Torpedo (self-propelling)
Robert Whitehead, UK *engineer with a works at Fiume, near Trieste, Italy, in conjunction with Captain Giovanni Luppis of the Austrian Navy. Originally carried 19 lb of guncotton, measured 11 ft long and travelled at 7 knots.*

Visible speech (lip-reading)
Alexander Melville Bell, *Scottish-American educationalist, working in London.*

Wood pulp (prototype sulphite process)
B C Tilghmann *and* R Tilghmann, *brothers at the mills of W W Harding & Sons, Manayunk, near Philadelphia, USA. Unable to overcome engineering problems associated with their process, the Tilghmanns never promoted the idea commercially.*

1867

Amphibious velocipede (lever-operated)
William George Crossley, *dentist of Cambridge, UK.*

Automatic electric block railway
Thomas S Hall *on the New York and Harlem Railroad, USA.*

Automatic printing telegraph
Professor Charles Wheatstone, UK

Barbed wire
Lucien B Smith *of Kent, Ohio, USA.*

'Bismarck' brown' (dye)
Carl Friedrich Philipp von Martius, *German naturalist.*

Double-decker railcars
Bidel, *France, for the Paris ring-line.*

Dynamite
Alfred Nobel, *inventor and manufacturer at Ammeberg, Sweden.*

Formaldehyde
August Wilhelm von Hofmann, *London-based German chemist.*

Mechanical 'writing ball'
Pastor Rasmus Hans Johan Malling Hansen, *Head of the Royal Deaf and Dumb Institute, Copenhagen, Denmark.*

Ring armature (generator)
Zénobe T Gramme, *Belgian scientist working in Paris.*
Time-switch
Dr Walter Christopher Thurger *of Norwich,* UK *– an eight-day clockwork 'gas controller'.*
'White coal' (hydro-electric power)
Aristide Bergès, *paper-maker, for his factory at Lancey, Isère, France.*

1868
'Air-hardening' tool steel
Robert Mushnet, UK.
Asbestos (industrial use)
200 tons of raw material were worked from an Italian mine, and it was discovered in Quebec, Canada. This heat-resistant, fibrous mineral was known and named by the ancient Greeks.
Boneshaker bicycle (steam-powered)
Sylvester Howard Roper *of Roxbury, Massachusetts,* USA, *and independently by* Pierre Michaux *and* L G Perreaux *of Paris.*
Compressed-air brake (railways)
George Westinghouse, US.

Bicycle tuition

Cryptograph
Sir Charles Wheatstone, UK *– an apparatus for writing in cipher.*
Elevated street railway
Charles T Harvey, *civil engineer, New York,* USA. *It was called the 'El'.*
Gas geyser (mechanical water heater)
Benjamin Waddy Maughan, *London decorator (see overleaf).*
Granny Smith apple
Miss Smith, *Australia.*
Helium (in sun's chromosphere)
Edward Frankland, *Professor of Chemistry, Royal Institute, London, with* Joseph Norman Lockyer, *astronomer.*
Leclanché battery
Georges Leclanché, *France, as a dry-cell non-rechargeable primary battery.*
Margarine
Hippolyte Megé-Mouriés, *at the Ferme Impériale de la Faisanderie, Vincennes, France. His was the only entry in prize competition organised by Napoleon III for 'a suitable substance to replace butter for the Navy and the less prosperous classes'.*
Mechanical sheep-shearing machine (prototype)
James Andrew Higham, *compositor, Melbourne, Australia.*
Naval searchlight
Commander Philip Howard Colomb. *It was installed on* 'HMS Minotaur', *for use in 'Colomb's flashing signals'.*
Nephoscope
Karl Braun *– to measure the velocity of clouds.*
Plywood systems
John K Mayo, *New York,* USA. *Mayo's various patents show applications for bridges, canal locks, sewer's, tables and doors. He even suggested wet and dry heating to make the veneers more pliable before their compression.*
Pocket calculating machine
Charles Henry Webb, US, *'Webb's Adder'.*

Maughan's gas geyser, 1868

Shellac mouldings
although recorded in very ancient writings, the first patents were taken out during this year.

Stapler
Charles Henry Gould *of Birmingham*, UK.

Steam man
New York – a figure constructed to drag a phaeton.

Street crossing signals
J P Knight, *railway signalling engineer, outside the Houses of Parliament, Westminster, London. The semaphore arms had red and green gas lamps for night-time. It was dismantled after it exploded, killing a policeman.*

1869
Aero-steam engine
George Warsop, *Nottingham*

mechanic, UK. *Compressed air was united with steam, saving 47 per cent fuel. A tug-steamer built for China was fitted with an aero-steam engine in 1870.*

Celluloid
John Wesley Hyatt *and his brother*, Isaiah Smith Hyatt, US. *Initially called collodion and used for coating billiard balls etc., it was very inflammable. It was also used for denture plates and toys – 'caveat emptor'!*

Chain-driven velocipede
Guilmet, *France*.

Colour photography (subtractive method)
simultaneously by Charles Cros *and* Louis Ducos du Hauron, *both of whom based their system on the Young-Helmholtz colour separation and mixing theory by superimposing three colour positive pictures on one another.*

Eugenics
Francis Galton, *scientist*, UK.

Gun (compressed steel)
Sir Joseph Whitworth, *industrialist and mechanical engineer, Manchester*, UK. *Its spiral polygonal bore gave it a considerably increased range.*

Hypodermic syringe (glass construction)
Luer, *France*.

Lithofracteur (stone-breaker)
Professor Engels *of Cologne, Germany, and manufactured by* Krebs. *A type of dynamite.*

Monorail-trolley 'dis-assembly' line
US *meat-packing firms in Cincinnati and Chicago.*

Nickel-plating.
Ominimeter
Eckhold, *German engineer, to replace chain measuring. It comprised a theodolite, a level, a telescope and a microscope.*

Pantograph
Francis Galton, *scientist for the Meteorological Office*, UK.

Periodic law
Dmitri Ivanovitch Mendeleyev,

1868

Professor of Chemistry at St
Petersburg, Russia.

Rayon acetate
Paul Schützenberger, *physician and
Director of the Sorbonne Chemical
Laboratory, Paris.*

Rickshaw
Reverend Jonathan Scobie, US
*Baptist minister, to transport his
invalid wife about the streets of
Yokohama.*

Suez Canal
Ferdinand de Lesseps, *French
engineer, after a project lasting nine
years.*

Vacuum cleaner
Ives W McGaffey. *The manually
operated sweeping machine
incorporated a suction device for
surface cleaning.*

Xylonite
Daniel Spill, *former partner of
Alexander Parkes, inventor of
Parkesine,* UK.

1870

Bucket-wheel water turbine
Lester Allen Pelton, US *engineer,
California,* USA.

Chewing gum (chicle-based)
Thomas Adams, *Staten Island
photographer – whilst researching a
substitute for rubber. He began a
factory two years later.*

Circular ship
Admiral Popoff, *Russian Navy. It
had six propellers and two 11-inch
turret guns and was 101 ft in diameter.*

Fireman's respirator
Dr John Tyndall, *Irish physicist.*

Goods-wagon (steam-engined)
John Yule, *Glasgow,* UK, *for
transporting large marine boilers from
his works at Rutherglen Loan to the
Glasgow docks.*

Helicopter (steam-engined)
Horatio Phillips *of Battersea,
London. It failed to take off.*

**Pasigraphy (international
communication by numbers)**
Anton Bachmaier *of Munich. He*

published a number-grammar of
German, French and English, enabling
4334 mental conceptions to be
communicated numerically.

Phenophthalmoscope
Franciscus Cornelis Donders, *Dutch
oculist and Professor of Physiology at
Utrecht, for investigating eye-ball
movements.*

**Pneumatic underground
railway**
New York, USA.

Stock-ticker
Thomas Alva Edison *and* Franklin
Leonard Pope, *electrical engineers at
a workshop in Newark,* USA. *Some
months before, Edison had repaired the
ticker of the New York Gold
Reporting Company and been made
foreman by the company's president,
Dr Laws.*

**Washout water closet (all-
earthenware)**
Thomas W Twyford *of Hanley –
'The National'. 100,000 were sold and
in use within two decades.*

**Wheel (threaded spoke with
flange-headed nut)**
William Henry James Grout,
engineer, UK.

1871

Automatic Coupler (railways).
Bulb horn.
**Dry-plate photograph (gelatine
silver bromide emulsion)**
Dr Richard Leach Maddox, UK
physician and microscopist.

**Halftone process (photographic
reproduction)**
Carl Gustaf Wilhelm Carlemann,
Swedish engraver.

**Ordinary ('Penny-farthing')
bicycle**
James K Starley *of the Coventry
Machinist Company, with* William
Hillman. *Called the 'Ariel'.*

Paranormal isolation cage
William Crookes *and* Dr William
Huggins, *Fellows of the London Royal
Society, to test the paranormal powers*

The 'Ariel' bicycle

of Daniel Home in laboratory conditions. Crookes observed how an accordion in a wire cage surrounded by an electrical current could be made to levitate and play tunefully without Home physically touching it.

Phenolphthalein dye
Johann Friedrich Wilhelm Adolf von Baeyer, *German organic chemist at the Berlin Technical Institute.*

Self-raising flour (chemical raising agent).

Toilet roll
Seth Wheeler, *New York,* USA.

Wind tunnel
Francis Herbert Wenham, UK *marine engineer, interested in aeronautics. It was built by Messrs Penn of London.*

1872
Cooke's circular dividing machine
Frederick Cooke, UK.

Hydroplane (boat)
Reverend Charles Meade Ramus, *Rector of Playden, near Rye, Sussex,* UK – *for designing steamships of great*

speed. *He also advocated the development of rocket rams (weapons).*

Passenger monorail (power-driven)
cable-drawn system at the Lyon Exposition, France.

Solar distillation system
Charles Wilson, UK *engineer at Las Salinas, Chile, to the designs of J Harding. It remained in operation for 40 years.*

Woodpulp (sulphite process)
Carl Daniel Ekman *at a mill in Bergvik, Sweden. By 1875 Ekman was producing 800 lb of pulp per annum. The process was soon adopted in the* USA *and Europe.*

1873
Barbed wire (mass-production machine)
Joseph Glidden *of de Kalb, Illinois,* USA.

'Burbank potato'
Luther Burbank, US *plant-breeder, Lunenburg.*

Daguerreotype disc
Pierre Jules César Janssen, *Head of the Astrophysical Observatory at Meudon, France – to photograph the movement of the stars.*

Electric motor (DC)
Zénobe Théophile Gramme, *former Belgian carpenter. His first, hand-driven generator was ready in 1871 and two years later he installed improved models in some French lighthouses.*

Electromagnetic radiation
James Clerk-Maxwell, *first Professor of Experimental Physics at Cambridge,* UK, *and organiser of the Cavendish Laboratory.*

Fire alarm
Professor Grechi. *A mechanical and chemical apparatus caused a bell to ring and a coloured light to flash when the temperature of the room was greatly increased.*

Oil (Russian)
Discovered at Baku.
Pyrophone
Frédéric Kastner *of Paris – a chemical harmonicon whose notes were produced by heating glass tubes of various lengths with little flames.*
Silo
Fred L Hatch, US. *He made the first attempt to build a silo above ground in McHenry County, Illinois, USA.*
Typewriter (mass-produced)
Christopher Latham Sholes, *Milwaukee senator and former postmaster/editor, with his friend* Carlos Glidden, *an attorney. Since 1867 Sholes and Glidden had built as many as 40 prototypes, finally managing to sell a much improved machine to the Ilion Arms Manufactory, who called their model 'the Remington' and built the first thousand machines in one year.*

1874
Abecedarium
William Stanley Jevons, UK *economist and logician, while Professor of Logic at Owen's College, Manchester.*
Aircraft (steam-engined)
Félix due Temple, *France. Ramp-launched, it made a very short hop.*
Austria (asteroid)
Johann Palisa, *astronomer, Director of the naval observatory, Pola, Austria. The first of 27 such asteroids he discovered.*
Boyton's life-preserving dress
Captain Paul Boyton, *using Merriman india-rubber as the constructional material for this inflatable suit. Boyton demonstrated his suit on the waterways of the USA, the UK and Europe.*
Cotton wool
Dr Percy, *to purify the air for ventilating the Houses of Parliament, London.*
Electric car
Sir David Salomons *at Tunbridge Wells, Kent, UK. His 1 hp three-wheeler was powered from Bunsen cells.*
Hydrokimeter (boiler-water-circulator)
Weir Ltd, UK.
Logograph
W H Barlow, *for giving graphic representation to the vibratory motions of the airwaves of speech, like a telegraphic message.*
Merchant ship (triple-expansion engine)
SS *'Propontis', built on the Clyde,* UK.
Pianoforte (sostenuto pedal)
Heinrich *and* Theodore Steinway, US *piano-makers of New York.*
Quadruplex telegraphy
Thomas Alva Edison, US.
Solitons
John Scott Russell, *Scottish engineer.*
Sprinkler system (commercial)
Henry S Parmalee, *New Haven, Connecticut,* USA.
Swinging saloon (steamship)
Sir Henry Bessemer, *with* E J Reed, *naval architect,* UK. *It was incorporated in the 'Bessemer' cross-Channel steamer as an anti-seasickness device, but was a complete failure.*
Tangent-spoke wheel
James Starley *of Coventry,* UK. *The spokes were set aslant so that each pair crossed each other at an angle, giving great strength and rigidity. It has been in popular use ever since.*
Tubular steel bridge
Captain James Buchanan Eads, *engineer, and* Charles Shaler Smith, *design consultant, at St Louis, Missouri, over the Mississippi River. It incorporated corrosion-resistant chromium steel in cantilever design, using compressed air caissons.*
Umbrella (curved steel rib frame)
Samuel Fox, *wire-drawer from Deepcar, near Sheffield,* UK.

1875
Canned baked beans
Burnham & Morrill Co. *of Portland,*

Maine, USA, for their local fishermen.
Compound locomotive
Anatole Mallet *for the local Bayonne-Biarritz line, France.*
Cube sugar manufacturing process
Eugen Langen, *Cologne, Germany.*
Electric dental drill (battery-operated)
George F Green, *Kalamazoo, Michigan, USA.*
Eosin scarlet dye
Adolf von Baeyer, *Professor of Chemistry at Munich, Germany.*
Gallium
Paul Emile Lecoq de Boisbaudron, *physical chemist, France.*
Magazine rifle
Benjamin Berkeley Hotchkiss, US, *manufacturer resident in France.*
Measurement of wave power
R S Deverell, *Australia.*
Mechanical ship's telegraph
William Chadburn, *Liverpool, UK, for transmitting orders from bridge to engine room.*
Milk chocolate
Daniel Peter, *F L Cailler's son-in-law at Vevey, Switzerland.*
Mimeograph
Thomas Alva Edison, *US, while experimenting with paraffin paper for possible use as a telegraphy tape.*

Edison mimeograph

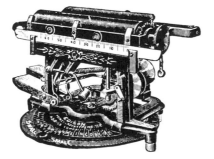

Phenyl-hydrazine
Emil Hermann Fischer, *chemist, Germany.*

Refrigerator freight wagon (railways).
Snooker
Lieutenant Neville Chamberlain, *while serving as subaltern with the Devonshire Regiment at Jubbulpore, India: a variation of 'Black Pool'.*
Solar-collector (axicon-type)
August Mouchot, *French engineer, using truncated cone reflector.*
Tide-calculating machine
William Thomson, *Professor of Natural Philosophy, Glasgow.*
Tinplated food cans (steel).
Xylotechnographica
A P Brophy – *a process for staining wood various colours.*

1876
'Aerotherapy'
Dr Carlo Forlanini, *Pneumatic Health Institute, Milan, Italy.*
Articulating telephone
Alexander Graham Bell, *Scottish-American Professor of Vocal Physiology, Boston University, USA – with his assistant* Thomas A Watson. *This invention was also patented by* Elisha Gray *for the Western Telegraph Co., but after Bell.*
Carburettor (surface type)
Gottlieb Daimler *of Cannstatt, Germany, and further modified by* Karl Benz *from 1884 for the 2000 motor cars he sold.*
Carpet sweeper
Melville R Bissell *of Grand Rapids, Michigan, USA. As a china shop owner, Bissell suffered terribly from headaches. Convinced the cause was the dusty straw in which the china was packed, Bissell developed a sweeper.*
Cream separator
Gustav de Laval *of Blasenberg. His mechanically whirling arm turbine completely displaced the skimming of cream by hand. His invention was shown at the Royal Show, Kilburn, London, in 1879.*
Crematoria
simultaneously by Julius le Moyne *in*

his own grounds at Washington, Pa, USA, and by Dr Pini of the Milan Cremation Society, Italy.

Dewey decimal system (library cataloguing)
Melvil Dewey, acting librarian at Amherst College, Massachusetts, USA.

Dietheroscope
Giuseppe Luvini of Tunis, for geodesy and teaching optics.

Electric candle
Paul Nickolaievitch Jablochkoff, Russian electro-technologist working in Paris. An electric current was passed through two carbons side by side with a strip of kaolin between them producing a steady, soft, noiseless light – the carbons burning like wax.

Erythrosin and phloxin
Adolf von Baeyer, organic chemist, Munich, Germany.

Glaciarum (artificial ice-skating rink)
King's Road, Chelsea, London, constructed by Dr John Gamgee. Freezing was accomplished by Raoul Pictet's liquefaction process and W E Ludlow's rotary engine and pump.

Hygeiopolis (city of health)
Dr Benjamin Ward Richardson. A company was proposed for its erection, but with no result.

'Phase rule'
Josiah Willard Gibbs, mathematical physicist, US.

Phenylethyl alcohol (synthetic perfume)
B Radziszewski. It had the scent of roses.

Plimsoll mark and line
Samuel Plimsoll, MP for Derby, UK and social reformer, 'the sailors' friend'.

Radiometer
William Crookes FRS, UK.

Rotary-foil hydrofoil craft (steam-engined)
M de Sanderal, France.

Stump-jump plough (three-furrow model)
Robert Bowyer Smith, farmer-engineer, Southern Australia. It was called the 'Vixen'.

Tomato ketchup (bottled)
Henry John Heinz & Co., Pittsburgh, Pennsylvania, USA.

Twine-binder
John F Appleby, US, enabling one farm labourer to do the work of eight. It also incorporated a device called a knotting-bill, invented in 1864 by Jacob Behel of Illinois, USA.

1877

Battery (solid depolariser)
Georges Leclanché, French railway engineer, after ten years' research.

Cyclometer (distance recorder for 'Ordinary' bicycles)
James C Thompson of Ealing, UK.

Dayyum
Vincenz von Kern.

Differential gear
James Starley of Coventry, UK, patented for tricycle transmission to overcome the problem of travelling round bends, when the outer wheel has to run faster than the inner one. This was later adapted for motor vehicles.

Friedel-Crafts Reaction
Charles Friedel, chemist and curator of the mineral collections at the School of Mines, Paris, with James Mason Crafts, US chemist resident at the school.

Ilmenium
Hermann.

Liquid oxygen
Louis-Paul Cailetet, ironmaster of Châtillon-sur-Seine, France, and independently by Raoul-Pierre Pictet, physics professor at Geneva University, Switzerland.

Microphone (dynamic)
simultaneously by Charles Cuttris with Jerome Redding, US, and E W Siemens, Germany.

Otheoscope
William Crookes, UK, for studying molecular motion.

Otto cycle
Nikolaus August Otto, German

engineer, *who developed a four-stroke internal combustion engine, some ten years after his first mains gas engine, based on the principle proposed by* Alphonse Beau de Rochas, *France.*

Photography (electrically lit).

Resistance welding
Elihu Thomson, *London-born chemistry teacher in Philadelphia,* USA.

Serial photography
Eadweard Muybridge, *with* John D Isaacs, *railway engineer, as commissioned by Californian Governor Leland Stanford, race-horse owner. Using 24 electrically activated cameras lined up beside Palo Alto race-course, their zoogyroscope was a direct forerunner of the modern motion picture.*

Spinthariscope
William Crookes, UK *chemist, for observing ionising radiation.*

Switchboard
Edwin T Holmes. *Installed for Boston's first six telephone subscribers, it served as a telephone system by day*

Telephone exchange

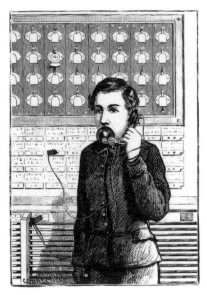

and a burglar-alarm by night.

1878

Cathode ray tube
William Crookes, UK *chemist.*

Electric alternator
Zénobe Gramme *and* Hippolyte Fontaine, *France, and independently by* Werner von Siemens, *Germany.*

Electrically synchronised city clocks
Barraud *and* Lund *for the City of London circuit of 108 clocks.*

Harmonic analyser
William Thomson, Lord Kelvin *and his brother,* Professor James Thomson, *for the Meteorological Office,* UK.

Helical logarithmic calculator
J E Fuller *of New York.*

Lenses (apochromatic objective)
Ernst Abbe, *Professor of Physics and Mathematics at Jena University, together with* Otto Schott, *glass chemist, and* Carl Zeiss, *industrial socialist.*

Megaphone
Thomas Alva Edison, USA, *for use by the deaf.*

Milking machine
L O Colvin, *and manufactured by Albert Durand of Auburn, New York,* USA.

Philippium
Marc Delafontaine, *chemist, Switzerland.*

Phonograph
Thomas Alva Edison and his team of mechanics *at Menlo Park Laboratory, New Jersey,* USA. *The first recording replayed was 'Mary had a little lamb. Its fleece was white as snow.'*

Phonometer (engine powered by human voice)
Thomas Alva Edison *et al,* USA.

Praxinoscope theatre (photographic optical toy)
Emile Reynaud, *France.*

Sailing railway car
C J Bascom *of the Kansas Pacific*

Edison's first phonograph

Railroad. *It was capable of 33 mph
and was used for three years to
transport repair gangs to pumps,
telegraph lines, etc.*

Salvo quad tricycle
James Starley *of Coventry,* UK.
*Following the delivery of two vehicles
to Queen Victoria, it was re-named the
'Royal Salvo'.*

Solar-powered printing press
Abel Pifre *at the Paris Exhibition.*

Theatre (electrically lit)
the Gaiety Theatre, London.

**'Willesden' rot-proof paper and
multi-ply green canvas**
*proofing was by immersion in copper
and ammonia, based on the pioneer
work of* Dr John Scoffern, George
Tidcombe *and* Elkanah Haley. *The
factory was at Willesden Mills,
London.*

Ytterbrium
Jean Charles Galissard de Marignac,
chemist, Geneva, Switzerland.

1879
Arc lighting system
Edwin James Houston, US *electrical
engineer, with his partner,* Elihu
Thomson. *A similar system was
devised by* Charles F Brush *of Ohio,*
USA.

1879

**Audiometer (hearing test
device)**
Professor David Edward Hughes,
London, UK. *It consisted of a battery of
two Leclanché cells, connected with a
simple microphone and telephone.*

Benday process
Benjamin Day, *West Hoboken, New
Jersey,* USA. *A mechanical method of
producing a tinted, shaded or stippled
background on a line plate, for the
reproduction of maps and illustrations.*

Blasting gelatine (explosive)
Alfred Nobel, *Swedish inventor and
manufacturer, and later modified by*
Professor Frederick Augustus Abel,
UK *War Department chemist.*

'Bordeaux mixture' (fungicide)
M Millardet, *Professor of Botany at
Bordeaux, France, to prevent an
American mildew disease ruining
the grapes in countless vineyards.*

**Carbon filament lamp (light
bulb)**
simultaneously by Joseph Wilson
Swan *at Newcastle,* UK, *and* Thomas
Alva Edison *at Menlo Park
Laboratory, New Jersey,* USA. *Whilst
Swan was assisted by high vacua
expert Stearn and glass-blower
Topham, Edison had a whole team of
experts. The major difficulty for both
inventors was to find the right material
for carbonisation as a filament. Those*

*Edison's and Swan's first incandescant
lamps*

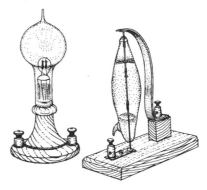

1879

tested included paper, bamboo and human hair.

Cash register
James J Ritty, US café owner, inspired by a ship's propeller revolution counter. It was built at Dayton, Ohio, but was commercially unsuccessful. Ritty sold his patent for a mere $1000 to what became known as the National Cash Register.

Coumarin and heliotropin (synthetic perfumes)
having the scent of tonka-beans and heliotrope respectively.

Decipium
Marc Delafontaine, chemist, Switzerland.

Domestic refrigerator (steam-pumped)
Karl von Linde, engineer, Germany.

Electric streetcar line (temporary)
Werner von Siemens and Johann Halske for the Berlin Industrial Exhibition.

'Five and Ten Cent Store'
Frank Winfield Woolworth, US merchant in Lancaster, Pennsylvania, USA.

'Gabardine'
Thomas Burberry, draper, of Basingstoke, Hampshire, UK, with Doctor Syme, medical reformer of Edinburgh.

Holmium
Per Theodor Cleve, chemist, Uppsala University, Sweden.

Milk bottles
Echo Farms Dairy Co., New York, USA.

Mosandrum
John Lawrence Smith, chemist, US.

Norwegium
Telef Dahl.

Rifle magazine box
James Lee, Scottish watchmaker and US citizen.

Saccharin
Constantine Fahlberg and Professor Ira Remsen at Johns Hopkins University, Baltimore, USA.

Samarium
Paul Lecoq de Boisbaudron, physical chemist, France.

Scandium
Lars Fredrik Nilson, chemist, Uppsala University, Sweden.

Steamship (Siemens-Martin steel-construction)
340 ft 'Pizarro', launched at Napier's Yard, Govan, near Glasgow, UK.

Submarine (steam-powered)
Reverend George W Garrett. Built at Liverpool, the 'Resurgam' had a two-man crew. Enough steam could be generated to drive her 12 miles after furnaces and chimney had been sealed off for diving. 'Resurgam' was sunk after being towed off Anglesey, UK.

Temnograph
A M Rymer-Jones, to plot to any accurate scale a section of ground over which the instrument has travelled.

Thulium
Per Theodor Cleve, chemist, Uppsala University, Sweden.

1880

Audiophone (hearing aid)
R G Rhodes, Chicago, USA, and improved by M Colladon, Geneva, Switzerland. A thin sheet of hard ebonite rubber or cardboard was placed against the teeth, through which vibrations could be conveyed to the auditory nerve.

Blowlamp
C R Nyberth of the Max Sievert Co., Stockholm, Sweden – petrol-burner. The paraffin blowlamp was developed by the Lindqvist brothers, also of Sweden.

Chinese 'alphabet'
Bishop Eligius Cosi of Canton.

Cigarette-making machine
James A Bonsack.

Electric clothes iron
France.

Electrically lit town
Wabash, Indiana, illuminated by 4000 candle-power Brush arc lights – in the business quarter.

Electrophone (piped music)
Dr Thaddeus Cahill. *This system whereby subscribers could dial in and listen to performances from theatres and concert halls, via a central switchboard, was also called the 'Telharmonium' or 'Dynamophone'.*

Gadolinium
Jean Charles Galissard de Marignac, *chemist, Geneva, Switzerland.*

Inoculation
Louis Pasteur, *Professor of Chemistry, La Sorbonne, Paris, whilst investigating chicken cholera. Having isolated the germ, he cultivated an attenuated form of it and injected fowls with progressively more virulent suspensions of the culture, so rendering them immune. Following immensely successful inoculation programmes for both anthrax and rabies, the Pasteur Institute was founded in 1888.*

Miniature electric motor
Thomas Alva Edison, *Menlo Park, New Jersey,* USA, *for an electric pen he had designed for producing punctured copying stencils. It was powered by a tiny two-cell battery. 60,000 such miniature motors and pens were manufactured by Western Electric Co.*

Photophone
Alexander Graham Bell *and* Sumner Tainter *of Washington,* USA. *This was a mirror-optical telephone.*

Pre-stressed concrete
P H Jackson, *San Francisco,* USA.

Refrigeration ship
the steamer 'Strathleven' brought to the UK *the first successfully carried refrigerated meat cargo.*

Rubber diaphragm contraceptive (metal-spring rims)
Professor W P J Mensinga.

Solar-powered 2-kW pump
W Adams *for his bungalow at Middle Colaba, Bombay, India.*

Solar spectrum map (infra-red region)
Captain William de Wiveleslie Abney, UK *chemist.*

Submarine-hulled steamship
Sergei Apostolov, *Russia – as a defence against a potential torpedo attack.*

1881
Bolometer
Samuel Pierpont Langley, US *astronomer, to measure radiant heat.*

Compressed-air clock
Victor Popp *of Vienna.*

Cyclostyle pen
David Gestetner, *Hungarian inventor. A spiked-wheel pen which could be used to make a series of perforations in a duplicating waxed stencil.*

Elektromote
Werner von Siemens, *German electrical engineer, and* Johann Halske *– as a trackless car. A line 540 metres long was erected in Halensee and the first demonstration took place in 1882. This was the prototype trolleybus.*

'Farthing-penny' (reversed Ordinary bicycle)
the Smith Machine Co., *New Jersey,* USA. *Called 'the Star', it had the smaller wheel at the front.*

Hydro-electric power station
Messrs Calder & Barrett's Central Power Station at Godalming, Surrey, UK, *for Pullman's leather mill on the River Wey. The principal customer was the Godalming Town Council.*

Interferometer
Albert Abraham Michelson, *ex-*US *Navy academician and physicist. It was used in 1887 for the critical Michelson-Morley experiment.*

Lawn-mower (tricycle-powered)
R Kirkham *of Chicago, Illinois,* USA.

Ocean liner (steel construction)
Canadian Allen Line's *'Buenos Ayrene'.*

Otto dicycle
E C F Otto, *and manufactured by the Birmingham Small Arms Co.,* UK.

Railway (electric)
Werner von Siemens *and* Johann

Halske, *German electrical engineers at Lichterfeld, Berlin – it travelled at 18½ mph.*

Rechargeable battery (lead lattice-work grid)
Camille Faure, *France.*

Telautograph
Elisha Gray, US.

Telephotography
Shelford Bidwell, UK *barrister and physicist – a process of transmitting to a distance images of objects by the agency of electricity and selenium – 'the wirephoto'.*

Télétroscope
Constantin Senlecq, *solicitor of Ardres, Pas-de-Calais, France.*

Tricycle (steam-powered)
Sir Thomas Parkyns *and* William Pateman, UK.

1882

Electric fan (commercial)
Dr Schuyler Skaats Wheeler, *Chief Engineer of the Crocker and Curtis Electric Motor Co. of New York, USA. It was a two-bladed desk fan.*

Electric flat iron
Henry W Seely *of New York*, USA.

Electric mine railway
Werner von Siemens *and* Johann Halske, *electrical engineers at the Zauckeroda coal mine, Saxony.*

Fusil photographique (photographic pistol)
Dr Jules Etienne Marey, *French physiologist. It took twelve consecutive pictures per second. Three years later he had developed his 'chambre chrono-photographique' and reduced exposure time to 1/25,000 of a second to photograph the flight of insects.*

Gas-heated bath tub
Ewart Co's 'General Gordon', Dudley, UK.

Induction coil system
Lucien Gaulard, *France, and* John Gibbs, UK. *They called it a 'secondary generator'.*

Judo
Jigoro Kano, *Japan.*

Oximes
Viktor Meyer, *German chemist at Zurich Polytechnic.*

Skyscraper
William Le Baron Jenny, US *architect, with the Home Insurance Company's ten-storey building in Chicago, Illinois,* USA.

Tuberculosis germ
Robert Koch, *German physician at the Medical School Faculty in Berlin. The following year he discovered the cholera bacillus.*

1883

Automatic temperature control
Albert M Butz, *Minneapolis,* USA.

Automatic vending machine (commercial)
Percival Everitt, *London.*

Diphtheria bacillus
Edwin Krebs *of Zurich, and the following year Friedrich Loeffler succeeded in cultivating the bacillus.*

Dustbin
Eugene Poubelle, *Paris Prefect of Police. It was a galvanised iron portable container.*

Electric blanket
first seen at an exhibition in Vienna.

Electric hair brush
Dr Scott. *'Warranted to cure nervous and bilious headaches, neuralgia and to arrest falling hair and premature greying', it was a complete hoax.*

Gasogen
Emerson Dowson, UK *chemist. It was used by the Crossley brothers to fuel their motor cars.*

Manganese steel
Robert Abbott Hadfield, *Sheffield-born* UK *metallurgist. He also developed silicon steel and stainless steel.*

Narrow-gauge railway (electric)
Magnus Volk. *It ran along the sea-front at Brighton,* UK.

Plastids-from-plastids
Andreas Franz Wilhelm Schimper, *botanist at the Bonn Botanical*

Institute, Germany.
Spark plug (stationary gas engine)
Jean Joseph Etienne Lenoir, *Paris*.
Suspension bridge (long span)
John Augustus Roebling, *German-born US engineer. Brooklyn Bridge, 1600 ft between its towers.*
Thiophene
Viktor Meyer, *German chemist at Zurich Polytechnic.*

1884
Aircraft (steam-engined with triple propellers)
Feodorovitch Mozhaiski, *Russia. Ramp-launched, it made one, very short hop.*
Carburettor (float-feed spray)
Edward Butler *for his 'Petrocycle', UK, then re-invented by both* Wilhelm Maybach *in Germany and* Charles E Duryea *in the USA, some ten years later.*
Cemented bifocals (spectacles).
Evaporated milk
John Mayenberg, *St Louis, USA, and manufactured by Mayenberg's Helvetia Milk Condensing Co., Highland, Illinois.*
'Farthing-Penny' (steam-powered bicycle)
Lucius D Copeland *of Phoenix, Arizona, USA – a converted 'Star' bicycle.*
Fountain pen
Lewis Edson Waterman, *US insurance salesman. Its capillary feed system made for a much freer flowing reservoir pen than those produced since the early 1800s.*
Greenwich Mean Time.
Kangaroo (bicycle)
E C F Otto *and* J Wallis, *UK, and manufactured by Hillman, Herbert & Cooper of Coventry.*
Local anaesthetic (cocaine)
K Köller, *Austrian ophthalmologist.*
Machine-gun (water-cooled and belt-loaded)
Hiram Stevens Maxim, *US inventor of*

Huguenot descent, who had settled in the UK (see overleaf).
Nipkow scanning disc
Paul Nipkow, *German engineer.*
Oil well (Sumatra)
Aeilko Jans Zijlker, *Dutch engineer.*
Phillips dipping entry wing
Horatio Phillips, *UK, who discovered that an aerofoil will provide lift if the upper surface is deeply curved and the lower surface is shallowly curved.*
Rollercoaster
L N Thompson, *founder of Luna Park, Coney Island, New York, USA. It was called the 'Switchback'.*
Steam-powered turbine engine
the Hon Charles Algernon Parsons, *UK engineer at a dynamo-manufacturing firm, Gateshead, County Durham.*
Synchronous multiplex telegraphy
Patrick Delany, *New York.*
Test for bacteria
Hans Christian Joachim Gram, *Danish bacteriologist.*
Transformer (practical commercial)
Max Deri, Otto Blathy *and* Karl Zipernowsky *of Hungary; also* Lucien Gaulard *of France.* William Stanley, *electrical engineer for the Westinghouse Company, US, invented his version of the transformer less than a year later.*
Washing machine (octagonally shaped and hand-rotated)
Thomas Bradford, *UK engineer.*

1885
Ammonium picrate (explosive)
Eugène Turpin, *France. Called 'Mélinite'.*
Commercial plywood factory
Christian Luther, *in Revel, Estonia, manufacturing 3-ply chairseats.*
Electric toothbrush
Dr Scott, US.
'Flying Bedstead'
W O Ayers, *US. Hand and foot pedals and compressed air motors drove seven*

Maxim beside his machine gun, (1884)

propellers – six for lift and one for
propulsion. It never flew as it was
never built.

'Golden syrup' (canned treacle)
Abram Lyle III, *Scottish sugar
refiner, with his sons, at Plaistow, East
London.*

Holiday caravan (custom-built)
Dr Gordon Stables RN, *and built by
the Bristol Wagon Co. Fully equipped
and horsedrawn, 'The Wanderer'
made a 1300-mile tour of the UK with
Dr Stables's valet, Foley, travelling in
advance on a tricycle to warn other
road-users of its approach.*

Humber tandem tricycle
Thomas Humber *and* T H Lambert,
Beeston, Yorkshire, UK.

Incandescent fabric gas mantle
Karl Auer, *Vienna-born chemist and
former pupil of Robert Bunsen.*

Moray car
G A Thrupp, *London coachbuilder.*

Motor-bicycle (petrol-engined)
Gottlieb Daimler *and* Wilhelm
Maybach *at Cannstatt, Germany. It
was the first to incorporate 'twistgrip'
handlebar acceleration.*

Motor car (petrol engined)
simultaneously by Gottlieb Daimler
with Wilhelm Maybach, *engineers at
Cannstatt; and* Karl Friedrich Benz,
*engineer at the Rheinisch Gasmotoren-
fabrik, Mannheim, Germany.*

Neodymium
Karl Auer, *Vienna-born chemist.*

Petrol pump
Sylvanus F Bowser, *Fort Wayne,
Indiana,* USA, *to enable local store-
keeper Jake Gumper to supply his
customers with kerosene lamp oil in
given quantities.*

Praseodymium
Karl Auer, *Vienna-born chemist.*

Safety bicycle
John Kemp Starley, *whose Rover*

Mark III with equal-sized wire-spoked wheels and diamond frame was the final design which ousted the highly unstable Ordinary (penny-farthing) of the 1870s. That Starley's Rover had its crank and pedals in the centre with a chain-drive to the back wheel was thanks to H J Lawson's 'bicyclette' of 1879. The only feature it still lacked was an efficient tyre.

1886

Aluminium (by electrolysis)
simultaneously by Paul Louis Toussaint Héroult, *aged 23, in France, and by* Charles Martin Hall, *aged 22, of Oberlin, Ohio,* USA, *working without collaboration. Both students made aluminium by the electrolysis of a solution of aluminium oxide in fused cryolite. Soon after, a water-powered aluminium factory was formed at Neuhausen on the Swiss-German frontier, with Martin Kiliani as its chief engineer. In 1889, the first 40 tons of pure aluminium were electrolysed.*

Automatic air brake (railways)
George Westinghouse, *Pittsburgh, Pennsylvania,* USA.

Boat (internal combustion engine)
Gottlieb Daimler *of Cannstatt, Germany. Secretive dawn trials were held along the local River Neckar.*

Canal rays
Eugen Goldstein, *German physicist.*

Celesta
Auguste Mustel, *Paris instrument-maker, as a keyboard glockenspiel.*

Coca Cola
Dr John Pemberton, *Atlanta, Georgia; marketed as 'Esteemed Brain Tonic and Intellectual Beverage'.*

Coin-operated ticket machine
James Longley, *and installed at Leamington Athletic Ground, Leeds,* UK.

Comptometer (key-driven calculating machine)
Dorr Eugene Felt, *Chicago, Illinois,* USA.

Diapason clock
Louis Breguet, *French clockmaker.*

Dinner jacket
Griswold Lorillard, *at the Autumn Ball held at the Tuxedo Park Country Club, New York. It became known as 'Lorillard's Tuxedo'.*

Dysprosium
Paul Emile Lecoq de Boisbaudron, *France.*

Electric light bulb (bayonet cap)
Edison & Swan United Electric Light Co., *Newcastle-on-Tyne,* UK.

Germanium
Clemens Alexander Winkler, *chemist, Freiburg, Germany.*

Graphophone
Chichester Bell *and* Charles Sumner Tainter, US *scientists, using a wax cylinder and a gouge-shaped sapphire stylus.*

Hatchet planimeter
Prytz.

Linotype machine
Ottmar Mergenthaler, *German-born engineer working in Baltimore, Maryland,* USA. *Following eight years of prototypes, Mergenthaler's 'blower machine' enabled type to be set mechanically, using a keyboard. The first 30 machines were installed in the 'New York Tribune' building.*

Mechanical potato digger.

Motorboat (paraffin vapour)
William *and* Samuel Priestman *of Hull,* UK.

Photographic hat
Lüders Brothers, *Gorlitz, Germany.*

Sound signal cyclometer
Messrs C V Boys *and* M D Rucker.

Submarine (electric)
Lieutenant Isaac Peral, *Spanish officer. The Peral was powered by 480 storage accumulators powering two 30 hp motors turning two propellers.*

Tintometer
Joseph William Lovibond, *brewer, maltster and spirit merchant of Salisbury,* UK.

1887

Automatic page-turner (piano music)
Augustin Lajarrige, *mechanic of Marseilles, France.*

Collapsible pocket umbrella.

Contact lenses
Dr Eugen A Frick *of Zurich.*

Electric elevator
Werner von Siemens *and* Johann Halske, *electrical engineers, for the Industrial Exposition at Mannheim, Germany.*

Electric heater
Dr W Leigh Burton. *Commercially manufactured by the Burton Electric Co. of Richmond, Virginia,* USA, *it was first used on electric tramcars.*

Electrocardiogram
Augustus Désiré Waller, *Parisian-born physiologist at London University.*

Electro-magnetic waves
Heinrich Rudolf Hertz, *while physics professor at the Karlsruhe Polytechnic and aged only 30. Hertz (Hz) remains the unit of measurement.*

Esperanto (auxiliary language)
Dr Lazarus Ludwig Zamenhof, *Polish oculist.*

Gas meter (coin-in-the-slot)
R W Brownhill *of Birmingham,* UK. *Only five years later did the South Metropolitan Gas Co. agree to instal 100 experimental meters.*

Gold extraction (cyanide process)
John Stewart Macarthur, *and the* Forrest brothers.

Gramophone record
Emile Berliner, *German immigrant living in Washington* DC, USA. *He also introduced the copying of flat disc recordings by making a wax disc from which a 'negative' metal matrix was made for producing endless positives. Although the first commercial manufacture was by Kammerer and Rheinhardt of Waltershausen, Germany, greater success was achieved by the Gramophone Co. formed eleven years later.*

Hydrazine
Theodor Curtius, *German organic chemist.*

Impact turbine (steam)
Carl Gustav Patrik de Laval, *Swedish engineer.*

Mach scale and angle
Professor Ernst Mach *of the Physics Faculties of Prague and Vienna, while studying supersonic projectiles and the flow of gases.*

Malted milk
William Horlick, *Racine, Wisconsin,* USA.

'Parlour' hansom cab
Joseph Parlour.

Phonograph doll
Thomas Alva Edison, *Menlo Park, New Jersey,* USA.

Pocket typewriter
Messrs Dobson & Wyn, *London.*

Rhodamine
Johan Friedrich Adolf von Baeyer, *Professor of Chemistry, Munich, Germany.*

Rotary aero engine (experimental)
Lawrence Hargrave, *Australia.*

1888

Ballistite
Alfred Nobel, *Swedish inventor and manufacturer,*

Cellulose film
John Carbutt, UK *photographer working in the USA. Carbutt persuaded a celluloid manufacturer to produce sufficiently thin sheets, coated with gelatine emulsion, for him to roll through a camera instead of the cumbersome glass-plate negative system.*

Compressed air tramcar
tested by the London Steam Tramway Company on its Caledonian route between King's Cross and the Holloway Road. It was found unable to climb hills.

Crude oil refining process
Hermann Frasch, *German-born*

Alfred Nobel, as depicted on the Nobel Prize medal

pharmacist, employed by the Standard Oil Co. of Ohio, USA.

Electric induction motor (AC)
simultaneously by Nikola Tesla, *Croatian-born engineer, at his laboratory in New York and by* Professor Galileo Ferraris *in Turin, Italy. Both men demonstrated the practical applications of alternating electrical current.*

Electric switch socket
Sir David Salomons, *Tunbridge Wells, Kent,* UK.

Electric tramway (regular service)
Frank J Sprague, US. *The route, in Richmond, Virginia, was 12 miles long.*

Electric vehicle disc brakes
Elmer Ambrose Sperry, *inventor, Chicago, Illinois,* USA.

Flexible roll-film camera
George Eastman, *dry-plate manufacturer from Rochester, New York,* USA. *Called the 'Kodak', it was given the slogan 'You press the button, we do the rest.'*

Flexible, unbreakable beer glass
G W A Kahlbaum, *Swiss chemist, using methylacrylate polymers.*

Monorail
Charles Lartigue. *The 9-mile line ran between Listowel and Balybunion,*

Ireland.

Monotype
Tolbert Lanston, *a Washington civil servant. The type-forming and composing machine was in production by 1897.*

Motor tricycle (production model)
Karl Benz, *as marketed by Roger in Paris.*

Phonograph (commercial model)
Thomas Alva Edison, *Menlo Park, New Jersey,* USA. *It had a clockwork motor and wax cylinders.*

Pneumatic bicycle tyre
John Boyd Dunlop, *prosperous Scottish veterinary surgeon living in Belfast, to help his ten-year-old son, Johnnie, win a race over the bumpy cobblestones of the city on his Edlin Quadrant tricycle. Hosepipe sections were filled with air, largely on the advice of the Dunlops' family doctor, Sir John Fagan, from experience he had gained with medical air beds and inflatable cushions. Following a patent, Dunlop went into partnership with Harvey DuCros, Irish industrialist, to form a manufacturing company.*

Recording adding machine (full keyboard)
William S Burroughs, *American Arithmometer Co., St Louis, Missouri,* USA.

Revolving door
Theophilus van Kannel, USA, *for use in the new skyscrapers, where air pressure outside and suction from elevators inside made conventional doors difficult to open.*

Solar thermocouple
E Weston, *who designed a machine whereby the solar energy, instead of focusing on mirror or plates, heated up a thermocouple so as to create a flow of electricity.*

Submarine (battery-electric motor)
Gustav Zédé's *'Gymnote', built to*

plans of H Dupuy de Lôme, France.

1889

Active molecules
Svante August Arrhenius, *Swedish scientist.*

Agricultural tractor (petrol-engined)
Charter Engine Co. *of Chicago, Illinois. It was used on wheat farms in the Dakotas,* USA.

Arithmographe (pocket calculator)
Troncète, *France.*

Artificial silk
Hilaire, Comte de Chardonnet, *French chemist. His silk was very inflammable so that his workmen called it 'mother-in-law' silk, since the present of a dress made of it was an effective means of disposing of a troublesome relative.*

Automatic telephone exchange
Almon B Strowger, *Kansas City, undertaker.*

'Bird Flight as the Basis of Aviation' (book)
Otto Lilienthal, *German civil engineer and man-carrying-glider innovator.*

Celluloid roll film
Henry Reichenbach, *at George Eastman's Kodak factory, Rochester, New York.*

Cordite
James Dewar *and* Frederick Augustus Abel. *Abel was chemist to the British War Department and Ordnance Committees.*

Cotton-picker
Angus Campbell, *pattern-maker from Chicago,* USA. *For the next 30 years, with financial backing from Theodore H Price, a wealthy cotton dealer, Campbell continued to develop and improve his spindle-type picker including equipping it with a petrol engine.*

Data-processing computer
Dr Herman Hollerith *of New York, for the* US *Census Bureau. It used punched cards containing information submitted by respondents to the census questionnaire. Seven years later, Hollerith established his Tabulating Machine Co.*

Dish-washing machine (power model commercially produced)
Mrs W A Cockran *of Shelbyville, Indiana,* USA, *after ten years of prototypes.*

Eiffel Tower
Gustave Eiffel, *French engineer, based on experience gained whilst building iron bridges at Oporto and Garabit. In Eiffel's office, 40 draughtsmen turned out 5000 drawings for the 985-ft high tower. The sections were prefabricated at his factory. The completed tower cost £260,000.*

Electric oven (experimental)
Hotel Bernina in Samaden, Switzerland, which had its own electric power supply, generated from a dynamo, driven by a waterfall.

Electric tricycle
John Kemp Starley *of Coventry,* UK.

Fruit machine
Charles Frey, *German immigrant at San Francisco,* USA. *It was called the 'Liberty Bell'.*

Kinetoscope (peep-show machine)
Thomas Alva Edison *and his assistant,* William Kennedy Laurie Dickson, *at Menlo Park Laboratory, New Jersey,* USA. *It used 158 glass plates, but later progressed to Eastman's and Goodwin's roll film.*

Life boat (steam-powered)
the British water-jet-propelled 'Duke of Northumberland'.

Lysine (amino acid)
Edmund Drechsel, *physiological chemist at Carl Ludwig's Institute, Leipzig.*

Northrup loom
James H Northrup, *English mechanic working for George Draper & Sons of Hopedale, Massachusetts,* USA. *The first commercially produced machine went into action at Seaconnet Mill,*

Fall River.

Plankton
Viktor Hensen, *Professor of Physiology, Kiel University, Germany.*

Pneumatic yacht (gas-jet-propulsion)
'Eureka' of Brooklyn, New York, USA.

Recording adding machine
Felt & Tarrant Manufacturing Co. *of Chicago,* USA. *It was called the 'Comptograph'.*

Skyscraper (riveted skeleton)
William Holabird *and* Martin Roche, US *architects, with the Tacoma Building, Chicago,* USA.

Table-tennis
James Gibb, *engineer and distinguished amateur athlete of Croydon, Surrey,* UK. *It was manufactured nine years later by John Jaques & Son Ltd and marketed by Hamley Brothers of Regent Street, London, first as 'Gossima' and soon after as 'Ping Pong'.*

Telephone call-box (coin-operated)
William Gray. *Following the installation of a prototype at Hartford Bank, Hartford, Connecticut,* USA, *the Gray Telephone Pay Station Co. was formed to rent out coin-operated telephones to store-keepers.*

Water closet (washdown flushing)
David Bostel, *plumber, Brighton,* UK.

1890

Aeroplane (man-carrying)
Clément Ader, *whose steam-engined 'Eole' 'hopped' 50 metres at Armainvilliers, France, with its inventor at the controls.*

All-steel oil rig.

Aluminium saucepan
Henry W Avery, *Cleveland, Ohio,* USA.

Babcock test
Professor S M Babcock *of the University of Wisconsin,* USA, *by which at a glance a farmer could see the*

Clement Ader's 'Eole'

percentage of butter fat in milk or cream on a graduated container. Although this test revolutionised dairy manufacture and the breeding of dairy animals, Professor Babcock refused to patent this or any of his other inventions, preferring that the world might benefit from them.

Compression-ignition engine
Herbert Akroyd Stuart, *and produced by Richard Hornsby & Sons of Grantham, Lincolnshire,* UK, *two years later. It was called the 'hot bulb' engine.*

Cuprammonium rayon
Louis Despeissis, *France.*

Diphtheria antitoxin
Emil Adolf von Behring, *German bacteriologist, and* Shibasaburo Kitasato, *Japanese investigator at Robert Koch's Institute for Infectious Diseases, Berlin. It was placed on the market in 1892.*

Electric bicycle
G P Hachenberger, *Austin, Texas,* USA. *It was to be powered and supported by electric telegraph wires.*

Electric chair
Harold P Brown *working with Thomas Alva Edison's chief electrician,* Dr E A Kennelly. *Following experiments with many animals, William Kemmler, convicted murderer, was executed at Auburn Prison, New York,* USA. *He took 8 minutes to die.*

Electric-generator windmill
P la Cour *in Denmark, who mounted patent sails and twin fantails on a steel tower.*

Elevated and surface railroad system
T C Clarke, *consulting engineer, New York*, USA.

High-frequency generator
Elihu Thomson, US *engineer and inventor, with* Edward James Houston.

Ice-cream sundae
Jed Smithson, *Wisconsin*, USA.

Kinematograph camera and projector
William Friese-Greene, *Bristol-born photographer and self-taught scientist, working in Holborn, London. His fourth camera, able to take pictures on modified celluloid film at 50 frames per second by means of parallel sprocket holes, was publicly demonstrated in Chester Town Hall during this year.*

Machine-gun (gas-operated)
Baron Adolf Odkolek von Augezd, *Austrian army officer.*

Pure nickel process
Monde, Langer *and* Quincke.

Radio conducteur (coherer)
Edouard Branly, *Physics Professor at the Catholic University, Paris.*

'Schizophone'
Captain Louis de Place, *France, to detect internal cracks or holes in metals.*

Underground electric railway
London. The City and South London Railway ran from Arthur Street to Stockwell, a distance of 3 miles, passing under the River Thames in separate tunnels, with one class and one fare, turnstiles instead of tickets, and hydraulic lifts or winding stairs. The initial idea of traction by an endless cable was abandoned in favour of electric power as the motive force.

Wells hive (two-queen system of bee-keeping)
George Wells, *bee-keeper, Kent*, UK.

1891

Aluminium boat
Escher, Wyss & Co. *of Zurich, Switzerland. Even its smoke stack and rigging were of aluminium. It was powered by a 2 hp naphtha engine, had a speed of 10 km per hour, weighed only 1000 lb and carried eight passengers. Soon after, the same company built the 43 ft naphtha-engined yacht, 'Mignon', whose machinery was of aluminium except for the cranks and shafting.*

Bazeries cylindrical cipher device
Captain Etienne Bazeries, *French cryptologist.*

Crankshaft compression phase (two-stroke engine)
Joseph Day.

Electric oven (commercially manufactured)
Carpenter Electric Heating Manufacturing Co., *of St Paul, Minnesota*, USA.

Electric torch
Bristol Electric Lamp Co., *whose 2-candle-power bull's eye lantern was used by ticket inspectors of the Bristol General Omnibus Co.*, UK.

Motor car (electric)
William Morrison, US *engineer. The vehicle was powered by 24 electric storage batteries placed under the seats giving it a total running time of thirteen hours with a passenger load of twelve people.*

Musk xyol and musk ambrette (synthetic perfumes)
A Baur.

Pneumatic typewriter
Marshal A Weir, *London.*

Premature baby incubator
Dr Alexandre Lion *of Nice, France. 137 out of 185 children reared in the incubators at Nice during the first three years survived infancy. Dr Lion introduced his 'Couveuses' to London in 1897.*

Silicon carbide
Eduard Goodrich Acheson, US. *By 1893 it had come into production as the abrasive, carborundum.*

Spectroheliograph
George Ellery Hale, US *astronomer at*

Kenwood Observatory, Chicago, Illinois, USA.
Telephoto lens (camera)
Thomas R Dallmeyer, *lens manufacturer*, UK. *Also* Dr Miethe *of Germany and* A Dubosq *of France.*
Traveller's cheque
American Express Co.
Zip fastener
Whitcombe L Judson, US *mechanical engineer. It was first exhibited at the Chicago Exposition two years later. Initially the Universal Fastener Co. produced the slide-fastener by hand, but following a re-design by Judson in 1905, the Automatic Hook and Eye Co. of Hoboken, New Jersey began machine-producing them under the trade name of 'C-Curity'.*

1892
Cellulose-fibre manufacture (viscose process)
Charles F Cross, UK *consulting chemist, and his partner* Edward John Bevan. *It went into commercial manufacture by C H Stearn and Topham as an artificial silk.*
Cholera vaccine
Waldemar Mordecai Wolff Haffkine, *Russian bacteriologist and physician working in India.*
Crown top and cast-iron bottle opener
William Painter, *Baltimore, Massachusetts*, USA.
Crum Brown's Rule
Alexander Crum Brown, *Professor of Chemistry, Edinburgh*, UK.
Diesel engine
Rudolf Diesel, *German engineer, subsidised by Krupps, after fourteen years' research on a compression-ignition engine. After a further six years' developmental work at Augsburg, manufacturing rights were acquired by Russia, Sweden and France.*
Electric fire (domestic)
Colonel Rookes Evelyn Bell Crompton *and* Herbert John

Dowsing, UK.
Iodoso compounds
Viktor Meyer, *German chemist at Heidelberg.*
Magnetic controller (for electric trolley cars)
Elihu Thomson, *Swampscott, Massachusetts*, USA.
Mega-globe
Alberto di Palicio, *Spanish architect, to be erected on San Salvadore to commemorate the 400th anniversary of Christopher Columbus's discovery of America. The globe was to be 100 ft in diameter, on a concrete base 250 ft high and topped by a huge statue of Columbus. Its concrete base was to contain museums, lecture halls, public meeting places and a planetarium. It was never built.*
Motorcycle (rotary-engined)
Félix Theodore Millet, *France.*
Outboard engine (petrol)
Alfred Seguin *of France, at the SNG garage in Geneva, Switzerland. It was called the 'Motogodille'.*
Phagocytes (cells)
Ilya Mechnikov, *Russian biologist, working at the École Normale, Paris.*
Portable typewriter
George C Blickensderfer, US. *Called the 'Featherweight Blick', it was based on an aluminium chassis.*
Toothpaste tube (collapsible metal)
Dr Washington Sheffield, *dentist of New London, Connecticut*, USA, *and later manufactured by his Sheffield Tube Corporation. Beecham's Tooth Paste was retailed in tubes the same year.*
Vacuum flask
Sir James Dewar, *Cambridge*, UK, *for laboratory work. It was commercialised in 1904 by Reinhold Burger, Germany, as the 'Thermos' flask.*
Virus
Dmitri Iosifovich Ivanovsky, *botanist, St Petersburg, Russia.*

1893

Aerograph air-brush
Charles L Burdick, US *artist working in London, who founded the Fountain Brush Co. in Clerkenwell Green.*

Breakfast cereal
Henry D Perky, *lawyer of Denver, Colorado,* USA – '*Shredded Wheat'.*

Electric toaster
Crompton & Co., UK.

Ionone (synthetic perfume)
Johann C W F Tiemann, *chemistry professor at the University of Berlin, Germany, with* P Kruger – *having the scent of violets.*

Kinetographic theatre (film studio)
W K L Dickson *at Menlo Park, New Jersey,* USA – *the studio revolved to follow the sunlight.*

Kirby grip hairpin
Hindes.

Language-teaching course (phonographs)
Dr Richard Rosenthal *of the International Institute of Languages, London and New York, who produced 50 wax cylinders with accompanying text.*

Lawn-mower (steam-powered)
James Sumner *of the Lancashire Steam Motor Co.,* UK. *150 1½-ton machines were sold, many for large country estates.*

'Model electric kitchen'
Chicago World's Fair, including both electric saucepan and electric kettle.

Photo-electric cell
Julius Elster *and* Hans F Geitel. *Geitel was Professor of Physics at the Brunswick Technical Institute, Germany.*

Solar-electric engine
M L Severy, *whose engine operated in conjunction with wet storage batteries to enable the user to have 24-hour electric power.*

1894

Aircraft (twin steam-engined)
Sir Hiram Maxim. *Unmanned, powered by two 180 hp engines, with wind-tunnel-tested aerofoils, it failed to fly.*

Argon gas
John William Strutt, UK *physicist, and* William Ramsay, *Scottish chemist at University College, London.*

Chromatic harp
Gustave Lyon *of Paris.*

Electric arc furnace
Ferdinand Henri Moissan, *Professor of Chemistry, University of Paris, France.*

Employee's time-recorder (clock-in/clock-out)
Daniel M Cooper, US. *The operation of a lever printed the time on a card divided by lines into seven equal spaces for the days of the week.*

Escalator
Jesse W Reno, *New York entrepreneur, as a novelty ride on Coney Island Pier, New York. Another designer of 'inclined elevators' was* Charles Seeberger. *Before long escalators had been installed in department stores and railway stations.*

Helium
William Ramsay, *Professor of Chemistry, University College, London.*

Lorry (petrol-engined)
René Panhard *and* Émile Levassor, *coachbuilders, D'Ivry, Paris.*

Meteorograph-carrying kite
W A Eddy, *above the* US *Blue Hill Weather Observatory, Massachusetts.*

Motor barge
Delamare-Debouteville *and* Malandrin. *'L'Idée' went on successful trials on the Tancarville Canal, Le Havre, France.*

Motorcycle (pneumatic-tyred)
Heinrich *and* Wilhelm Hildebrand *and* Alois Wolfmüller, *German engineers.*

Wireless telegraphy
Guglielmo Marconi, *twenty-year-old*

Italian at *Villa Griffone, his family home outside Bologna. By 1897, the Marconis had formed their Wireless Telegraph and Signal Co. in London.*

1895

Aerial
Professor Aleksandr Stepanovich Popov *of Kronstadt.*

Amphibious passenger train (steam-engined)
Magnell. *Built by Ljunggreen of Christianbad, Sweden, it ran on water rails between Farum and Fredericksdal, near Copenhagen, Denmark. By 1897 'Swan' had carried 40,000 passengers.*

Baked beans in tomato sauce (canned)
H J Heinz Co., *Pittsburgh, Pennsylvania,* USA.

Chiropractic (spinal re-adjustment)
Daniel David Palmer, *osteopath,* USA.

Ciné film show
Louis *and* Auguste Lumière, *French brothers. A sequence of films was shown in a Parisian café: a train entering a station, a rowing boat leaving a harbour and workers coming out of the Lumière factory at Lyons.*

'Daddy Longlegs'
Magnus Volk – *an electrically-driven pier-boat on stilts.*

Electric hand drill
Wilhelm Fein, *Stuttgart, Germany.*

Epicyclic gearbox
Dr Frederick William Lanchester, *engineer, Birmingham,* UK.

Fish paste (in jars)
Charles Shipham, *high-class pork butcher, Chichester, Sussex,* UK.

Flaked breafast cereal
Dr John Kellogg *of the Battle Creek Sanatorium, Michigan,* USA, *with 'Granose Flakes'. 'Corn Flakes' were first manufactured three years later by Kellogg's brother William and marketed by Sanitas Food of Battle Creek.*

Glass-blowing machine
Michael J Owens *of Toledo, Ohio,* USA. *His production model was ready by 1904.*

'Invar' (nickel-steel alloy)
Charles Edouard Guillaume, *Swiss metrologist, International Bureau of Weights and Measures, Sèvres, near Paris.*

'Latham Loop' (motion picture camera device)
Enoch J Rector, US, *working for the Latham brothers – to prevent the snapping of film. Exactly the same device was invented the same year by* Thomas Armat *for his Vitascope 'Beater' Movement film projector.*

Microtones (musical notation)
Julián Carillo, *Mexican composer.*

Milking machine (pulsator model)
Dr Alexander Shields *of Glasgow,* UK, *with the Thistle Mechanical Milking Machine.*

Motor ambulance
a Daimler-engined Panhard and Levassor.

Motor bus (petrol-engined)
Netephener Omnibus Co.'s *5 hp Benz single-decker enclosed landau, for a 15-km route in the North Rhineland, Germany.*

Motorcycle (four-cylinder)
Captain Capel Holden, UK *civil engineer.*

Motorcycle (quantity-produced)
Marquis Albert de Dion *and* Georges Bouton, *France.*

Photographic portrait machine
Ferrer.

Photographic typesetting
William Friese-Greene, *kinematography inventor of London, who took out a patent, although his system was never built for commercial use.*

Pneumatic-tyred automobile
André *and* Edouard Michelin, *brothers, on a Peugeot during the Paris-Bordeaux-Paris race. They were forced to retire after innumerable*

punctures and bursts, despite carrying 22 spare inner tubes on board.

Rotogravure
Karl Klietsch *in Vienna.*

Talking watch (phonographic)
Sivan, *watchmaker, Geneva, Switzerland.*

Time machine
Herbert George Wells, *fictional writer,* UK – *published as a short story, still to be built as a practical reality.*

Triplet hydrocycle
Messrs Keen, Marriott & Cooper *of Holborn Viaduct, London. It was successfully tested on the River Thames.*

Tse-tse fly (as sleeping sickness carrier)
David Bruce, *Australian-born physician, Royal Army Medical College, Millbank,* UK.

X-rays
Wilhelm Könrad von Röntgen, *Director of the Physical Institute, Würzburg, Germany.*

One of Röntgen's cathode-ray tubes

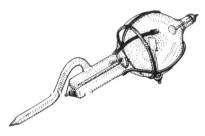

1896
Armoured car
E J Pennington, US *at Motor Mills, Coventry,* UK. *The four-man, 16 hp Pennington Military Autocar had Harveyised steel armour and twin machine guns.*

Automatic oyster-picking machine
modified diving bell.

Automatic rotary stencil machine
Lowe *in the* USA *and* Ellams *in the* UK.

Blériot acetylene headlamps
Louis Blériot, *Paris.*

Centrifugal-type speedometer (motor car)
Edward Prew, *schoolmaster, Yorkshire,* UK.

Dial telephones
Private automatic exchange, City Hall, Milwaukee, USA.

Direct top gear (motor car)
Dr Frederick William Lanchester, *engineer, Birmingham,* UK.

Electrons
Joseph John Thomson, *Cavendish Professor of Experimental Physics, Trinity College, Cambridge,* UK.

Europium
Eugène Anatole Demarçay, *French chemist.*

Four-cylinder motor car engine
René Panhard *and* Émile Levassor, *France.*

Histidine (amino acid)
Albrecht Kossel, *professor of biochemistry at Heidelberg, and* Sven Anders Hedin, *professor at Uppsala, Sweden.*

Ice-cream cones
Italo Marcioni, *Italian immigrant to New Jersey,* USA.

Ion-radiated food (theory of)
both Wilhelm Könrad von Röntgen, *German physicist, and* Antoine Henri Becquerel, *French physicist.*

Mechanically-tipping hat
James Boyle *of Washington,* USA.

Motor car (production model)
René Panhard *and* Émile Levassor, *former coachbuilders of Paris.*

Periscope (submarine device)
Gustav Zedé, *France, for use on the 'Morse'.*

Portable propeller pitchometer
J Chapman *and* J Hunter, UK.

Rear-view vehicle mirror
John William Cockerill, *Captain Surgeon, Army medical staff, London.*

Septic tank (sewage treatment)
Donald Cameron *of Exeter,* UK. *The effluent from the septic tank was treated on contact-beds as developed*

several years before by W J Dibden.

Sphygmomanometer
Dr Scipione Riva-Rocci, *Italy.*

Steamroller-ship
Ernest Bazin, *St Denis, France. 180
ft, with a carrying capacity of 300
passengers, it would roll over the
waves. Following satisfactory trials
under Vice-Admiral Coulembeaud,
the project was abandoned.*

Visible typewriter
Herman L Wagner, US. *It was called
the 'Underwood Number One'. With
previous machines the typist was
unable to see what he/she was typing
from word to word.*

1897

Agrimotor (petrol-engined)
Messrs Hornsby *with their Hornsby-
Ackroyd Patent Safety Locomotive of
18 bhp for working threshing
machinery, road haulage and other
tasks.*

Bismuth effect (radiology)
Walter Cannon, *physiology student at
Harvard University,* USA.

Cathode-ray tube
Ferdinand Braun, *Professor of
Physics, Strasbourg, Germany.*

Cell-free fermentation
Eduard Buchner, *chemist, Tübingen,
Germany.*

Decumtyplet bicycle
Waltham Manufacturing Company,

USA, *for ten riders as a pacing machine.
Soon after, a Quindicuplet was
constructed for fifteen riders.*

Digestion physiology
Ivan Petrovic Pavlov, *Professor of
Physiology at the Military Medical
Academy, St Petersburg, Russia.*

Electric omnibus
one tested in London.

**Filaria (malaria-carrying
parasite)**
Ronald Ross, *officer in the Indian
Medical Service, and* C Finlay *in
Cuba. Work completed by* Patrick
Manson, *Scottish doctor.*

Foot accelerator
Dr Frederick William Lanchester,
Birmingham, UK.

**Galalith and Erinoid (casein
plastics)**
Krische *and* Spitteler, *to meet the
unusual demand in Germany for white
blackboards made from milk.*

Hydrofoil (steam-powered)
the Comte de Lambert, *working with
Englishman,* Horatio Phillips, *on the
River Seine, France.*

Oscillograph
William du Bois Duddell, UK
engineer.

Pianola
Edwin S Votey, *Farrand & Votey
Organ Co., Detroit, Michigan,* USA.
*Perforated paper-roll-automata-
piano.*

1897

Plasticine
manufactured by William Harbutt,
Bath, UK.

Scooter
Walter Lines, *fifteen-year-old
London schoolboy who, when he grew
up, founded the Triang Toy Company.*

**Simms-Bosch low tension
magneto**
Frederick R Simms, UK;
*manufactured in Stuttgart by Robert
Bosch.*

Taximeter motor cab
Friedrich Greiner, *Stuttgart,
Germany.*

Turbine-driven steam yacht
the Hon. Charles Algernon Parsons,
Newcastle-upon-Tyne, UK. *The* 100-
ft *'Turbina' achieved the unheard-of
speed of 34.5 knots.*

Umbrella boat
Percy Pilcher *and* W G Wilson, UK.
*They rigged up a 'cyclone sail' for a
17-ft boat and tested it on the Solent.
Her 360 sq ft of umbrella gave her a
stable swiftness, while only 200 ft of
conventional canvas proved too much.*

Wildlife-preserving park
*Umfolozi Game Park, Natal, South
Africa.*

Worm gear
Dr Frederick William Lanchester,
engineer, Birmingham, UK.

1898

Anti-neuritic Vitamin B
Christiaan Eijkman *and his assistant*
G Grijns *at Batavia in the
Netherlands Indies (present-day
Indonesia).*

**'Bellowing telephone'
(loudspeaker)**
Sir Oliver Lodge, *Professor of
Physics, University College,
Liverpool,* UK.

Direct-drive motor car
Louis Renault, *France. An adapted
De Dion Bouton tricycle, it led to a
production model.*

Electric torch (tubular)
Electric and Novelty Manufacturing
Co. *of New York – later the Ever-
Ready Company.*

Garden cities
Sir Ebenezer Howard,
*Parliamentary shorthand writer and
visionary.*

**Impermeable millboard
(commercial building material)**
D M Sutherland, *former employee at
a firm of Edinburgh paper mill
engineers. In 1882 he had established a
factory at Sunbury-on-Thames,* UK, *to
produce millboard as a backing for
Lincrusta-Walton. The millboard was
made by hot-pressing waste paper.*

Krypton
William Ramsay, *Professor of
Chemistry, University College,
London, with* Morris William Travers.

Loudspeaker
Horace Short *of London. The
compressed-air 'Auxetophone' was
first used atop Blackpool Tower,* UK,
*and the Eiffel Tower, Paris, to
broadcast phonograph records of
operatic arias.*

Motor car (enclosed body)
Renault's $2\frac{1}{2}hp$ 2-seater, *France.*

**Motor fire engine (self-
propelled)**
Cambie et Cie *of Lille, France,
demonstrated at the French Heavy
Autocar Trials held at Versailles.*

Neon
William Ramsay *and* Morris William
Travers, *University College, London.*

Osmium Glühlampe (light bulb)
Karl Auer, Freiherr von Welsbach,
Austrian physicist and chemist.

Parabellum (automatic pistol)
Georg Lüger, *Germany.*

Pedal lawn-mower
William Burnet, US.

**Pneumatic tyre valve (one-piece
replaceable core)**
George H F Schrader, *New York,*
USA.

Polonium
Pierre *and* Marie Curie, *France.*

Remote-radio-controlled boat
Nikola Tesla, *Yugoslav-born* US

inventor. It was tested on the lake in Central Park, New York City.

Submarine (petrol-electric)
John Philip Holland US, *immigrant Irish schoolteacher. His Mark VIII craft was driven by a petrol engine on the surface and an electric motor when submerged, both installed on the same shaft. It carried both a guncotton torpedo tube and a submarine gun.*

Thermit process
Hans Goldschmidt, *industrial chemist, Essen, Germany.*

Xenon
William Ramsay *and* Morris William Travers, *University College, London.*

1899

Actinium
André Louis De Bierne, *chemist, France.*

Artificial parthenogenesis
Jacques Loeb, *German-American biologist.*

Aspirin (drug)
Dr Felix Hoffman *of Bayer* AG *of Leverkusen, Germany, based on the synthesization of acetylsalicyclic acid by* Karl Gerhardt, *Alsatian chemist, some 46 years before.*

Catalytic hydrogenation of oils
Paul Sabatier *and* Jean Baptiste Senderens, *French chemists.*

Electric car (high speed)
Camille Jenatzy *of Belgium, who achieved 66 mph on a road outside Paris, beating his arch electric-car rival, Count Gaston de Chasseloup-Laubat of France.*

Electric-pneumatic signalling
Bishopsgate Station, London.

Electric-wave wireless telephone
A Frederick Collins, US.

Foot-and-mouth disease vaccine
Friedrich August Johannes Löffler, *Professor of Hygiene, University of Griefswald, Germany.*

High-speed steel
Frederick Winslow Taylor *and* Maunsel White, US.

Inboard-outboard marine engine drive
Société du Propulseur Universel Amovable *of Neuilly, France. It was called the 'Lautonautile'.*

Integral body-and-chassis construction (motor car)
Dr Frederick William Lanchester, *engineer, Edgbaston, Birmingham,* UK.

Life-saving trunk
Herr Bunse, *Leipzig commercial traveller.*

Oil-engined omnibus
London.

Pupin coil
Michael Idvorsky Pupin, *for extending the range of long-distance telephone calls. It was patented by the American Telephone and Telegraph Company in 1901.*

RDX (explosive)
Hans Henning, *German chemist; fully developed 45 years later as 'Hexogen' or 'Cyclonite'.*

'Robin starch'
Reckitt & Sons Ltd, *Dansom Lane, Hull,* UK.

Tape-recorder (magnetisable steel tape)
Valdemar Poulsen, *Danish electrical engineer, who demonstrated his 'Telegraphone' the following year at the Paris Exhibition.*

Thoron isotope
Ernest Rutherford, *New Zealand-born Professor of Physics, McGill University, Montreal, Canada.*

Tielocken coat
Thomas Burberry *for* UK *generals during the Boer War.*

1900

Alkaline storage cells
Thomas Alva Edison, *Menlo Park, New Jersey,* USA.

'Cineorama'
Raoul Grimoin-Sanson *at the Paris Exposition. A battery of ten projectors showed an elaborately hand-coloured film on a completely circular screen*

330 ft in circumference. The audience was seated in the middle.

Clementine (fruit)
Father Clément *of Oran, France.*

Cord tyre (motor cars)
The India Rubber, Gutta Percha and Telegraph Works Co. Ltd of Essex, UK. *Based on the invention of* John Fullerton Palmer *seven years before.*

Heavy-production grinding machine
the Norton Co., *Worcester, Massachusetts,* USA, *using artificial abrasives.*

Kala-azar parasite
William Boog Leishman, *assistant Professor of Pathology, Army Medical School, Netley, Hampshire,* UK.

Kodak 'Brownie' camera
5 shillings (25p) each.

Metal nuts and bolts (mass-produced)
Guest, Keen & Co.

Paper clip
Johann Vaaler, *Norwegian working in Germany.*

Photo-electric cell (practical)
Julius Elster *and* Hans Friedrich Geitel, *German physicists.*

Quantum Theory (general)
Max Karl Ernst Planck, *ordinary Professor of Theoretical Physics, Berlin, Germany.*

Radon (radiation emanation)
Friedrich Ernst Dorn, *chemist, Halle, Germany.*

Rigid airship (petrol-engined)
Lieutenant-General Count Ferdinand von Zeppelin, *Friedrichshafen, Germany. It was 420 ft long.*

Rolling pavement (variable speed)
Paris Universal Exposition. It transported 6.7 million visitors.

Submarine (petrol-electric)
Maxime Laubeuf, *France: the 'Narval'.*

Tryptophan (amino acid)
Frederick Gowland Hopkins, *biochemist, Prelector in Physiological*

Chemistry, Trinity College, Cambridge, UK, *and* S W Cole. *Six years later, Hopkins carried out his memorable experiment on animal nutrition.*

1901

Blood groups (four primary)
Karl Landsteiner, *Austrian scientist working in Switzerland, who classified them as A, B, AB and O.*

Diesel engine (reversible gear)
Frederic Dyckhoff *of Bar-le-Duc, France.*

Double-edged safety razor
King Camp Gillette, *former travelling hardware salesman of Fond du Lac, Wisconsin,* USA. *By 1906, with modifications by* William Nickerson, *Gillette's American Safety Razor Co. had sold some 90,000 razors and 12,400,000 blades.*

Gillette's safety razor

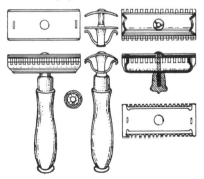

Electric typewriter
Dr Thaddeus Cahill *of Washington* DC. *Only 40 machines were built before the Cahill Writing Machine Co. ceased production.*

Fire engine (petrol-propelled)
Adler. *Water was discharged by gas pressure generated chemically.*

'First car radio'
Guglielmo Marconi, *Italian engineer working in London. A fully equipped Thornycroft steam wagon used for wireless telegraphy experiments.*

Grignard reagents
François Auguste Victor Grignard, *chemistry lecturer at Nancy, France.*

Hearing aid (electric)
Miller Reese Hutchinson *of New York. It was called the 'Acousticon'.*

Meccano
Frank Hornby, *Liverpool*, UK.

Motorcycle engine (centre-frame location)
Eugene *and* Michel Werner, *Russian émigrés working in Paris.*

Motor car mileometer
Bell Odometer by S H Davis Manufacturing Co., *Portland, Massachusetts*, USA.

Motor car speedometer
Thorpe & Salter Ltd, *Clerkenwell, London; calibrated from 0 to 35 mph.*

Multi-storey car park
City and Suburban Electric Carriage Co., *Piccadilly Circus, London – for owners of vehicles supplied by this company.*

Pre-selector gearbox
Dr Frederick William Lanchester, *Sparkbrook, Birmingham*, UK.

Vacuum cleaner
Hubert Cecil Booth, UK *civil engineer. His first machine was electrically powered and housed inside a horse-drawn carriage complete with 800-ft hose to go inside the house being spring-cleaned.*

Valine and proline (amino acids)
Emil Hermann Fischer, *German organic chemist in Berlin. He discovered oxyproline the following year.*

Wireless telegraphy (long-distance)
Guglielmo Marconi, *Italian engineer working in London. The letter 'S' was transmitted in Morse Code (. . .) from Poldhu, Cornwall, 2170 miles across the Atlantic to an aerial suspended from a kite in St John's, Newfoundland.*

1902
Air-conditioning
Willis H Carrier, US. *A 30-ton unit was installed in the Sackett-Wilhelms Lithographing & Printing Co's plant at Brooklyn, New York*, USA.

Air-cooled automobile
J Wilkinson *for* H H Franklin, US.

Alarm-clock tea-maker (spirit stove)
Frank Clarke, *gunsmith of Birmingham*, UK, *manufactured in 1904 by the Automatic Boiler Co., also of Birmingham.*

Disc brake (motor car)
Dr Frederick William Lanchester, *engineer, Sparkbrook, Birmingham,* UK.

Hormones
William Maddock Bayliss *and* Ernest H Starling, *his brother-in-law, physiologists at University College, London. They called their discovery 'secretin' but it was only the first of many hormones to be discovered in the human body.*

Indanthrene (non-fading dye)
James Morton, *textile manufacturer. He procured hundreds of samples of dyed fabrics and exposed them to light in a greenhouse, selecting those that did not fade.*

Intravenous anaesthetic
E Fischer, *biochemist, Germany. It was called 'Veronal'.*

Ivel agricultural motor
Dan Albone, *cycle manufacturer of Biggleswade, Bedfordshire*, UK, *after five years' development. Named after the local river, it remained in production until 1916.*

Kennelly-Heaviside layer (ionosphere)
discovered almost simultaneously by Arthur Edwin Kennelly, *Professor of Electrical Engineering, Harvard University*, USA, *and* Oliver Heaviside, *physicist working for the Great Northern Telegraphy Co., Newcastle-upon-Tyne*, UK.

Lawn-mower (petrol-engined)
James Edward Ransome, *Ipswich,* UK. *The 42-inch chain-driven machine*

was powered by a 6 hp Simms petrol motor and equipped with a passenger seat.

Mercalli Scale of felt intensity (earthquakes)
Giuseppe Mercalli, *Italian geologist and seismologist.*

'Mischmetal' (cerium-iron alloy)
Karl Auer, Freiherr von Welsbach, *chemist, Austria.*

Motorboat (transatlantic)
New York Kerosene Co., USA.
Crewed by Captain William C Newman and son, 'Abiel Abbot Low' took 36 days to motor from New York to Falmouth, UK.

Motor car (all-metal body)
the aluminium Napier 9, manufactured in London.

Motor caravan
Panhard *and* Levassor, *Paris, for Dr E E Lehwess, Germany, for an unsuccessful attempt to make the first motorised circumnavigation of the world.*

Motorcycle engine (V-twin, two-stroke)
Bichrone, *France.*

Overhead camshaft
Reginald W Maudslay *with* Alex Craig, *Coventry, UK – for both the first Maudslay car and commercial vehicle.*

Passenger excursion ship (steam-turbine)
'King Edward', launched on the Clyde.

Polygraph
James Mackenzie, *cardiologist, Burnley, Scotland; to measure arterial and jugular pulse simultaneously.*

Radium
Pierre *and* Marie Curie, *husband and wife, physicists at the School of Physics, La Sorbonne University, Paris, based on work by* Antoine Henri Becquerel *who had discovered 'Becquerel rays' in 1896. Having coined the term 'radioactivity', Marie died from its harmful side-effects.*

Solar engine (flat-plate/two-fluid system)
H E Willsie, *Olney, Illinois,* USA.

Straight-eight motor car engine (adapted)
CGV (Charron, Girardot and Voigt) *of Paris, using two of their four cylinder engines coupled together.*

Sulphur extraction process
Herman Frasch, *German-born chemist working for the Standard Oil Company at Louisiana,* USA.

Synchronous electric motor
Ernst Danielson, *Swedish engineer.*

Teddy bear
Simultaneously by Morris Mitchom, *Russian immigrant in a small sweet shop in Brooklyn, New York, with permission from* US *President Theodore ('Teddy') Roosevelt, and* Richard Steiff *at Giengen, Swabia.*

Track-circuit-controlled railway systems
London & South Western Railway Co., UK.

Aeroplane (controllable wing-warping)
Orville *and* Wilbur Wright, *former bicycle engineers at Kill Devil Hill, Kitty Hawk, North Carolina,* USA. *Their first practical aeroplane was the 16 hp 'Flyer III' of 1905, which made 40 flights, some for over half an hour, at 35 mph.*

Barbiturate drugs
Emil Hermann Fischer *(Germany) and* Emil Adolf von Behring, *who was Professor of Hygiene at Marburg University, Germany.*

Box-kite boat
Colonel Samuel Franklin Cody, US, *who had a kite tow his boat across the English Channel.*

Cross-ply tyre
Christian Hamilton Gray *and* Thomas Sloper, *at the India Rubber, Gutta Percha and Telegraph Works Co., Ltd, Essex,* UK. *They also patented, but never marketed, a*

radial-ply tyre in 1913.

Instant photographic printing
G C Beidler, *and marketed by the* US
Rectigraph Co. in 1907.

Micro-cinematography
Martin Duncan *with* Charles Urban,
London, UK. *'The Unseen World' as
filmed by the Urban-Duncan Micro-
Bioscope was first shown at the
Alhambra Theatre, London.*

Motorboat (diesel-engined)
Rudolph Diesel, Frederick Dyckhoff
and Adrien Bochet. *'Petit Pierre' was
a canal barge belonging to Hachette
and Driout, French iron-founders.*

Motor-chemical fire engine
Messrs Merryweather *in close
collaboration with the* Tottenham
District Council, *London.*

**Motorcycle engine (in-line, air-
cooled, four-cylinder)**
FN armaments firm, *Belgium.*

'N-rays'
René Blondlot, *Professor of Physics at
Nancy University, France.*

Nitrogen-from-air process
Kristian Olaf Birkeland *and* Samuel
Eyde *of Kristiania, Norway.*

Postage-franking meter
Karl Uchermann, *Norwegian
inventor, manufactured by Krag
Maskinfabrik of Kristiania.*

Saloon car (enclosed)
Duryea Co. *of Coventry*, UK.

Seamless bust support
Kate Morgan *of London.*

**Skyscraper (reinforced
concrete)**
the Ingalls Building, Cincinnati, USA,
with sixteen storeys.

**Steam-turbine ship
(transatlantic)**
the small yacht, 'Emerald'.

Straight-six motor car engine
Jacobus Spyker, *De Industrieele
Maatschappij, Trompenburg,
Netherlands.*

**String galvanometer
(electrocardiograph)**
Willem Einthoven, *Professor of
Physiology at Leyden, Holland.*

Swinging axles
Edmund Rumpler, *engineer,
Germany.*

**Trimaran (steam-engined/
aerially propelled)**
Lawrence Hargrave *of Stanwell
Park, Australia.*

Ultramicroscope
Richard Adolf Zsigmondy, *Austrian
chemist.*

V-8 motor car engine
Clément Ader, *France.*

1904

**Air-cushion hydrofoil (swing-
wing)**
Clément Ader, *France. It worked well
but was too complicated and unwieldy
for practical use, so was presented by
its inventor to the Conservatoire des
Arts et Metiers, Paris.*

Caterpillar/crawler tractor
Benjamin Holt, US, *and* David
Roberts *with* Ruston Hornsby, UK.

Double-sided gramophone discs
International Talking Machine Co.,
*Weissensee, near Berlin – called
'Odeon' records.*

'Emanium X'
Friedrich Oskar Giesel, *chemist,
Germany.*

Forced lubrication (motor car)
Dr Frederick William Lanchester,
engineer, Birmingham, UK.

Heckelphone
Wilhelm Heckel, *musical instrument-
maker, Biebrich-am-Rhein, Germany.*

Ivel armoured car
Dan Albone *of Biggleswade,
Bedfordshire*, UK. *With steel armour-
plating, supplied by Cammel, Laird &
Co. of Birkenhead, to cover the three-
wheeled vehicle, it was a development
of the Ivel tractor.*

Kapok lifebelts
*made from a vegetable fibre found
around Java in the Dutch East Indies
that was impervious to water and
highly buoyant.*

Motor crash helmet
'casques', used during a motor race at

the Parc des Princes, France.
Motorcycle engine (parallel vertical twin)
Eugene *and* Michel Werner, *France.*
Novocaine (local anaesthetic).
Parabolograph
H Payne *of London University, with a grant from the Drapers' Company. It was designed and engineered by Swiss instrument-maker, Coradi.*
Pinacyanol (infra-red sensitising agent)
Homolka.
Rolling-road testbed dynamometer
Serge Berditschewsky Apostoloff, *Russian immigrant working in London.*
Rubber engine mountings
Dr Frederick William Lanchester, UK *engineer, Sparkbrook, Birmingham,* UK.
Ship (diesel-engined)
Messrs Nobel Brothers *at the Kolomna Works, St Petersburg, Russia. The 'Wandal', an oil tanker, was used on the Caspian Sea.*
Slow-motion film
Lucien Bull *and* Henri Nogues *at the Nancy Institute, France.*
Snow-chains (car tyres)
Harry D Weed *of Canastota, New York.*
Submarine (diesel-engined)
the French Navy's 'Aigrette' and 'Cicogne'.
Tantalum light filament (incandescent lighting)
W von Bolton, USA.
Telescopic motorcycle fork
simultaneously by Terrot *of France and* Alfred A Scott, *dyer's technician, Yorkshire,* UK.
Thermionic valve (vacuum tube)
John Ambrose Fleming, *Professor of Electrical Engineering at University College, London. This made possible long-wave radio transmissions. Fleming also invented the potentiometer.*

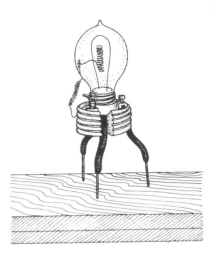

Fleming's thermionic valve

1905
Amphibian vehicle (petrol-engined)
Fournier, *France.*
Capillary motor-regulating mechanism (medical)
Schack August Steenberger Krogh, *marine biologist, University of Copenhagen, Denmark, following a study of the respiration of frogs.*
Dérailleur gear (bicycle)
Paul de Vivie, *France.*
Detachable wire wheel (motor car)
J V Pugh *of the Rudge-Whitworth Co.,* UK.
Diesel engine (reversible two-stroke)
Sulzer Brothers, *Sweden.*
Electric motor horn
Miller Reese Hutchinson *of New York.*
Electric strip lighting
D McFarlan Moore *of Moore Electric Co., London – used for advertising.*
Electromagnetic seismograph
Prince Boris Borisovich Golitsyn, *physicist, Russia.*
Engine bearings (tapered oil film)
A G M Michell, *to avoid mechanical*

friction.

Fire-extinguisher (chemical foam)
Professor Alexander Laurent, *St Petersburg, Russia.*

Float-mounted glider (man-carrying)
Gabriel Voisin, *France. It took off from the River Seine, towed by 'La Rapière', a racing motorboat.*

Fluid coupling (hydraulic centrifugal clutch)
Hermann Fottinger; *developed by Vulcan of Hamburg for marine and diesel railway use.*

Gas turbine (workable)
René Armengaud *and* Charles Lemale, *Société des Turbomoteurs, Paris.*

Hydrofoil ladder (hull attachment)
Enrico Forlanini, *Italy. In 1906 Forlanini tested his hydrofoil boat on Lake Maggiore. It lifted clear of the water and reached a speed of 38 knots using a 75 hp engine.*

Hydroxy citronellal (synthetic perfume)
having the scent of lilies of the valley.

Juke-box (pre-selective)
John C Dunton *of Grand Rapids, Michigan,* USA. *It gave the listener a choice of 24 different Edison phonographic recordings.*

Liquid metal polish
W H Slack *for Reckitt & Sons Ltd, Hull,* UK – *'Brasso'.*

Pneumatic bumpers (motor car)
Frederick R Simms *at his Kilburn factory, North London, for the 20 hp Simms-Welbeck car.*

Pneumatically operated semaphore indicator arms (motor car)
Fritz Berger *of Berlin. They were self-cancelling.*

Pressure cooker (aluminium)
Presto Co. *of Eau Claire,* USA.

Safety glass
John Crewe Wood, *solicitor of Swindon, Wiltshire,* UK, *who fitted it to his Peugeot Bébé.*

Silicones
Frederic Stanley Kipping, *Professor of Chemistry, University College, Nottingham,* UK.

Special theory of relativity
Albert Einstein, *German-born, Swiss-Jewish mathematical physicist, whilst examiner at the Swiss Patent Office.*

1906

Alexanderson high frequency alternator:
Ernst Frederik Werner Alexanderson, *Swedish-born engineer, General Electric Co. Schenectady, New York,* USA.

Animated cartoon film
simultaneously by James Stuart Blackton *for the Vitagraph Co. of New York,* USA, *and* Walter Booth *for the Charles Urban Trading Co., London.*

Automatic flashing apparatus (lighthouses)
Nils Gustaf Dalén, *engineer-manager, Swedish Carbide and Acetylene Co. He also invented 'Solventil' and 'Agamassin'.*

Battleship (steam-turbine)
HMS *'Dreadnought'.*

Carbon suboxide
Otto Paul Hermann Diels, *Professor of Chemistry, Berlin University, Germany.*

'Cherry Blossom' (shoe polish)
Chiswick Soap & Polish Co., *London.*

Chromatography
Mikhail Tswett, *Russian botanist, Warsaw University.*

Constant-volume gas turbine
Dr Hans Holzwarth. *The first machine was built in Hanover, Germany, two years later.*

Crystal radio detectors (silicon, galena etc.)
H H C Dunwoody *and* G W Pickard, US.

Electric motor horn
United Motor Industries Ltd,

Sherbourne Works, Coventry, UK – *called the Wagner Electric Motor Horn.*

Electric washing-machine
Alva J Fisher, *Chicago, Illinois,* USA. *It was marketed in 1910 by the Hurley Machine Co. as the 'Thor'.*

Ellipsoidal-wing floatplane
Gabriel Voisin *and* Louis Blériot, *France. Trials on Lake Enghien, Belgium, were disastrous.*

Growth-stimulating vitamins
Frederick Gowland Hopkins, *biochemist, Prelector in Physiological Chemistry, Trinity College, Cambridge,* UK.

Heat theorem (third law of thermodynamics)
Hermann Walther Nernst, *Professor of Physical Chemistry, Berlin.*

Hookless zip-fastener (matching metal locks on a flexible backing)
Dr Gideon Sundback, *Swedish émigré electrical engineer, working for the Judson Company, Chicago. In 1913 he further developed a zip-fastener manufacturing machine.*

Humidity control and dust filter (air conditioning)
Stuart Warren Cramer, *mill-engineer, Charlotte, North Carolina,* USA.

Ionium
Bertram Borden Boltwood, *Professor of Physics, Yale University,* USA.

Kinemacolor (motion pictures)
George Albert Smith *of Brighton,* UK, *sponsored by businessman Charles Urban. Regular public screenings were held three years later at Urbanora House, Wardour Street, London.*

Lutetium
Georges Urbain, *assistant Professor of Analytical Chemistry, la Sorbonne. Latin 'Lutetia' is the ancient name of Paris, where Urbain was working.*

Marine outboard engine (mass-produced)
Cameron B Waterman, George Thrall *and* Oliver E Barthel, *Detroit, Michigan,* USA.

Nichrome
Albert Marsh *of Lake County, Illinois,* USA. *This was the ideal alloy for electric-fire elements.*

Permanent waving (hairdressing)
Karl Ludwig Nessler, *German-born hairdresser working in London.*

Pneumatic-tyred undercarriage (aircraft)
Trajan Vuia, *Italian, with his tractor-propeller monoplane, the Vuia No. 1.*

Public radio broadcast
Reginald Aubrey Fessenden, *Canadian-born professor at his Brant Rock Station, Massachusetts,* USA.

Röntgen radiation of the elements
Charles Glover Barkla, *research physicist and chorister, University of Liverpool,* UK.

Sound-on-film motion pictures
Eugen Augustin Lauste, *French engineer working in Brixton, London. Four years later he produced a workable system using an electromagnetic recorder and string galvanometer.*

SOS (distress signal)
British Marconi Society *and* German Telefunk *organisation at the Berlin Conference. Formally introduced in 1908.*

Steamcar (high-speed)
F O *and* F L Stanley, *whose vehicle, driven by Fred Marriott, reached 127.5 mph on Ormonde Beach, Daytona, Florida,* USA.

Telephone amplifier valve
Robert von Lieben, *Vienna, Austria.*

Triode amplification tube
Professor Lee de Forest *of New York,* USA. *It was called the 'Audion'.*

Wheeled undercarriage (aircraft)
Alberto Santos-Dumont, *Brazilian engineer-aviator working in Paris, for his '14-Bis'.*

1907

Banked motor-racing circuit (custom-built)
Messrs Donaldson *(railway technician)*, Holden *(colonel, Royal Engineers)* and L G Mouchal, *as commissioned by Hugh Fortescue Locke-King of Brooklands Estate, Weybridge, Surrey,* UK.

Cinema projection system
The *'Kinedrome'* – D J Bell *and* A S Howell, USA.

Detergent (household)
Henkel et Cie, *Düsseldorf. Based on the dry soap powder process of* Professor Hermann Geissler *and* Dr Hermann Bauer, *it was called 'Persil'*.

Helicopter (petrol-engined)
independently by Frenchmen Paul Cornu, *with a twin-rotor design, and* Louis Breguet, *with four rotors on outriggers. Cornu and his passenger hovered at 5 ft for over a minute.*

Hydroplane (aerially propelled)
Comte de Lambert *of France and* Horatio Phillips, UK. *Powered by a Serpollet steam engine with a flash boiler, it was tested on the River Seine, France.*

Monoplane
Louis Blériot, *France, whose 50 hp Antoinette-engined Type VII made a long enough flight to be deemed airworthy. Blériot's 1909 cross-Channel flight underlined the monoplane's stability. Blériot also invented the aileron.*

Multiplane (venetian-blind type)
Horatio Phillips, UK. *This, his second attempt at a multiplane machine, had 50 horizontal winglets and was powered by a 22 hp engine. It flew to 500 ft during a flight test over Norbury,* SE *London – according to Phillips.*

Ocean liners (steam-turbine engines)
Cunard Shipping Co.'s *'Lusitania' and 'Mauretania'.*

Spangler's upright vacuum cleaner, trade-named 'Hoover'

Rotary aero engine (production model)
Laurent Seguin, *France. It was marketed as the 50 hp 'Gnome'.*

Supercharged motor car
Lee S Chadwick and his assistant *at Pottstown, Pennsylvania,* USA – *'The Great Chadwick Six'.*

Upright vacuum cleaner (dustbag attached)
J Murray Spangler, *janitor in a department store at New Berlin, Ohio. Spangler's patent rights were purchased by former harness-maker,* W H Hoover, *who produced a commercial model the following year.*

1908

Automatic typewriter
T A McCall *of Ohio,* USA. *It was controlled by a pianola-type roll.*

Bakelite
Dr Leo Hendrik Baekeland, *Belgian-born inventor working in Yonkers, New York. At temperatures of over 100°C, Baekeland experimented with formaldehyde and phenolic bodies to produce a synthetic shellac from simple molecules. After developing his process into a three-stage reaction, he took out patents and in 1912 became President of the Bakelite Coporation with the slogan 'the material of a thousand uses'.*

1908

BCG (tuberculosis) vaccine
Leon Charles Albert Calmette, *Director of the Pasteur Institute, Lille, France, with* Dr Camille Guérin.

Cellophane
Dr Jacques Edwin Brandenberger, *Swiss-born French dye-chemist at Thaon-les-Vosges – a transparent wrapping material. The Comptoir de Textiles Artificiels, then the largest rayon producer, agreed to finance him. A company called La Cellophane was formed and machine production began in 1912.*

Crankshaft torsion dampers
Dr Frederick William Lanchester, UK *engineer, Sparkbrook, Birmingham,* UK.

Diesel engine (reversible four-stroke)
Nobel Brothers, *oil distributors, St Petersburg, Russia.*

Filter coffee machine
Frau Melitta Bentz, *Dresden, Germany.*

Gyroscopic compass
Dr Hermann Anschütz-Kämpfe *and* Dr Max Schuler, *Kiel, Germany, for polar exploration. The Sperry-Ford version followed in 1910.*

Hydroplanes (multi-step/petrol-engined)
William Henry Fauber, *former bicycle engineer of Chicago, Illinois* USA.

Language-teaching system
Thódore Rosset *of Grenoble, France. Using gramophone discs and ciné films to teach French to foreigners.*

Liquid helium
Heike Kamerlingh-Onnes, *Dutch physicist at the Cryogenic Laboratory, Leyden.*

Nickel-cadmium storage cells
Jungner, *Sweden.*

Nickel-iron alkaline storage battery
Thomas Alva Edison, *Menlo Park, New Jersey,* USA.

'Noiseless' typewriter
W P Kidder *and* C C Colby *of Quebec, Canada.*

Oil (south-west Persia)
William Knox d'Arcy, *wealthy Englishman, who had made a fortune in gold-mining in Australia. The Anglo-Persian Oil Company was formed in 1909, following the strike at Masjid-i-Sulaiman.*

Paper cups (for drinking)
Public Cup Vendor Co., *New York.*

Poliomyelitis
Karl Landsteiner, *pathologist, Austria.*

Pre-transfusion blood tests
Dr Reuben Ottenberg, *New York.*

Safety film (cellulose acetate base)
Kodak, *Rochester, New York,* USA.

Sedimentation equilibrium
Jean Baptiste Perrin, *Professor of Physical Chemistry, Paris University.*

Ship-steadying gyroscope
Dr E O Schlick *and* Pierre Schlowsky.

Silencer (firearm)
Hiram Stevens Maxim, *London.*

Solar thermoelectric device
W Zerassky, *Russian engineer.*

Thought machine (needle-deflecting-scale)
Edmonde Savary d'Odiardi, *Paris.*

Tungsten filament (ductile)
William D Coolidge, *General Electric Co., Schenectady,* USA.

V-Eight motorcycle
Glen L Curtiss, US.

1909

Barany indication test
Robert Barany, *Austro-Hungarian lecturer in Otology, the Ear Clinic, Vienna University.*

Doppler effect in canal rays
Johannes Stark, *physicist at the Technische Hochschule, Aachen, Germany.*

Double-decker bus (enclosed)
Widnes Corporation, UK.

IUD (intra-uterine contraceptive)
Dr R Richter, *German physician. He devised a ring-shaped* IUD *out of silkworm gut.*

Salvarsan (compound 606)
Dr Paul Ehrlich *and* Hideyo
Noguchi, *bacteriologists, Frankfurt-
am-Main, Germany. Nicknamed 'the
magic bullet', this arsenic compound
was the first chemical cure for disease
in this case, early syphilis. It was
manufactured by Hoechst of Germany.*
Synthetic ammonia process
Fritz Haber, *Professor of Physical
Chemistry, Karlsruhe, Germany;
industrially exploited in 1913 by Carl
Bosch at the Badische Anilin und
Sodafabrik's pioneer factory.*
Typhus fever body louse
Charles Jules Henri Nicolle, *Director
of the Pasteur Institute, Tunis.*

1910
Aero-engined car
Fiat *of Turin, Italy. A 300 hp airship
engine was installed in a motor car,
giving it a speed of 132.37 mph at
Ostende, Belgium.*
Duralmin (metal)
Durener Metallwerke AG, *Germany.*
Gene theory
Thomas Hunt Morgan, *Professor of
Experimental Zoology, Columbia
University, New York, during his
study of thousands of generations of the
fruit-fly Drosophila.*
Hydro-aeroplane (seaplane)
Henri Fabre *of Marseilles, France. In
his first flight from Martigues, Fabre
was airborne at 6 ft for 500 yards.*
Neon lighting
Georges Claude, *French physicist. It
was first used to illuminate the
peristyle of the Grand Palais during
the Paris Motor Show. Within four
years, some 150 neon signs had been
installed on buildings in Paris alone.*
Powers' accounting machine
James Powers, US – *automatic card
punch, sorter and tabulator.*
Radio-direction finder
Ettore Bellini *and* Alessandro Tosi,
*Italy; acquired by the Marconi Co.,
UK, and first installed at Boulogne,
France.*

Spring-operated mouse-trap
James Henry Atkinson, *Leeds,* UK.
Tanker ship (diesel-engined)
*'Vulcanus', built by the Netherlands
Shipbuilding Co. of Amsterdam for
operation by the Anglo Saxon
Petroleum Co.*
Triplex laminated glass
Edouard Benedictus, *Paris.*
Tumour-inducing virus
Francis Peyton Rous, *associate of the
Rockefeller Institute for Medical
Research, New York,* USA.

1911
Aircraft carrier (adapted)
the US *Navy's cruiser, 'Pennsylvania',
with a platform built over her stern.
The plane was landed by Eugene Ely.*
Air-raid
Lieutenant Giulio Gavotti, *Italy, in
an Etrich monoplane, dropped several
4½ lb Citelli bombs on both the Turkish
position at Ain Zara and the oasis at
Tagiura.*
**Airship-tethering mast
(offshore)**
H B Pratt *of Vickers-Armstrong Ltd
for the Cavendish Dock, Barrow-in-
Furness,* UK, *to secure the 'Mayfly'
airship.*
**Amphibious biplane (canard
configuration)**
Gabriel Voisin. *Piloted by Maurice
Colliex, it took off from Issy
aerodrome, France, landed on the
River Seine and returned to Issy. A
dozen models were acquired by both the
French and Russian Navies soon after.*
Binet intelligence scales
Alfred Binet, *Director of
Physiological Psychology at the
Sorbonne, France, with* Theodor
Simon.
Blood groups A1 and A2
Karl Landsteiner, *Professor of
Pathological Anatomy, Vienna.*
**Calculating machine (full
automatic multiplication and
division)**
Jay R Munroe, *New Jersey,* USA.

1911

Cloud chamber
Charles Thomson Rees Wilson, *Scottish physicist at the Cavendish Laboratory, Cambridge,* UK.

Cosmic radiation
Victor Franz Hess, *Assistant at the Institute of Radium Research, Vienna Academy of Sciences.*

Diesel ship (transatlantic)
'Toiler', fitted with a Swedish engine.

Electric frying pan.

Harmonic balancers
Dr Frederick William Lanchester, *engineer, Sparkbrook, Birmingham,* UK.

Hydrofoil biplane
A Guidoni, *Italy, using a modified Farman F.1.*

Intertype (printing process)
Hermann Ridder.

Isotopes
Sir Joseph John Thomson *at the Cavendish Laboratory, Cambridge,* UK.

Machine-gun (air-cooled)
Isaac Newton Lewis, US *artillery officer. It was the first machine-gun to be fired from an aircraft – in 1912.*

Molecular beam
Louis Dunoyer de Segonzac *at Marie Curie's Laboratory, La Sorbonne, France.*

Motion-picture camera (electric drive)
E F Moy *of London. With a gyroscope to stabilise it and a displacement film-magazine, it replaced hand-cranked machines.*

'Nivea' cream
Beiersdorf GmbH, *Hamburg, Germany.*

Oil (underwater drilling)
Gulf Refining Co., *in Caddo Lake on the Louisiana-Texas border,* US.

Powerboat (aero-engined)
DeKorwin hydroplane, 'Soulier-Volant', powered by a US Wright aero-engine. It raced at Monaco.

Retractable undercarriage
Wiencziers *monoplane, Germany.*

Self-starter (automobile)
Charles F Kettering *at Dayton, Ohio,* USA. *The first 4,000 'Delco' starters were sold to Cadillac Motors.*

Theory of atomic structure
Ernest Rutherford, *New Zealander, and his Danish assistant,* Niels Henrik Bohr, *at the University of Manchester,* UK.

'US Patent No. 1,000,000'
Francis H Holton, *Akron, Ohio – described as 'vehicle tyre improvements'.*

1912

Activated sludge process (sewage)
Arden *and* Lockett, *Manchester,* UK.

Aerograph (pneumatic-drive cine camera)
Arthur Samuel Newman, *inventor, Highgate, London, for use on horseback by Count Proszynski of Russia.*

Cabin biplane
Igor Sikorsky, *Russia, aged 23. With four 100 hp engines, a 92 ft wingspan, an enclosed cockpit for pilot and co-pilot with dual controls, a luxuriously decorated cabin for sixteen passengers and even a toilet, this was the prototype of the modern airliner, even if it only made a few flights. A cabin biplane was also built by* Alliot Verdon Roe, UK, *during this year.*

Diffraction of X-rays by crystals
Max Theodor Felix von Laue, *Professor of Theoretical Physics at Munich, with his assistants,* Friedrich *and* Knipping.

Electric fire (fire-clay frame)
Charles Reginald Belling, *Enfield, Middlesex,* UK. *It was called the 'Standard'.*

Hydraulic hoist (device for dumping trucks)
Garfield Arthur Wood, *mechanic and boatbuilder, St Paul, Minnesota,* USA.

Liquid silver polish
Reckitt & Sons Ltd, *Hull,* UK – *'Silvo'.*

Locomotive (diesel-engined)
North British Locomotive Co. –
*1000 hp diesel-mechanical unit with a
direct drive Klose-Sulzer engine.*
Monocoque monoplane
Louis Becherau, *32-year-old French
engineer working for Armand
Deperdussin, wealthy silk broker. It
was called the 'Deperdussin'.*
Motor-narrowboat (custom-built)
Fellows, Morton & Clayton *of
Birmingham*, UK.
**Regenerative, oscillating
features of the triode**
simultaneously by Lee de Forest, US;
and Edwin Howard Armstrong, US;
and Fritz Löwenstein, US, Siegmund
Strauss, *Austria, and* A Meissner,
Germany.
Self-service grocery store
*Alpha Beta Food Market at Pomona,
and also Ward's Groceteria in Ocean
Park, California*, USA.
**Stainless steel (chromium
alloys)**
simultaneously by Harry Brearley,
self-taught metallurgist, UK*; and*
Elwood Haynes, *automobile inventor
of Kokomo, Indiana*, US; *and* Edward
Maurer *and* Benno Strauss *of
Krupps, Germany.*
Synthesised butadiene
Chaim Weizmann, *German-born
chemist at Manchester University*, UK
– bacterial fermentation culture.
Synthetic wood glue
Dr Leo Hendrik Baekeland, *Belgian-
born chemist, General Bakelite
Corporation*, USA.
Vest pocket Kodak camera
USA*; almost 2 million sold by 1926.*
Vitamin B1 (thiamine)
Casimir Funk, *Polish-born biochemist
working at the Lister Institute,
London.*

1913
Autopilot
Lawrence Sperry, US, *on a Curtiss
floatplane.*

Bicycle (semi-bodyshelled)
Etienne Buneau-Varilla, *France. It
was called the 'Vélo Torpille' (torpedo
bicycle).*
Crossword puzzle
Arthur Wynne, *Liverpool-born. It
was first published in the weekend
supplement of the 'New York World'.*
Diesel-electric railcar
*Mellersta Sodermanlands Railway,
Sweden.*
**Dumet (iron-nickel alloy wire
for incandescent lamps)**
Dr Colin Garfield Fink,
*electrochemist, Head of the Research
Laboratory, Chile Exploration Co.,
New York City*, USA.
**Electric self-starter
(motorcycle)**
Indian Co., US. *Called the
'Dynastart', it was used on their
Hendee Special.*
Gas-filled tungsten lamp
Irving Langmuir, *Research
Laboratory, General Electric Co.,
Schenectady, New York*, US.
Hydrocycle (aerial propeller)
Charolais-Favelier, *France.*
Isotopic labelling technique
Georg von Hevesy (*Hungary*) *and*
Friedrich Adolf Paneth (*Austria*)
whilst staying in Vienna.
**Moving assembly line for
automobile magnetoes**
Henry Ford *of Ford Motor Company,
Detroit*, USA, *for the mass-production
of his Model T utility automobile, 'Tin
Lizzie', alias 'Flivver'. By 1914 the
system had been extended to the
assembly of the entire car. By 1927
over 15 million 'Tin Lizzies' had come
off the production line and been sold.*
Protactinium
Frederick Soddy, *Lecturer in
Physical Chemistry at Aberdeen, with*
John A Cranston, UK.
Quantum theory (sub-atomic)
Niels Bohr, *Danish physicist,
following work with* Thomson *at
Cambridge and* Rutherford *at
Manchester – applying the theory to*

sub-atomic physics.

Synthetic detergent
A Reychler, *Belgian chemist. It was marketed in 1917 as 'Nekal'.*

Vitamin A
simultaneously by Thomas B Osborne, *biochemist, US, with* Lafayette B Mendel, *Professor of Physiological Chemistry, Yale University, and* Elmer V McCollum *with* M Davis. *at the Wisconsin Agricultural Experiment Station, USA.*

Wireless communication to an aeroplane
Received by Elmo Pickerill in a Curtis biplane, up to a height of about 1 mile.

Woolworth Building (60 storeys)
Cass Gilbert, *American architect. Broadway, New York City, USA.*

1914

Aluminium-foil bottle caps
Josef Jonssen *of Linköping, Sweden.*

Brassière
Miss Mary Phelps Jacob *(later glamorous heiress, Mrs Caresse Crosby), whilst a débutante in New York, USA. Miss Jacob's patent for the 'backless brassière' was eventually acquired by the Warner Brothers Corset Company for a mere $15,000.*

Curved windscreens (automobile)
Kissel, *USA.*

Dye-coupler colour process
Dr Rudolph Fischer *and* Siegrist, *Berlin, Germany – one of the factors which ultimately made colour photography a practical reality. The weakness in the Fischer-Siegrist process was the lack of diffusion control between the layers of sensitising dyes.*

Electric traffic light (red)
Alfred A Benesch, *Cleveland, Ohio, USA, who formed the Traffic Signal Co.*

'Fléchettes' (steel darts)
jettisoned by Royal Flying Corps pilots from their aircraft over enemy lines.

Giant mortar (gun)
'Big Bertha', built by the Skoda

Works in Austria-Hungary and used by the Germans in battering the forts at Liège and Namur, Belgium. Capable of firing a 1980 lb shell, it was nicknamed after Frau Bertha von Bohlen, head of the Krupp armaments factory.

Grand Prix engine (desmodromic valve gear)
Automobiles Delage, *Courbevoie, France.*

Insulating board
Karl Munch, *Germany.*

Leica (35 mm film still camera)
Oskar Barnack, *microscope designer at Leitz in Wetzlar, Germany. It went into production ten years later.*

Oskar Barnack's original 35mm Leica camera

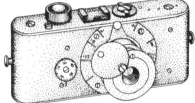

Multi-engined heavy bomber
Igor Sikorsky's *'Ilya Mourometz', for the Imperial Russian Air Corps.*

Panama Canal
John F Stevens, *US engineer, and* Lt Colonel George W Goethals, *in charge of the US Army engineering detachment.*

Passenger-carrying aeroplane (commuter)
Benoist Flying Boat, carrying passengers between St Petersburg and Tampa, both in Florida, USA.

Tear gas (xylyl-bromide)
Dr von Tappen, *Berlin. It was used in 1915 by the Germans against the Russians at Bolimow, Poland.*

1915

Airship air-raid
'Zeppelin L-3,' over Great Yarmouth, UK. It killed two people.

All-metal aeroplane (cantilever wing)
Hugo Junkers, Dr Mader *and* Otto Reuter *at Dessau, Germany. The Junkers J.1 was nicknamed the 'Tin Donkey'.*

Amplitude modulation (AM)
Hendrick Johannes van der Bijl, *South African physicist, and* Raymond A Heising, *electrical engineer,* US, *at the Western Electric Co., New York.*

Bacteriophage
Frederick William Twort, *Professor of Bacteriology, University of London.*

Carrel-Dakin treatment
Alexis Carrel, *French Army Surgeon, and* Henry Drysdale Dakin, UK *chemist, as a solution for treating wounds received in combat.*

Chlorine gas (weapon)
suggested by Sergeant Fritz Haver, *Jewish chemist. It was used by Germany against the Allied Forces at Ypres, West Belgium, so was nicknamed 'Ypérite' by the French troops, who tried tying pads of cotton dipped in a chemical solution over nose and mouth to avoid harmful effects.*

Gas mask (veil mask respirator)
Robert Davis, Siebe Gorman & Co., *Lambeth, London.*

Equal-distance bomb sight (for aircraft)
Warrant Officer F W Scarff, *Royal Navy, and soon adopted by the Royal Naval Air Service,* UK.

Motor scooter
Auto-Ped Co. *of New York,* USA.

Periscope rifle
Lance-Corporal Beech *from Sydney, Australia, whilst fighting at Gallipoli, Turkey.*

Propeller-synchronised machine-gun
Anthony Herman Gerard Fokker, *Dutch engineer working for the Germans. This enabled the pilot to fire his machine-gun without destroying the wooden blades of his propeller. The scourge of the Fokker Eindecker*

monoplane with its synchronised gun lasted nine months before a comparable device was produced by George Constantinesco for the Allied Forces.

Radiotelephone (long-distance)
Western Electric Telephone and Telegraph Co., USA, *with long-distance calls from Arlington, Virginia to San Francisco and Hawaii, then three weeks later across the Atlantic to Paris.*

Tank (tracked, armoured vehicle)
RNAS Lieutenant Walter Gordon Wilson, *Armoured Car Squadron, working with* William Tritton *of Fosters in Lincoln,* UK. *Following their prototype, 'Little Willie', the lozenge-shaped boiler-plated 'Mother' was so successful that 100 ironclads were ordered. For secrecy's sake, they were called 'Water Carriers', hence the word 'tank'. By 1917 some 450 improved British Mark IV tanks attacked the Hindenburg Line at Cambrai, capturing 7500 prisoners and 120 guns. Each tank had an eight-man crew.*

'Tank' – tracked armoured vehicle

Tank (French design)
Jean-Baptiste Estienne, *French artillery officer.*

Thermostatic oven regulator (gas stoves).

1916
Concrete shell structure
Eugène Freyssinet, *French architect.*

Albert Einstein

Two giant airship hangars at Orly Airport, France, were completed by 1924.
General theory of relativity
Albert Einstein, *Director of the Kaiser Wilhelm Physical Institute in Berlin.*
Long-wave M-lines (X-ray spectroscopy)
Karl Manne Georg Siegbahn, *Deputy Professor of Physics at Lund University, Sweden.*
Mechanical windscreen wipers (automobile)
Willys-Knight, *New York City*, US.
Microphone (condenser type)
E C Wente.
Nissen hut
Peter Norman Nissen, *Canadian-born officer commanding 29 Company Royal Engineers. A prototype was erected at Hesdin, France.*
Open-air public address system
Bell Telephone Co., *Staten Island, New York*, USA.
Organic micro-analysis
Fritz Pregl, *Professor of Applied Medical Chemistry, Innsbruck, Austria.*
Radio tuning device (selective)
Ernst Frederik Werner Alexanderson, *General Electric Co., Schenectady, New York*, USA.
Vibraphone (musical instrument)
Hermann Winterhoff, *of the Leedy Drum Company*, US.

1917
Chernikeef ship's log
Captain B Chernikeef *of the Russian Naval Hydrographic Department.*
Echo-sounder
Paul Langevin, *French physicist, and* Professor Robert W Boyle. *Called* ASDIC *(Anti-submarine Detection Investigation Committee) it was ready for use by the Royal Navy in 1918.* SONAR, *the US version, was developed concurrently.*
General treatment for paralysis
Julius Wagner von Jauregg, *Austrian neurologist and psychiatrist.*
Machine-gun (water-cooled/recoil-operated)
John Moses Browning, *inventor, Ogden, Utah*, USA.
Mustard gas
German scientists. *In 1918, during ten days of the Somme offensive, the Germans used some 500,000 mustard gas shells. The newly developed, unwieldy gas masks used by the Allied Forces were of little help.*
Roll-towel cabinet
George Steiner *of the American Linen Supply Co., Salt Lake City, Utah,* USA.
Superheterodyne radio receiver
Lucien Lévy, *France.*
Tractor (mass-produced)
Ford Motor Co., *Detroit*, USA. *By 1926, 70,000 'Fordsons' were coming off the assembly line per year, 75 per cent of all tractors made in the* USA.
Very High Frequency (VHF)
Guglielmo Marconi, *Italian electrical engineer, experimenting near*

Caernarvon, Wales, UK.
Windscreen wipers (electric)
Ormond Edgar Wall, *Hawaiian dentist.*

1918
Anti-tank gun
German armaments engineers.
Domestic refrigerator (electric-power)
Nathaniel Wales *and* E J Copeland, US. *It was called the 'Kelvinator'.*
Food mixer (electric)
Universal Co., US.
Non-rigid airship (helium-filled)
US Navy.
Practical electric clock (AC)
Henry Ellis Warren, *electrical engineer, Ashland, Massachusetts,* USA.
Sheet glass
Emile Fourcault *in Belgium and* Irving Colburn *in the* USA.
Traffic lights (red-green-amber)
New York City, USA. *They were*

The 'Universal' electric food-mixer

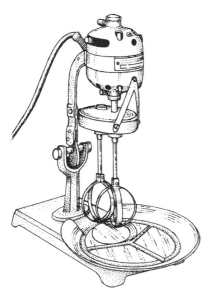

electrically powered but worked manually from a 'crow's nest'.
Ventriculography
Walter Edward Dandy, *Professor of Neurological Surgery, Johns Hopkins Hospital,* USA.
Vitamin D (antirachitic factor)
Edward Mellanby, *Professor of Physiology, University of London,* UK.

1919
Aircraft (transatlantic)
Vickers Vimy, piloted by Captain John Alcock *and* Lieutenant Arthur Whitten-Brown *from Newfoundland to Ireland, taking almost sixteen hours.*
All-welded ship (offshore)
Cammell Laird *of Birkenhead,* UK, *whose 'Fullager' was built for the Anchor Brocklebank Line of Liverpool.*
Atomic nuclei bombardment
Sir Ernest Rutherford, *Cavendish Professor of Physics at Cambridge University,* UK.
Compound epicyclic gear train
Walter Gordon Wilson, UK. *First tried experimentally with a 1927 Vauxhall, it was later used in his pre-selector gearbox.*
Electrets (permanent electrostatic fields)
Mototaro Eguchi, *Japan.*
Flying boat (transatlantic)
Glenn Curtiss's *NC-4 from Long Island,* USA *to Lisbon, Portugal.*
Grease-filled 'nipple'-gun
Edward Coe Critchlow, *production superintendent of the Union Oil Co. of California,* USA.
Hydrofoil (high-speed)
Alexander Graham Bell. *His 60 ft 'Hydrodome IV' achieved a world water speed record of almost 71 mph.*
Mass spectrograph (double-focusing)
Francis William Aston, *scientist at the Cavendish Laboratory, Cambridge,* UK, *to investigate isotopic structures of elements.*

1919

Microfilm-reading machine
Rear-Admiral Bradley A Fiske, US *Navy, New York. It was called the 'Fiskeoscope'.*
Neutrodyne receiver
Alan Hazeltine, *Professor of Electrical Engineering at Stevens Institute, Hoboken, New Jersey*, USA.
Parachute (manually operated)
Leslie 'Ski-Hi' Irvin *and* Floyd Smith, *former* US *circus acrobats and balloonists. The first successful jump was from a de Havilland 9 biplane from a height of 1500 ft. The Irvin Air Chute Co. was formed in Buffalo soon after.*
Servo-assisted, four-wheel brakes
innovated for the Type H6 Hispano-Suiza automobile.
Straight-eight automobile engine
Fred Duesenberg, *Indianapolis*, USA.
Thyroxine
Edward Calvin Kendall, US *chemist. It was synthesised seven years later by* Charles Robert Harington, UK *chemist.*
'Trigger relay' (electronic valve as switch)
William Henry Eccles *and* Richard Jordan, *physicists at the Imperial Wireless Telegraphy Commission*, UK.

1920

Airscrew-propelled motor car
Leyat, *France.*
Autopilot equipment
the Aveline-Stabiliser, as used on the Handley-Page 0/10, built and flown from Cricklewood, North London.
Catgut (surgical thread).
Electrical recording process
Lionel Guest, *stockbroker and* H O Merriman *of Paddington,London.*
Fire-fighting engine (foam-equipped)
Foamite Childs Corp., US.
Light bulb and vacuum tube mass-production machine
Corning Glass Works, Illinois, USA.

Capable of making 20-30,000 glass bulbs per day.
Microphone (double-button type)
Bell Telephone Laboratories, Holmdel, New Jersey, USA.
Radio station (regular broadcasts)
Westinghouse's KDKA, *Pittsburgh, Pennsylvania*, USA.
Rotary hoe (steam-powered)
Cliff Howard, *farm-engineer, Gilgrandra, New South Wales, Australia. It was later called the 'Rotovator'.*
Tea-bags
Joseph Krieger, *San Francisco*, USA.
X-ray analysis of crystal structure
William Henry Bragg, *Quain Professor of Physics, University College, London, and his son,* William Lawrence Bragg *at Manchester University – using their spectrometer specially developed for this purpose.*
Windscreen wipers (vacuum-operated)
William Mitchell Folberth, *auto-mechanic, Cleveland, Ohio*, USA; *manufactured by the Trico Corp. of Buffalo.*

1921

Aerial crop-dusting
Lieutenant John B Macready, *piloting a Curtiss JN6 light aircraft over a catalpa-infested grove in Troy, Ohio*, USA. *It had been suggested by C R Neillie, Cleveland entomologist.*
Aerocycle
Edouard Nieuport, *former bicycle racer and aircraft manufacturer, France. Pedalled by Gabriel Poulain, the Aviette flew 10.54 metres at a height of 1.5 metres above the Longchamps racecourse, so winning the Peugeot Prize.*
Automobile reversing light
Wills-Sainte Claire Co., *Marysville, Michigan*, USA.
Glutathione
Frederick Gowland Hopkins,

Professor of Biochemistry, Cambridge,
UK.
Hydraulic four-wheel brakes
Duesenberg Motor Co.,
Indianapolis, USA.
Insulin (animal)
Frederick G Banting *and* Charles H
Best *at J J R Macleod's laboratory,
Toronto University, Canada.*
Ion-radiated food (practical use)
Schwarz, US, *to kill 'trichinella
spiralis' bacillus in meat.*
Lie detector
John Larsen, *medical student,
California*, USA.
Ley lines
Alfred Watkins, *mill owner and
photographer, Herefordshire*, UK, *who
claimed that the pre-Roman trackways
of Britain were constructed in straight
lines, marked out by mystical and
religious temples and monoliths.
Following publication of his 'Ley
Hunter's Manual' in 1927, the
Straight Track Postal Portfolio Club
was formed to re-plot the network of
'leys'.*
Magnetron
Dr Albert Hall, *General Electric Co.,
Schenectady, New York.*
'Neracar'
J Neracher, *Syracuse, New York,*
USA.
'Robot'
the word was used by Karel Capek,
Czech playwright, in his play 'RUR'
*(Rossum's Universal Robots) – from
the Czech 'robota' (work).*
**Slotted aircraft-wing (anti-
stall-and-spin device)**
simultaneously by Gustav Viktor
Lachmann, *engineer at Professor
Ludwig Prandtl's Aerodynamic
Institute, Göttingen, Germany and*
Frederick Handley Page, *aircraft
manufacturer, Cricklewood, London.*
**Tetraethyl lead (anti-knock
petrol compound)**
Charles Franklin Kettering *and*
Thomas Midgley, *General Motors
Research Laboratory*, USA. *A practical*

*manufacturing process for 'premium'
gasoline was developed by* Doctors
Charles A Kraus *and* Conrad C
Callis *for the Standard Oil Company.*
**Triple-hydro-triplane (giant
flying-boat)**
Gianni Caproni *of Italy. With eight
aero-engines, 'Capronissimo' only
made one successful test flight, of less
than 2 km.*
Wirephoto
Western Union Cables, USA.

1922
Aircraft carrier (custom-built)
*the Japanese Navy's 'Hosho', followed
nine months later by* HMS *'Hermes'.*
**Automatic doors (underground
railway carriages)**
Cammell Laird *of Birkenhead*, UK –
for London's Piccadilly Line.
Clavilux
Thomas Wilfrid, *Danish-born singer
and scientist working in the* USA.
*Rather than make sound, the Clavilux
had its own colour orchestration with
works composed and performed by
Wilfrid such as his Opus 39
Triangular Etude. In 1926 Leopold
Stokowski conducted the Philadelphia
Orchestra playing Rimsky-Korsakov's
'Scheherezade' to a synchronous
performance by Wilfrid on his 'colour-
organ'.*
**Continuous hot strip rolling
(steel)**
John B Tytus, *working for the
American Rolling Mill Co., with
assistance from* Charles Hook,
*chairman of that company. By 1923,
the first continuous mill was in
production. The rolling of wide strips
several hundred feet in length was the
achievement of* H M Naugle *and* A J
Townsend *working at the Columbia
Steel Co. in Butler, Pennsylvania,*
USA.
**Electrolux absorption
refrigeration system**
Carl Munters *and* Baltzar von
Platen, *undergraduates at the Royal*

Institute of Technology in Stockholm, Sweden. The patent was purchased by Dr Alex Wenner Gren of AB *Electrolux, Sweden.*

Film (3-dimensional motion pictures)
Perfect Pictures, US. *Viewers were given spectacles with one red and one green lens to watch 'The Power of Love'.*

Flettner rotorship
Anton Flettner *of the University of Göttingen, Germany. It derived its power from the lift and drag forces developed in large, rotating cylinders mounted above the hull. In 1925 a Flettner rotorship, 'Buckau', 680 tons, crossed the Atlantic Ocean.*

Football pools
John Jervis Barnard, *ex-Coldstream Guards officer, in Birmingham,* UK. *Called the 'Pari-Mutual Pools'.*

High-frequency induction furnace (steel)
Dr Edwin Fitch Northrup, *of both Princetown University and the Ajax Electrothermic Corp., Trenton, New Jersey,* USA.

Infra-red grill (sandwich toaster)
UK.

Motorised perambulator
Dunkleys *of Birmingham,* UK.

Patterned concrete building blocks
Frank Lloyd Wright, US *architect.*

Phase shift distortion (radio waves)
Dr Hoyt Taylor *and* Leo C Young, US *Navy scientists.*

Radio advert
WEAF Station. *10 minutes were sold to a furniture promoter of Jackson Heights, New Jersey,* USA.

Radio-broadcasting station (UK)
a Marconi team *at their laboratory in Writtle, near Chelmsford, with weekly half-hour broadcasts.*

Self-winding wrist-watch
John Harwood, UK *watchmaker and repairer. Unable to interest Swiss manufacturers, Harwood and Cutts formed a company which employed Swiss firms to make their watch. Between 1928 and 1931, 30,000 Harwood self-winders were made before the business was forced into liquidation.*

Sky-writing (smoke-trails)
Major J C Savage *of Hendon, London, using* SE5 *aeroplanes aerobatically to write messages in the sky.*

Tank (revolving turret)
Vickers Medium Mark I, UK.

Theory of polymers
Hermann Staudinger, *German chemist at the Federal Institute of Technology, Zurich.*

Tractor (diesel-engined)
Benz *of Stuttgart, Germany. A two-cylinder, three-wheel machine was soon followed by a four-wheel version.*

Tri-Ergon sound film
Joseph Engl, Joseph Massolle *and* Hans Vogt. *The first film, based on a Hans Andersen tale, was made by* UFA, *Germany's leading film company.*

Tubular immersion heater unit (kettle)
Arthur Leslie Large; *manufactured by Bulpitt & Sons Ltd of Birmingham,* UK.

Vitamin E
Herbert McLean Evans *and* K 3 Bishop, *physiologists,* US.

Water-skiing (powered)
Ralph W Samuelson *at Lake City, Minnesota,* USA. *By 1925 he was able to go over a ski ramp lubricated with cooking lard, and also took tows from Curtiss flying-boats.*

1923
Autogyro
Juan de la Cierva. *Following early and largely unsuccessful experiments in his native Spain, from 1926 the Cierva Autogyro Co., privately financed by Lord Weir, was based in the* UK. *Cierva was killed in a commercial air-crash in 1936.*

Coding machine (production model)
German Cipher Machines Co. *It was called the 'Enigma', as developed by* Dr Arthur Scherbius, *engineer, Berlin.*

Compton effect
Arthur Holly Compton, *Professor of Physics, Chicago University,* USA.

Frozen food
Clarence Birdseye, *former field naturalist and fur trader,* USA, *who established a company to industrialise his process in 1924 at Gloucester, Massachusetts. The first retail sales of packaged frozen peas, etc. were at Springfield in 1930. Pre-cooked frozen foods were introduced nine years later.*

Hafnium
Georg von Hevesy, *Hungarian-Swedish chemist at the Institute of Physics, Copenhagen, with* Dirk Coster, *Dutch physicist.*

Hearing aid (valve system)
Marconi Ltd, *London. Called the 'Otophone', it weighed 16 lb.*

Hectographic (or spirit) duplicating
Ormig Gesellschaft, *Germany.*

Iconoscope (electronic camera tube)
Vladimir Kosma Zworykin, *Russian émigré to the* USA. *He also developed his picture tube display, the kinescope, during the same year.*

In-flight refuelling (aircraft-to-aircraft)
US Army Air Service *with de Havilland four-day bombers.*

Lorry (diesel-engined)
Benz *of Stuttgart, Germany.*

Microphone (ribbon-type)
W H Schottky *and* Erwin Gerlach, *Germany.*

Low-pressure automobile tyre
Simultaneously by the Firestone Tire & Rubber Co., Akron, Ohio, USA, *as the 'Balloon'; and by Michelin et Cie, Clermont-Ferrand, France, as the 'Confort'.*

Phonofilm
Professor Lee de Forest, US, *who showed a number of singing and musical shorts using his process, at the Rialto Theater, New York.*

Planetarium (geocentric type)
Dr Walther Bauersfield *of the Carl Zeiss Optical Company, Jena, Germany, for the German Museum in Munich.*

Sound film on motion pictures (experimental):
Joseph Tykocinsky Tycocyner, *University of Illinois,* USA.

Short-wave radio transmitter
Dr Frank Conrad *of Westinghouse Electric Corp. from the* KDKA *radio station in Pittsburgh, Pennsylvania to* KDPM *radio station, 100 miles away in Cleveland, Ohio.*

Window polish
Reckitt & Sons Ltd, *Hull,* UK – *'Windolene'.*

Whooping cough vaccine
Thorvald Madsen, *Danish bacteriologist.*

1924

Aerial archaeology
O G S Crawford, *former* RFC/RAF *observer, was sponsored by A Keiller to fly over Hampshire, Wiltshire and Dorset,* UK. *The results were published in a book, 'Wessex from the Air'.*

Aetherophon
Professor Thérémin *of Leningrad,* USSR: *an electric musical instrument where the player regulated the pitch simply by movements of the hand towards and away from an upright rod connected with a valve.*

Aga kitchen range
Nils Gustav Dalén, *blind Swedish physicist.*

Automatic post office
installed in Bath, UK. *The kiosk combined posting box, stamp-vending machine and pay phone.*

'Beetle' (thermosetting plastic)
Fritz Pollak, *Vienna, Austria. Instead of heating phenol-formaldehyde resin*

under pressure, he heated urea. Unlike Bakelite, Beetle was odourless and tasteless, and much brightly coloured tableware came to be made in this material.

Cellulose paint finish
Oakland Cars, US.

Chanel No. 5 (perfume)
Mademoiselle 'Coco' Chanel, *Paris.*

Disposable paper handkerchiefs
Kimberley-Clark Co. *of Nennah, Wisconsin,* USA: *'Kleenex'.*

Fast rotary press (punched cards)
International Business Machines, US.

Gas chamber (execution)
Major D A Turner, US *Army Medical Corps. It was used at Nevada State Prison, Carson City, for the execution of a convicted murderer, Gee Jon. He took 6 minutes to die.*

'Harpic' (lavatory cleaning agent)
Harry Pickup, *Harpic Manufacturing Co., London.*

International chess match by radio
Between Haverford College, Philadelphia, USA, *and Oxford University,* UK. *Facilitated by five and a half hours' clear transmission.*

Language laboratory (gramophone-based)
Ralph Waltz *at the University of Utah,* USA, *was the first to use this phrase.*

Motorway
Italian Autostrada between Milan and Varese.

Pauli exclusion principle
Wolfgang Ernst Pauli, *Austrian-Swiss theoretical physicist, Federal Institute of Technology, Zurich.*

Pilotless aircraft flight
N-9 aircraft, as equipped by the US *Naval Research Laboratory, College of Ordnance and Naval Proving Grounds.*

'Presdwood' (prototype hardboard)
William Horatio Mason *in a sawmill*

shed in Laurel, Mississippi, USA, *while attempting to make a by-product out of sawmill waste, using a self-designed steam press. He began a factory in 1926 and trade-named his board 'Masonite'. Through one of his associates, Arne Asplund, the technology spread to Sweden in 1929.*

Undulatory theory of matter
Prince Louis Victor Pierre Raymond de Broglie, *French physicist, whilst investigating electron beams at La Sorbonne.*

Variable-pitch airscrew
Dr H S Hele-Shaw *and* T E Beacham, UK. *After convincing tests, the British Air Ministry ordered a dozen of these airscrews, but there was a general lack of interest until 1937 when 'Rotol' Airscrews Ltd began serious production.*

Variable-propellor water turbine
Viktor Kaplan, *Austrian engineer at Brunn, Czechoslovakia.*

Wash/spin-dry machine (electric-powered)
Savage Arms Corp., *New York,* USA.

Aerial crop-dusting service (commercial)
C E Woolman *of Huff-Daland Dusters Inc., flying a Petrel aircraft to discharge calcium arsenate over a Georgia cotton plantation.*

Bélinographe
Édouard Belin, *French researcher.*

Cinemascope
Dr Henri Chrétien, *French physicist: wide-screen cinema.*

Continuously variable friction gear
Frank Hayes *of New Jersey,* USA.

Electrodynamic loudspeaker (high-fidelity)
C. W. Rice *and* E. W. Kellogg, *General Electric Co., Schenectady, New York,* USA.

Gramophone (all-electric)
Brunswick Co. *of Dubuque, Iowa,* USA

– called the 'Brunswick Panatrope'.

Masurium and rhenium
Ida Eva Noddack *and her husband*,
Walter Karl Friedrich Noddack,
*chemists, National Physical
Laboratory, Germany.*

Motor car chassis (all-steel)
André Citroën, *France: the Citroën
B10 and B12.*

Petrol pump clock dial
S F Bowser and Co., *Fort Wayne,
Indiana,* USA.

**'Pulse ranging' (ionospheric
measurement)**
Dr Gregory Breit *and* Dr Merle A
Tuve, *Carnegie Institution of
Washington* DC, USA.

Quantum mechanics
Max Born, *Professor of Physics,
Göttingen, Germany, and* Werner
Karl Heisenberg, *German theoretical
physicist in Copenhagen, Denmark.*

Rigid airship (helium-filled)
US Navy, *Lakehurst, New Jersey,* USA.
It was designated the Z-R1.

Teaching machine
Sidney L Pressey *of Ohio State
University,* USA.

1926
Aerosol can
Erik Rotheim, *Norway.*

Automatic traffic lights
London.

Bulldozer
Robert Gilmour Le Tourneau,
*muck-shifting contractor, while
building a highway between Oakland
and Stockton, California,* USA. *He
fitted a steel scraper blade in front of a
Best tractor and equipped it with an
electric cable.*

'Buna S' (synthetic rubber)
I G Farben, *Germany.*

Cellophane (moisture-proof)
William Hale Charch *and* Karl
Edwin Prindle *of the Du Pont
Cellophane Co.,* USA.

**Crossbar (electro-mechanical
telephone exchange)**
Sweden.

**Electro-mechanical television
systems**
John Logie Baird, *Scottish inventor
in London and* Charles Francis
Jenkins *in Washington and* D Mihaly,
*Hungarian inventor in Munich. They
used neon lamps, Nipkow discs and
photoelectric cells.*

**Lawn-mower (electric-mains-
powered).**

Liquid-fuelled rocket
Robert H Goddard, *Professor of
Physics at Clark University,
Massachusetts,* USA. *It was fired to 41
ft altitude from nearby Auburn.*

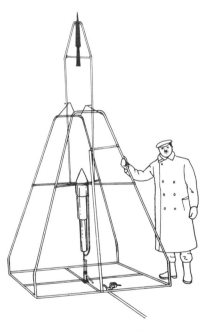

Goddard beside his liquid-fuelled rocket

Macromolecular chemistry
Hermann Staudinger, *Professor of
Organic Chemistry at Freiburg,
Germany.*

Power steering (motor vehicles)
Francis W Davis, *former chief
engineer of the Pierce-Arrow Motor
Car Co., at a small engineering shop in
Waltham, Massachusetts, with the
assistance of* G W Jessup, *tool-maker*

1927

and engineer.
Propeller-rudder (marine)
E Schneider *of Vienna. It was first
constructed by Messrs J M Voith of
Heidenheim for an experimental
launch, 'Torqueo'.*
Technicolor (cinematography)
Herbert Thomas Kalmus, D F
Comstock *and* W B Wescott *at the
Massachusetts Institute of
Technology, Cambridge,* USA.
**Tungsten carbide (industrial
quantities)**
Krupp Gesellschaft, *Germany, as*
'Widia', and General Motors, US, *as
'Carboloy'. It was originally
discovered twenty years before by*
Henri Moissan, *French chemist.*
Urease (crystalline enzyme)
James Batcheller Sumner,
*biochemist, Cornell University
Medical School,* USA.
Waterproof watch
Rolex, *Switzerland; called the
'Oyster'.*

1927
**Amyl cinnamic aldehyde
(synthetic perfume)**
having the scent of jasmine.
Continuous casting of metals
Dr Siegfried Junghans, *at the
brassworks he ran for his family firm of
clock and watchmakers in Germany.
He later progressed to other metals,
including steel, and his process was
commercially exploited by* Irving
Rossi *of New York.*
Electron diffraction by crystals
simultaneously by Clinton Joseph
Davisson *with* L H Germer *at Bell
Telephone Laboratories, and* George
Paget Thomson, *Professor of Physics
at Aberdeen,* UK.
Ethylene glycol
*'anti-freeze' liquid for motor car
radiators.*
Feature-length sound film
Warner Brothers *of Hollywood,
California,* USA. *Called 'The Jazz
Singer', it starred Al Jolson.*

Frogman flippers
Louis de Corlieu, *France.*
Genoa jib (sail)
Sven Salen, *Swedish helmsman, whose
boat 'Lilian' had a 'balloon staysail' at
the International 6-Metre Class races
of the Genoa Regatta, Italy.*
Iron lung (medical)
Professor Philip Drinker *of Harvard
University. It was manufactured by
Warren E Collins Inc. of Boston,
Massachusetts as the 'Drinker
Respirator'.*
Jukebox (all-electric)
Automatic Musical Instrument Co.
of Grand Rapids, Michigan, and
Seeburg Co. *of Chicago, Illinois,* USA.
MNS blood groups
Karl Landsteiner *and* Philip Levine
*at the Rockefeller Institute of Medical
Research, New York,* USA. *They also
discovered the 'P' system the same
year.*
Motorboat (rocket-powered)
Steel hull by Mullins Manufacturing
Corp., *Salem, Ohio,* USA. *Modified in
Florida by Dick Pope and test-piloted
by his brother Malcolm, 'Pirate Kid'
was powered by 32 rockets fired from
its iron-plated hull, and went out of
control at 60 mph when its rudder
burnt through.*
Motor car bodywork cleaner
Reckitt & Sons Ltd, *Hull,* UK; *called
'Karpol'.*
**Pentode (five-electrode radio
valve)**
Dr Gilles Holst *and* Bernard
Tellegen, *Philips Research
Laboratories, Eindhoven, Netherlands
– called the 'B443'.*
Pop-up toaster
Charles Strite, *mechanic from
Stillwater, Minnesota,* USA – *called the
'Toastmaster'.*
Rocket-propelled car
Fritz von Opel, *Germany.* RAK *1 was
powered by two solid-fuel rockets and
fitted with aerofoils.* RAK *2 achieved
195 mph at the Avus Circuit, Berlin.*

Salad cream
H J Heinz Co. Ltd, *Harlesden, London*, UK.
Sex hormone
Bernhard Zondek *and* Selmar Ascheim.

1928

Catalytic cracking of petroleum
Eugene J Houdry, *French engineer, whose search for a catalyst that could be regenerated led him to examine over 1000 substances before he discovered activated clay. His process was perfected and made practical first by Vacuum Oil Co., then Sun Oil Co.*
Cloverleaf intersection
State Highway Routes 4 and 25, Woodbridge, New Jersey, USA.
Colour television transmission (electro-mechanical)
John Logie Baird *at the Baird Studios, Long Acre, London.*
The Diels-Alder reaction
Otto Paul Hermann Diels, *Professor of Organic Chemistry, Kiel, Germany, with his assistant* Kurt Alder.
'Elastoplast' (adhesive bandage)
T J Smith & Nephew Ltd, *Hull*, UK.
Four-cylinder, two-stroke outboard engine
Ole Evinrude, US – *called the 'Elto Quad'.*
Geiger counter
Hans Geiger, *Professor of Physics at Tübingen University, and his assistant* E W Müller, *for the purposes of investigating Beta-ray radioactivity in atoms. The Geiger-Müller tube is named after them.*
Hot-air hand-drying apparatus
Elektra Händetrockner GmbH, *Germany.*
'I Speak Your Weight' machine
Joseph Tripodi. *It was first installed in Genoa, Italy.*
Image dissector tube (television)
Philo Taylor Farnsworth, *engineer, Utah*, USA.
Ondes musicales
Maurice Martenot, *French composer*

in Paris. The electric musical instrument was played using a ring on the finger fastened to a cord, producing a wailing glissando.
Pedal wireless
Alfred Traeger, *engineer, to assist* John Flynn *in his Flying Doctor Service in Australia.*
Penicillin
Alexander Fleming, *Scottish bacteriologist, St Mary's Hospital, Paddington, London.*
PVC (poly-vinyl-chloride)
simultaneously by Carbide and Carbon Chemical Corp. *and* Du Pont, *both* US, *and* I G Farben, *Germany.*
The Raman effect
Chandrasekhara Venkata Raman, *physicist, Calcutta University, India.*
Robot
Captain Rickards *and* A H Reffell *at Gomshall, Surrey*, UK. *It opened the Model Engineering Exhibition, London.*
Synthetic rubber (commercial)
J C Patrick. *Marketed as 'Thiokol A', it had remarkable oil-resistance, but its very unpleasant smell hampered its success.*
'Talking beacon' (lighthouse aid)
Charles Alexander Stevenson.
Teletype
Edward Ernst Kleinschmidt *of Morkrum-Kleinschmidt Co., Chicago,* USA.
Television recording system
John Logie Baird, *Long Acre, London. His 'Panovision' used aluminium gramophone records.*
Tomography
André Bocage, *France.*
Tubular-nibbed fountain pen
Rotring ('red ring'), *Hamburg, Germany. It was called the 'Tintenkuli'.*
Vitamin C (isolated from citrus fruits)
Albert von Nagyrapolt Szent-Györgi, *Hungarian biochemist at Cambridge University*, UK.

1929

Airship (diesel-engined)
Lieutenant-Colonel Victor Richmond, *leading a design team at Cardington, Bedford,* UK. *The five-Beardmore-engined* R101 *airship crashed at Beauvais, France, in 1930.*

Car radio (commercial manufacture)
Galvin Manufacturing Corp., *Chicago,* USA. *It was called the 'Motorola'.*

Coaxial cable
Bell Telephone Laboratories, US.

Electrical-mechanical flight simulator
Edwin Albert Link, US. *It was adopted by the* US *Navy in 1931.*

Electro-encephalogram (EEG)
Hans Berger, *Professor of Neuropsychiatry at Jena. It was popularised five years later by Nobel prize-winner* Lord Edgar Douglas Adrian, UK.

Electron microscope
Max Knoll *and* Ernst Ruska *in Berlin. Three years earlier they had set out to investigate the discovery made by* Dr Hans Busch *of Jena University that when a beam of electrons passes through a wire-coil, which acts as a magnet, the beam can be focused by a lens. Knoll and Ruska used a coil to form an image of a small aperture at a magnification of one; they then added a second stage which stepped up the magnification to seventeen. By 1933 Ruska's 'Supermicroscope', in achieving magnifications of up to 12,000, had gone well beyond the resolution of the optical microscope.*

Esterone
Adolf Friedrich Johann Butenandt, *German biochemist at Göttingen, simultaneously with and independently of* Edward Albert Doisy, *Director of the Department of Biochemistry at St Mary's Hospital, St Louis, Missouri,* USA.

Foam rubber
Dunlop Latex Development Laboratories, *Fort Dunlop, Birmingham,* UK. *A Dunlop scientist,* E A Murphy, *had the idea of beating latex into a foam with an egg-whisk. His colleague* W H Chapman *used a gelling process to produce a rubber foam capable of setting in moulds. 'Dunlopillo' was soon being used for all forms of upholstery.*

Giant flying-boat (twelve engines)
Dornier *of Germany. The 150-passenger DoX just managed a transatlantic crossing, but otherwise proved totally unmanageable.*

Hydroponics (growing plants in water)
William Frederick Gericke, *University of California,* USA.

Magnetic sound-recording tape (plastic)
Dr Fritz Pfleumer, *Austrian research worker with* I G Farben, *Allgemeine Electrizitäts Gesellschaft. Cellulose acetate tape was later replaced by 'Luvitherm' (*PVC*) tape, running at 30 inches per second.*

Oxygen-lance steel-making process
Dr Robert Durrer *and his assistant* Dr Heinrich Hellbrügge *at the Berlin Institute of Technology. The first experiments were at von Roll* AG, *Switzerland in 1948. A full scale operation was perfected by* Dr Theodro Eduard Suess *at the Austrian steelworks,* VOEST.

Quartz crystal clock
Warren Alvin Marrison, *clockmaker of Orange, New Jersey,* USA.

Radiotelephone service
set up on the steamship 'Leviathan'.

Retractable undercarriage (hydraulically operated)
George Messier, *France.*

'Scotch' Tape
Richard Drew, *research technician for the Minnesota Mining and Manufacturing Corp. (*3M*), at St Paul. It was introduced to Britain from France as 'Sellotape'.*

Submersible decompression chamber
Sir Robert H Davis *of Siebe-Gorman Ltd. It was successfully tested in Loch Long, Scotland.*
Synchromesh gearbox
Cadillac *and* La Salle, USA.
Synthesis of haemin
Hans Fischer, *organic chemist, Munich Technical High School, Germany.*
Television transmitting studio
John Logie Baird *at Long Acre, London. Sound and vision were transmitted in 1930 to the owners of his 'Televisor' receiving apparatus.*
Toughened plate-glass windscreen
St Gobain, *French glassmakers, based on discoveries by the* US *Corning Glass Company.*
Tune-playing motor horn
Sparks & Witherington Co.

1930
Bathysphere
Charles William Beebe *and* Otis Barton, US. *In 1934 Beebe recorded a depth of 3000 ft in their hollow steel globe off Bermuda.*
Cosmic ray magnetic cloud chamber
Carl David Anderson *and* Robert Andrews Millikan, *California Institute of Technology*, USA.
Cyclotron
Professor Ernest O Lawrence *and* N E Edlefsen *at the University of California, as a sophisticated particle accelerator to study the internal structure of disintegrated atoms. Lawrence had been inspired by a paper written by a German physicist called Wideroe.*
Differential analyser
Dr Vannevar Bush *and a team at the Massachusetts Institute of Technology, Cambridge*, US. *It was mechanically operated. Mark II was produced in 1942.*

Francium
Fred Allison, *physicist, Alabama Polytechnic Institute*, USA, *using his magneto-optic method.*
Hygienic tampon
Earl Hass, US. *His 'Tampax' Co. was formed in 1937.*
Ion-radiated food preservation
Wust, *France.*
Pepsin (crystalline enzyme)
John Howard Northrop, *chemist, Rockefeller Institute of Medical Research, New York*, USA.
Photo-flash bulbs (for photography)
General Electric Co., *Schenectady, New York*, USA.
Pluto (planet)
Clyde Tombaugh, *astronomer, at the Lowell Observatory, Flagstaff, Arizona*, USA, *using photographic methods.*
Polystyrene
I G Farben, *Germany, from benzine or crude oil. One of the first commercial products was 'Styron', marketed in 1937.*
Railway carriage air-conditioning
B & O Railway Co., USA. *For its dining-car 'Martha Washington', as tested behind 'The Columbian' between Baltimore and Cumberland*, USA.
Radiographic enlarger
Irving Langmuir, *physical chemist at General Electric's research laboratories, Schenectady*, USA.
Radio-pulse detection of ships (bounce-back system)
W A S Butement *and* P E Pollard, *Signals Experimental Establishment, Woolwich, London.*
Self-regulating hydraulic shock absorber
F G G Armstrong.
Supermarket
Michael Cullen *of Long Island, New York. It was called the 'King Cullen Food Stores'.*
Telescopic umbrella
Hans Haupt, *Berlin, Germany.*

Televisual telephone
'Ikonophone' – American Telephone & Telegraph Co. USA. A two-way telephone with two-way television transmission.

1931
'Alka-Seltzer'
Miles of Elkart, Indiana, USA.
Amphibious bicycle (inflatable floats)
Alfred Kreutz, forester from Sybba, East Prussia.
Amphibious tank
Vickers-Armstrong, UK. The Vickers-Carden Lloyd A4 was powered by a 50 hp Meadows engine.
Androsterone
Adolf Friedrich Johann Butenandt, privatdozent chemist, Göttingen, Germany.
Astatine
Fred Allison, Professor of Physics, Alabama Polytechnic Institute, USA.
'Blood bank'
Professor Sergei Sergeivitch Yukin at the Sklifosovsky Institute, Moscow's central emergency service hospital. The term 'blood bank' was coined six years later by Bernard Fantus at Cook County Hospital, Chicago, USA.
Caterpillar tractor (diesel-engined)
Caterpillar Tractor Co., Peoria, Illinois, USA.
Commercial electric guitar
Adolph Rickenbacker, Los Angeles-based machine and engineering contractor, with Messrs Beauchamp and Barth. Their Electro String Instrument Corp. produced an aluminium and steel banjo nicknamed the 'Frying Pan'.
Electric-generator windmill
G Darrieus, France. Three sails of aerofoil section like the blades of an aeroplane propeller were mounted on a pylon.
Electric kettle (resettable, self-ejecting safety plug)
Walter Henry Bulpitt of Birmingham, UK.
Empire State Building
New York City, USA. Messrs Shreve, Lamb and Harman, architects with N. G. Balcom Associates, engineers. It had 102 storeys and was 1250 ft high.
Exposure meter (photoelectric)
J Thomas Rhamstine of Detroit, Michigan. It was called the 'Electrophot'.
Glass-fibre
Owens Illinois Glass Company, US bottlemakers. Glass-fibre cloth was manufactured for heat and cold insulation. The Owens-Corning Fiberglass Corp. was formed in 1938.
Guitar pick-up
Horace Rowe, engineer, and De Armond, musician, US.
Iconoscope electronic TV camera
Vladimir Zworykin at the Westinghouse laboratories, Pittsburgh, Pennsylvania, USA.
Incompleteness Theorem
Kurt Gödel, German logician.
Microphone (crystal type)
C B Sawyer of Cleveland Heights, Ohio, USA.
Neutrino
Wolfgang Pauli, Austrian Professor of Physics, Zurich, Switzerland. Not until 1956 was a neutrino beam detected by F Reines and C L Cowan, Los Alamos, New Mexico, USA.
Radio astronomy
Karl Guthe Jansky, radio engineer at Bell Telephone Laboratories, Holmdel, New Jersey, USA. Using an improvised aerial, Jansky picked up radio emissions – as opposed to traditional light rays – from the Milky Way.
Railcar (pneumatic-tyred)
Edouard Michelin, tyre manufacturer, with Jacques Hauvette and Pierre-Marcel Bourdon, at the Michelin factory, Clermont-Ferrand, France. Called the 'Micheline'; by 1939, 140 such railcars were in use

throughout Europe.
Razor (electric)
Colonel Jacob Schick, US. *It was manufactured at Stamford, Connecticut,* USA.
Tailless aircraft
Dr Alexander Martin Lippisch, aerodynamicist, Germany.
TWX (teletypewriter exchange service)
Bell Telephone & Telegraphy Company, US. *The British Post Office's Telex system was operative by 1932.*
Tractor (rubber-tyred production model)
B F Goodrich, *Akron, Ohio,* USA.

1932
Car radio
Blaupunkt *five-valve AS5. It was first fitted in a* US *Studebaker automobile.*
Defibrillator
Dr William Bennett Kouwenhoven, *biomedical engineer, Dean of the Engineering school, Johns Hopkins University, Maryland,* USA.
Deuterium (heavy hydrogen isotope)
Harold Clayton Urey, *chemist, Columbia University,* USA.
Distant-reading compass
L C Bygrave, *Royal Aeronautical Establishment, Farnborough,* UK.
Gallup Poll
Dr George Horace Gallup, *Professor of Journalism at Northwestern University, Iowa,* USA.
Infra-red medical photography (subcutaneous analysis)
Professor Haxthausen *of Copenhagen, Denmark.*
Kort nozzle rudder (marine)
Ludwig Kort, *German aerodynamicist.*
Lethane (synthetic organic insecticide).
Lithium disintegration (proton bombardment)
John Douglas Cockcroft *and* Ernest

Thomas Sinton Walton, *nuclear physicists, Cavendish Laboratory, Cambridge,* UK.
'Mars Bar' (nougat and caramel-filled chocolate bar)
Forrest Mars, *Mars Confection Ltd, Slough,* UK.
Motion pictures (stereophonic sound)
Abel Gance *and* André Debrie, *film-makers, Paris. They made a re-edited version of Gance's 1927 eight-hour silent epic, 'Napoléon Bonaparte'.*
Neoprene
Arnold M Collins *and* Dr Wallace H Carothers *of Du Pont,* USA, *based on the twenty-year pioneer research of a religious Belgian-born pure scientist,* Father Julius A Nieuwland *and his 'divin-acetylene'.*
Neutron
James Chadwick, *physicist, Cavendish Laboratory, Cambridge,* UK.

Dr Wallace Carothers, inventor of Du Pont's nylon and neoprene

1932

Parking meter
Carlton C Magee, *editor of
'Oklahoma City Newspaper' and
chairman of the Businessmen's Traffic
Committee in that city. He further
refined his coin-operated machine
three years later, as manufactured by
his Dual Parking Meter Co.*

Positron
simultaneously by Carl David
Anderson, *California Institute of
Technology*, USA, *and* Patrick
Maynard Stuart Blackett, *Cavendish
Laboratory, Cambridge*, UK.

Synthesised Vitamin C
Walter Norman Haworth, *Professor
of Organic Chemistry at Birmingham
University*, UK.

**Thiram (dithiocarbonate
fungicide)**
Du Pont de Nemours, USA.

VHF radio-telephony
Marquese Guglielmo Marconi, G A
Mathieu *and* G A Isted, UK.

**Tractor tyre (pneumatic, low-
pressure)**
*Firestone Tire & Rubber Co., Akron,
Ohio,* USA. *First tested by Harvey S.
Firestone on his farm in Columbiana.*

**Wind tunnel (for mass-
production automobile design)**
Ford Motor Co., *Detroit, Michigan,*
USA – *for their Lincoln car.*

Zoom lens (cinematography)
Taylor Taylor & Hobson *of Leicester*,
UK, *with their 'Cooke' lens; then* Bell
& Howell, *Chicago,* USA. *The first
practical zoom lens was produced 21
years later by* Angenieux, *France.*

1933

Anchor (CQR)
Geoffrey I Taylor, *Cambridge*, UK.
*To enable flying boats/seaplanes to
moor on the water.*

Bicycle (supine-recumbent)
Charles Mochet, *France, ridden by
François Faure and called the
'Vélocar'.*

'Dayglo' fluorescent pigments
Joe *and* Bob Switzer *at their father's
drug store in Los Angeles, California,
*USA. *Together with chemist* Dick
Ward, *the Switzers began a
lithography company in Cleveland,
Ohio to produce fluorescent inks such
as 'Dayglo Orange'.*

'Dettol' (antiseptic liquid)
Reckitt & Sons Ltd, *Dansom Lane,
Hull,* UK.

**Flue-gas desulphurisation
process**
London Power Company's research
chemists, *to eliminate chimney gases
from the newly built Battersea 'A'
power station. It was known as 'lime-
scrubbing'.*

Frequency Modulation (FM)
Edwin Howard Armstrong, *Professor
of Electrical Engineering, Columbia
University*, USA.

**Microscope (miniature, hand-
held and versatile)**
John McArthur, *medical student and
Leica camera enthusiast, University
College Hospital, London. Although a
dozen instruments were manufactured
in 1936, it was not until 1959 that
Cooke, Troughton & Simms Ltd put it
into series production.*

**Million-volt electrostatic
particle accelerator**
Merle A Tuve, *at the Department of
Terrestrial Magnetism of the Carnegie
Institution of Washington*, USA. *It was
based on the work of* Robert van der
Graaff, *former Rhodes scholar from
Alabama*, USA.

'Monopoly' (board game)
Charles Darrow, US.

**Multiple-flash sports
photography**
Harold E Edgerton *and* Kenneth J
Germeshausen, *Massachusetts
Institute of Technology, Cambridge,
*USA.

Non-magnetic engine
Vickers-Petter Ltd, UK, *for
'Research', a motorboat to explore the
earth's magnetic field. The Atomic 160
hp engine was mainly of bronze alloy.*

Pantothenic acid
Roger J Williams, *biochemist*, US, *et al.*

Polythene (polyethylene)
Dr Reginald Gibson *and* E W Fawcett *at* ICI*'s Alkali Division research laboratories, Winnington, Cheshire,* UK. *The first ton of polythene was complete in December 1938, and a factory went into production less than a year later.*

Sodium vapour lamp
General Electric Co., Schenectady, New York, USA.

Stereophonic disc recording
Alan Dower Blumlein, UK *scientist working for Electric and Musical Industries (later* EMI*), London.*

Technicolour cartoon film
Walter Elias Disney, *producer*, US, *whose 'Flowers and Trees' achieved a tremendous success and was soon followed by 'Three Little Pigs'.*

Telephone speaking clock
M Esclangon, *Director of the Paris Observatory, for the French Ministry of Posts and Telecommunications.*

Tractor (three-point hydraulic plough hitching system)
Harry George Ferguson, *engineer, Northern Ireland.*

Vitamin B2 (riboflavin)
Richard Kuhn, *Austrian chemist,* Albert von Nagyrapolt Szent-Györgyi, *Hungarian biochemist, and* Julius Wagner von Jauregg, *Austrian neurologist. Synthesised 2 years later.*

1934
Automatic injection moulding press
for the various acetate and formaldehyde moulding powders recently developed.

Catseye
Percy Shaw, *road-repairer, Halifax, Yorkshire,* UK – *a reflecting road-stud to enable motorists to see in dark or foggy conditions. Shaw was particularly inspired by the glinting eyes of a cat sitting on a fence.*

Cerenkov Effect
Pavel Alekseevich Cerenkov, *physicist, Lebedev Physical Institute, Moscow. It was further developed during the next three years by* Ilja Michajlovic Frank *and* Igor Jevgenevic Tamm, *also of the Lebedev Institute,* USSR.

Colt chimney cowl
Herr Kuckuck, *to prevent downdraught and smoky fires.*

Diesel-electric railway traction
Charles Franklin Kettering, *General Motors Corporation, Detroit,* USA, *with* George Codrington (*ex-Winton Gas Engine Co.*) *and* Harold L Hamilton (*ex-Electromotive Company*). *The first* US *diesel-powered main line locomotive, 'General Motors 103', made its first 83,000-mile test run in 1939.*

Drive-in cinema
Richard Hollingshead, *on a 10-acre site at* Camden, *New Jersey,* USA.

Electromagnetic detection of aircraft (experimental)
Pierre David *at Le Bourget Airport, France.*

Launderette
J F Cantrell, *Fort Worth, Texas,* USA. *It was called the 'Washeteria'.*

Oxidation enzymes
Axel Hugo Theodor Theorell, *crippled Swedish biochemist, Kaiser Wilhelm Institute, Berlin.*

'Perspex' (methyl methacrylate polymer)
Dr Rowland Hill *and researchers at* Imperial Chemical Industries, UK. *Similar researches by* Dr Otto Rohm *and his colleague* Hass *of Darmstadt, Germany, produced 'Plexiglas' and by* Du Pont de Nemours, USA, *'Pontalite' then 'Lucite'.*

Progesterone (hormone)
Adolf Friedrich Johann Butenandt, *Professor of Organic Chemistry, Danzig Institute of Technology, Germany.*

Radio beam navigation aid
Dr Hans Plendl, *high-frequency radio*

expert from Rechlin, Germany. Called 'Knickebein' (bent leg), it was first used by the Luftwaffe in 1940.

Radio-synthetic organ
Abbé Pujet *in the Church of Notre Dame de Liban of Paris. The harmonics of the pipes were picked up by microphones, mixed synthetically into new tone colours and finally sounded over loudspeakers.*

Streamlined production automobile
Chrysler *'Air-Flow' Coupé,* USA, *and the 'Traction Avant' Citroën, France.*

Synthesised prontosil dye
Dr Gerhard Domagk, *pharmacologist employed by I G Farben, Germany.* Daniel Bovet, *Swiss-born scientist, identified prontosil dye's active compound as sulphanilamide. Sulphonamides became the latest generation of chemotherapeutic drugs.*

Tannoy portable loudhailer
Guy R Fountain *of the Tulsmere Manufacturing Co.,* USA. *The name was based on the words 'tantalum alloy'.*

Tone-wheel electric organ
Laurens Hammond *and* John Hanert, *engineers of Chicago, Illinois,* USA, *using the synchronous motor Hammond had developed for a non-ticking clock. The public début of the 'Hammond Model A' came in 1939 at the Industrial Arts Exposition in New York's Radio City* RCA *Building.*

Vitamin K
Carl Peter Henrik Dam, *Professor of Biochemistry at Copenhagen University, Denmark, simultaneously with* Edward Adelbert Doisy, *Director of the Department of Biochemistry at St Mary's Hospital, St Louis, Missouri,* USA.

1935
Aero-advertising
Richard Ormonde Shuttleworth, *wealthy young aviator of Old Warden, Bedfordshire,* UK. *He experimented with fitting a series of neon letters not*

exceeding 4 ft to the underwings of his de Havilland 84 Dragon biplane. BEER IS BEST *or* OXO *would appear in the night sky. Although a company was formed, the project failed when aircraft were banned from flying below a certain level.*

Aircraft (geodesic construction)
Barnes Neville Wallis, *designer and engineer, Vickers Aviation Ltd, Weybridge, Surrey,* UK – *the Vickers G4/31 monoplane.*

Colour film (monopak-still)
Leo Godowsky Jr *and* Leopold Mannes, *two music students and self-taught chemists in New York,* USA. *Following years of initial experiments in Mannes' kitchen, the two Leos were later employed by Dr C E K Mees of Eastman-Kodak and 'Kodachrome' was the commercial result. The following year the German 'Agfacolor' was marketed, also as a monopak-process.*

Dellinger Effect (sunspots on radio communication)
John Howard Dellinger, *physicist, Chief of the Radio Section,* US *National Bureau of Standards.*

Desk calculator (tabulator with independent cycles)
Compagnie des Machines Bull, *Paris, France.*

Electric typewriter (mass-produced)
International Business Machines, USA.

Fireboat (steel-hulled and diesel-engined)
J Samuel White Ltd, *Cowes, Isle of Wight,* UK. *The 78 ft 'Massey Shaw' for work on the River Thames, London.*

Fluorescent lighting (hot-cathode)
General Electric Co. *at Nela Park, Ohio,* USA.

Hearing aid (electronic)
Edwin A Steven's *'Amplivox'.*

Ink cartridge (pen)
M Perraud, *Director of Jiff-*

Waterman, France.
Mesotron
Hideki Yukawa, *physicist, Osaka University, Japan. Two years later* S Neddermeyer, J Street, E Stevenson *and* C Anderson *verified Yukawa's postulation about what later became known as the meson.*
Microphone (non-directional)
Bell Telephone Laboratories, Holmdel, New Jersey, USA. *Called the 'Western Electric Eight Ball'.*
Miniature radio valve (two-volt, multi-electrode type)
Stephen Philip de Laszlo, *High Vacuum Valve Co, Holborn, London,* UK. *Called the 'Hivac Valve'.*
Paperback book
Sir Allen Lane *of London, whose Penguin books could be bought for the same price as ten cigarettes.*
Quick-setting ink (printing)
Dr D Askew, *chief research chemist of Coates Brothers Inks Ltd of St Mary Cray, Kent,* UK. *Their 'liquid phase separation' considerably speeded up the drying process.*
Radiolocation of aircraft (reliable system)
simultaneously by Robert Alexander Watson-Watt, *Scottish physicist, Superintendent of the Radio Department of the National Physical Laboratory, Teddington,* UK, *and* Dr Rudolph Kühnold *of Kiel, Germany, and the* Société Française Radioélectrique, *Le Havre, France.*
Richter Magnitude Scale (earthquakes)
Charles Francis Richter, *seismologist, California Institute of Technology,* USA.
Semi-acoustic electric guitar
Lloyd Loar *for the Orville Gibson Guitar Co., Kalamazoo, Michigan,* USA. *It was called the 'Vivi-Tone'.*
Solar house heating
M Hottinger, *Zurich Institute of Technology, Switzerland.*
Strain seismograph
Hugo Benioff, *Seismological*

Laboratory, Carnegie Institution of Washington, Pasadena, USA.
Tape-recorder (commercially available)
AEG GnbH, Germany. Called the 'Magnetophone'.
US Patent No. 2,000,000
Joseph Ledwinka *for the Edward G Budd Manufacturing Co., Philadelphia, Pennsylvania: safer construction for pneumatic railcar wheels.*
VHF electronic television
Marconi-EMI, UK, *using the 'Emitron' electronic camera.*

'Emitron' electronic television camera

Vinyl polymers
B F Goodrich Co., *Akron, Ohio,* USA, *following some nine years extensive research. By 1936 polymers became available on a large scale.*

1936
Australopithecus
Robert Broom, *S. African morphologist at Sterkfontein, South Africa.*

Boulder Dam
Colorado River, USA.
Electric iron (thermostat-control)
Messrs Morphy *and* Richards, UK.
Electro-mechanical calculator
Konrad Zuse, *Germany. By 1941 he was successfully operating his third prototype, Z3, a program-controlled electro-mechanical calculator.*
Intertype Fotosetter
C E Scheffer, *Swiss engineer, whose prototype was taken up by the* US *Intertype Corp. who then took a further ten years to perfect it under the direction of* H R Freund.
Jet-engine
simultaneously by Frank Whittle, RAF *officer-engineer at Cambridge University,* UK *and* Dr Hans von Ohain *in Germany. Whittle's theoretical work had begun some ten years before.*
Plastic lenses
Arthur Kingston, UK *photographer and inventor, who chanced upon a discarded test sample of* ICI *methyl methacrylate polymer (later Perspex) on sale in the Caledonian Market, London. From further research, Kingston helped to form Combined Optical Industries Ltd to manufacture plastic spectacle lenses, camera lenses, magnifying glasses, etc.*
Plexiglass contact lenses
I G Farben, *Germany.*
'Polaroid' (polarised plastic material)
Edwin Herbert Land, *physics undergraduate, Harvard University,* USA.
Practical helicopter (contra-rotating rotors)
Professor Heinrich Focke, *of Focke-Wulf, Germany – the Fa-61.*
Private automobile (diesel-engined)
Mercedes-Benz, *Stuttgart, Germany – the 260D.*
Spitfire (fighter-aeroplane)
Reginald Joseph Mitchell, *Chief*

Designer, Supermarine Aviation Ltd, Southampton, UK *– the Supermarine Type 300. By 1946, over 22,000 Spitfires had been built and seen active service.*
Vitamin B6 (pyridoxine)
T W Birch *and* Albert von Nagyrapolt Szent-Györgyi, *Hungarian biochemist.*
Private automobile (diesel-Videophone service
The German Post Office, between Berlin, Nuremburg and Leipzig.
Volkswagen ('people's car')
Dr Ferdinand Porsche *of Stuttgart, Germany, at the behest of the Nazi Party. Factory built at Wolfsburg in 1938 for the 'KdF-Wagen', later the 'VW'.*

1937
Chairlift
Union Pacific Railroad engineers *in Chicago for the Dollar Mountain at Sun Valley, Idaho,* USA, *a resort famous for its ski-runs. It was 720 metres in length with a difference in level of 216 metres. Six other chairlifts were built for Dollar Mountain in quick succession. The first European chairlift was constructed in 1939 by Nevrly, an engineer at Pusterny, Czechoslovakia.*
Citric acid cycle
Hans Adolf Krebs, UK *biochemist and lecturer at Sheffield University.*
Drive-in-bank
Los Angeles, California, USA.
Electro-convulsive therapy
Doctors Ugo Cerletti *and* Lucio Bini *as a treatment for schizophrenia.*
Mirror reflex ciné camera
August Arnold *and* Robert Richter, *Munich, Germany. It was called the 'Arriflex 35'.*
Niacin
Conrad Arnold Elvehjem, *biochemist, University of Wisconsin,* USA, *and co-workers.*
Oil well (offshore)
Superior Oil Co. *and* Pure Oil Co., *in*

the Gulf of Mexico.

Photosensitive glass
R H Dalton *of the Corning Glass Works*, USA. *He found that if glass is exposed to ultra-violet radiation while it is still cold, the colour it eventually acquires is deeper.*

Radio telescope
Grote Reber, *of Illinois*, USA. *This was a dish, measuring 31 ft in diameter, for sending out and receiving back radio waves from outer space.*

Rayon yachting sail
used on the 'Ranger', UK.

Supermarket trolley (shopping cart)
Sylvan N Goodman *for the 'Humpty Dumpty' Supermarket, Oklahoma City*, USA.

Technetium
Emilio Gino Segré *and* C Perrier, *physicists, Palermo, Italy.*

Ticonal G (magnet type)
Philips Research Laboratories, *Eindhoven, Netherlands.*

Tinlet paintpot (model-making)
Douglas S Barton *of the Humber Oil Co., Hull*, UK – *to contain his Humbrol Art Oil Enamel paints.*

Titanium and zirconium
W J Kroll, *Luxembourg metallurgist, after seven years' research in his own private laboratory. The first manufacturing plant using the Kroll process, as developed by* F S Wartman *(*US *Bureau of Mines), was in operation at Boulder City, Nevada,* USA *by 1944.*

1938
Airliner (pressurised passenger cabin)
Boeing 307 Stratoliner, which went into service in 1940 for Transcontinental Airways.

Ballpoint pen
Ladislao J *and* Georg Biro, *brothers of Budapest, Hungary. Ladislao was a journalist and painter, whilst Georg was a chemist. On the outbreak of war,* the Biro brothers emigrated to Argentina where in *1943 with the help of* Henry Martin, UK *financier, they formed a company to perfect and produce the pen. It was at this stage that the familiar tube of ink and ball-bearing point were introduced. After the war, the Biro brothers' invention sold in its millions as the 'Eversharp' in the* USA *and ultimately as the 'Bic' in Europe.*

Bren gun
the ZB26 as developed at Brno, Czechoslovakia.

Cortisone
simultaneously by Edward C Kendall *with* Dr Philip S Hench *of the Mayo clinic in Rochester, Minnesota,* USA, *and* Tadeus Reichstein *of the Technische Hochschule in Zurich, Switzerland, after almost twenty years' concurrent research into a drug which would alleviate Addison's disease or rheumatoid arthritis.*

Domestic steam iron (thermostatic control)
Edmund Schreyer, *Ridgefield, Connecticut,* USA.

Ejection seat
Junkers Aircraft Factory, *Germany, although three other German aircraft works were also experimenting with the idea. The first aircraft to be fitted was a Junkers 88 in 1939. By 1940 more than 200 test ejections had been made using a compressed air gun to fire seats up rails from ground test rigs, some 70 of them with human subjects. The Luftwaffe began to use ejection seats in 1941 and by 1943 German aircrews were using the seat in battle emergencies. The* UK *Martin-Baker ejector seat was not tested from the air until 1946.*

Fluorescent lighting
Dr Arthur H Compton, *Professor of Physics, Chicago University and consultant to the* US *General Electric Co., with* Dr George Inman, *head of GEC's research team. Work began on the project in 1934. Unknown to GEC,*

1938

a similar lamp had been invented in
1926, by three Germans – Meyer,
Spanner *and* Germer *of the Rectron
Co. – but they had not developed their
invention.*

Flying-boat-launched seaplane
Short Brothers *of Rochester,* UK *– the
'Composite', with the 'Mercury'
taking off from the 'Maia', in flight.*

Folacin
P L Day.

**High definition colour television
(mechanical)**
John Logie Baird, *Long Acre,
London, who demonstrated his
'Telechrome' in a closed laboratory-
studio.*

Instant coffee
Nestlé *of Vevey, Switzerland,
following eight years of research after a
suggestion made by the Brazilian
Institute of Coffee. It was trade-
named 'Nescafé'.*

**Language laboratory (wire-
recorder based)**
*First tested at Middlebury College,
Vermont,* USA. *Based on the 'Sound
Mirror' process developed by* Marvin
Camras *at the Armor Research
Foundation, Illinois Institute of
Technology. Marketed by Brush
Development Corporation. Camras's
'record and erase' process was
developed three years later.*

Nuclear fission
Otto Hahn *and* Fritz Strassman,
*Kaiser Wilhelm Institute of
Chemistry, Berlin.*

Nylon
Dr Wallace Carothers, *leading a team
of organic chemists at the Eastern
Laboratory of Du Pont de Nemours,
Repauno, Delaware,* USA. *$27,000,000
was invested by Du Pont in nylon
research over thirteen years before
their factory at Seaford, Delaware
could mass-produce it. Although at
first twice as expensive as silk, it was
also twice as durable.*

Nylon toothbrush
Dr West's *'Miracle Tuft Toothbrush',*
USA.

Perlon (polyamide)
P Schalk, *Germany.*

PTFE (polytetrafluorethylene)
Roy J Plunkett, *research chemist at
the Kinetic Chemical Co.,* US.

**Pressure-cooker (self-sealing,
interlocking pan and lid)**
Alfred Vischer, *Chicago
draughtsman.*

Sulphonamide drugs
research team at May & Baker's
laboratories *in Dagenham, Essex,* UK.
*After considerable experimentation,
they produced M&B 693 or
'Sulphapyridine' to arrest the growth
of pneumococcus, the organism
responsible for pneumonia.*

**Television camera for outside
broadcasts**
Sidney Rodda *and* Hans Lubszynsky
of Marconi-EMI for the BBC, UK.
Called the 'Super Emitron'.

**Wein Bridge Oscillator
(stabilized version)**
William Hewlett, *electrical engineer,*
and David Packard, *formerly of the
Vacuum Engineering Department,
General Electric Co., Schenectady,
New York,* USA. *With an order for 8 of
their '200 A Audio Oscillators' from
Walt Disney, the Hewlett-Packard
Co. was formed in Palo Alto,
California.*

**Xerographic photo-copying
machine (prototype)**
Chester Carlson, US *physicist. The
first-ever copy read '10-22-38
Astoria'. It was perfected from 1944
by* Roland M Schaffert, *research
physicist for the Battelle Memorial
Institute, Columbus, Ohio,* USA.

1939

Betatron
Donald W Kerst, *Assistant Professor
of Physics, University of Illinois,* USA.

Binary calculator
John Atanasoff *and* George R
Stibitz, *mathematicians at Bell
Telephone Laboratories,* USA.

DDT (insecticide)
Paul Hermann Müller, *German chemist at J R Geigy, Basle, Switzerland. Trade-named 'Gesarol', it was to be used in large quantities to kill fleas and body lice amongst troops and refugees.*

'Electric aura'
Semyon Kirlian, *Soviet electrician, while repairing an electrotherapeutical instrument. For the next 30 years, Kirlian battled to perfect his apparatus, capable of 'photographing' the aura around the human hand.*

Jet-engined aircraft
Heinkel He 178, with a powerplant built to Dr Hans von Ohain's *design. By 1941* Frank Whittle's *W1 engine, as produced by Power Jets (UK), was installed in the successful Gloster E28/29. By 1944 the* RAF *had formed a squadron of jet aircraft and the Germans had one engine, the Junkers 004, in production.*

Gloster E28/29 aircraft

Microfilm camera (mass card-filing system)
Elgin G Fassel *of Milwaukee, Wisconsin. It was marketed by the Microstat Corporation of America.*

Nylon 'Supra' spectacles
Neville Chappell, *manufacturing optician, Amersham, Buckinghamshire,* UK – *the first spectacles to have a continuous top rim and bridge, and subsequently to use nylon.*

Oil (commercial quantities – UK)
British Petroleum Co. Ltd *at Eakring, Nottinghamshire.*

Polyurethane paint
Otto Bayer, *Germany.*

Practical helicopter (single main and small tail rotors)
Igor Sikorsky, *Soviet émigré at the Vought-Sikorsky Co. in Bridgeport, Connecticut,* USA – *the VS-300.*

Speech-simulating machine
Bell Telephone Laboratories, Holmdel, New Jersey, USA. *Exhibited at the New York and San Francisco World Fairs.*

Vaccine 17-D (yellow fever)
Max Theiler, *South African bacteriologist, International Health Division, Rockefeller Institute of Medical Research, New York,* USA.

1940

Amphibious Volkswagen
Dr Ferdinand Porsche, *Stuttgart, Germany – Type 128, called the 'Schwimmwagen'.*

Automatic gearbox
General Motors, *Detroit,* USA.

Cavity magnetron (centimetre radar valve)
John Turton Randall, *Professor of Physics, King's College, University of London, with* James Sayers *and a team at the Birmingham University Physics Laboratory.*

Deuterium (heavy-hydrogen isotope)
Hans von Halban *and* Lew Kowarski, *French physicists. It was called 'heavy water'.*

Extra-sensory perception
Joseph Banks Rhine, *psychologist and professor at Duke University,* USA.

'Flying wing' aircraft (tailless)
US Northrop Aviation's *N-1M.*

Helicopter (production model)
Focke-Aghelis-223, with a passenger capacity of six persons.

Jeep
Karl K Pabst, *consulting engineer of the Bantam Car Co., Butler, Pennsylvania, for the* US *Army. Powered by a 45 bhp, four-cylinder Continental engine, the first batch of 70 went into service the following year. 649,000 jeeps were produced during the war.*

'Mae-West' lifejacket
RAF *Mark I inflatable lifejacket.*
Neptunium
Edwin Mattison McMillan, *physical chemist, and* Philip Hauge Abelson, *Radiation Laboratory, University of California, Berkeley,* USA.
Orgone accumulator
Wilhelm Reich *of Long Island, New York,* USA – *a mini-greenhouse capable of capturing 'Orgone' or basic life energy and increasing the vitality of anyone sitting in it.*
Penicillin (stabilised and quantity-produced)
Dr Howard Walter Florey, *Australian pathologist, and* Dr Ernest Boris Chain, *German-refugee chemist, working at Oxford University,* UK: *produced in quantity by* US *drug firms.*
Plutonium
Glenn Theodore Seaborg *and* Edwin Mattison McMillan, *nuclear scientists, Radiation Laboratory, University of California,* USA.
Radar (radio detection and ranging)
word coined by Commander S M Tucker, US *Navy, to describe innovations made since 1936 by* Robert M Page, US *Naval laboratory.*
Radio-controlled tank
Marcel Pommelet, *French engineer.*
Rhesus blood group system
Karl Landsteiner *and* Alexander Solomon Wiener, *haematologists at the Rockefeller Institute of Medical Research, New York,* USA.
Rocket-powered glide bomb (radio-controlled)
Dr Herbert Wagner, *Germany – the HS293.*
Vitamin H (biotin)
Vincent du Vigneaud, *biochemist, professor at Cornell University Medical College,* USA.

1941
Aerially-launched fighter-planes
Soviet Air Force's Zveno 'parasite'. In experiments, three Polikarpov 1-5 biplanes were wing-launched from an in-flight Tupolev TB-3 heavy bomber.
Aerosol insect spray (carbon-dioxide propulsion)
L D Goodhue *and* W N Sullivan, US.
Anti-tank grenade projector
Watts *and* Jeffries, UK. *Based on the work of* Lieutenant Colonel Blacker, *it was called* PIAT *(projector, infantry anti-tank).*
'Bat bombs'
Louis F Fieser *and* nineteen university professors *on the* US *National Defense Research Committee. Thousands of artificially hibernated bats, an incendiary bomb attached to each one, would be released from aircraft bomb bays. Awoken by the warm air, they would fly into buildings where the incendiaries would ignite. Planned for Japanese cities, they were never used.*
Battery (silver-zinc)
H André, *France.*
Cardiac Catheter
Dick W Richards US, *and* André F Cournand, *France at Bellevue Hospital, New York. By developing a crude rubber tube originally used in 1929 by* Dr Werner Forssmann, *Germany, for investigating diseased hearts.*
Colour TV service (experimental)
Dr Peter Goldmark *of Columbia Broadcasting Service, New York.*
Electronic autopilot
Minneapolis-Honeywell Co., USA – *called the C-1.*
Electronic flight simulator
Dynatron Radio Ltd, *Hampton Court, Surrey,* UK.
Giant windmill (electricity-generating)
a 1250 Kilowatt windmill was erected at 'Grandpa's Knob' in the central Vermont mountains, USA. *It had twin blades, 175 ft in diameter. The experiment ended when a blade cracked*

after three years' service.

Locomotive (gas-turbine-engined)
Switzerland.

MCPA (herbicide)
Messrs Slade, Templeman *and* Sexton *at Jealott's Hill Research Station,* ICI, UK.

Microwave radar set
US *Radiation Laboratory.*

Particleboard (commercial)
Heseke *of Torfitwerker* AG, *Bremen, Germany – from wood chips and thermosetting resins. Worldwide manufacture did not begin until 1946.*

Portable two-way FM radio
Motorola, USA. *Called the 'Handie/Walkie-Talkie'.*

Shell-moulding ('C' process)
Johannes Croning, *proprietor of a Hamburg foundry, using a thin mould of sand held together by phenolic resins. By 1948 Ford of Detroit had a pilot line shell-moulding 1000 exhaust valves per day.*

Television advertisement
WBNT – *New York, for the Bulowa Watch Co.*

Terylene
John R Whinfield, *Cambridge chemistry graduate, and his assistant* James T Dickson *at the research laboratories of the Calico Printers' Association, Accrington, Lancashire,* UK. *It was marketed and mass-produced by* ICI *in the* UK, *and under licence by Du Pont de Nemours,* US *first as 'Fiber V' then as 'Dacron'.*

2,4-D (herbicide)
Messrs Nutman, Thornton *and* Quastel *at Rothamstead Experimental Station, Harpenden, Hertfordshire,* UK.

1942
Atomic pile reactor
Enrico Fermi, *Italian refugee physicist,* and a team *working at Stagg Field, University of Chicago,* USA *(known as the Manhattan Project).*

Bazooka gun
Lieutenant-Colonel Leslie A Skinner (US Army) *to fire 2 ft long ground rockets from a steel tube; the charge was set off by an electric battery. It was officially known as 'Launcher, Rocket,* AT, M-I*'.*

DUKW (amphibian)
Sparkman *and* Stephens, *together with* General Motors Corporation Truck and Coach Division, US.

GR-S (government rubber styrene)
US Government, *following the Japanese over-running of Malaya and then the Dutch East Indies to cut off main sources of rubber from the* USA. *As a substitute, this synthetic rubber was manufactured from butadiene and styrene.*

Napalm (incendiary)
Louis F Fieser *at the* US *National Defense Research Committee's laboratories. It was later used to saturate Japanese cities in the closing weeks of the war.*

Supersonic long-range war rocket
General Walter Dornberger, Eugen Sänger, Wernher von Braun, Hermann Oberth *and* 2000 *research workers at the Rocket Research Station, Peenemünde, on the Nazi Baltic Coast. Over 35,000 such V2 rockets were launched on London and Antwerp.*

1943
Aqualung
Commandant Jacques-Yves Cousteau, *French Navy, with* Emile Gagnan, *control-valve engineer.*

Bailey bridge (prefabricated steel bridge for rapid temporary use)
Donald Coleman Bailey, UK *engineer.*

Bouncing bomb
Barnes Wallis, *Vickers Aviation Ltd, Weybridge, Surrey,* UK. *Four years in development, and inspired by the memory that Admiral Horatio Nelson*

had had an idea of aiming his cannons at the sea, so as to bounce up the shot and hit the French warships, Wallis's bombs successfully destroyed the Mohne, Ader and Sorpe dams on the Ruhr.

LSD (drug)
Dr Arthur Stoll, *research director of the Swiss Sandoz pharmaceutical company, whilst synthesising the first natural ergot alkaloid. His assistant, Dr Albert Hofman, chemist, was unknowingly the first to undergo the hallucinatory effects of this drug. Although originally manufactured for use by qualified professionals investigating mental disease,* LSD *also fell into the hands of drug addicts and was soon known as a 'killer'.*

Programmable electronic computer
Professor Max H A Newman *and* T H Flowers *at Bletchley Park, Buckinghamshire,* UK. *Called 'Colossus', it used 1500 valves and was operated to decipher the output of the German coding machine, Enigma.*

Sailing trimaran
Victor Tchechett, *Soviet émigré to the* USA.

Streptomycin (antibiotic drug)
Dr Selman Abraham Waksman, *Russo-Jewish-born,* US-*trained bacteriologist and biochemist, with his students at Rutgers University, New Jersey,* USA. *By 1944, Merck & Co. Laboratories had developed a deep vat fermentation process for its mass-production.*

Teflon
Du Pont de Nemours, US.

1944
Air-to-air missile
Dr Herbert Wagner, *Germany. The* HS *298 was designed to be carried by either the Do 217 or FW 90 aircraft.*

Americium
Glenn T Seaborg *and* Albert Ghiorso, *University of California, Berkeley,* USA.

ASCC Mark I (automatic sequence-controlled calculator)
Professor Howard H Aiken *of Harvard University,* US *in association with* International Business Machines (IBM) engineers *at Endicott, New York.*

Curium
Glenn T Seaborg *and* Albert Ghiorso, *University of California, Berkeley,* USA.

Decca navigator-receiver QM1
used by the Royal Navy for mine clearance.

Forward-swept-wing aircraft (jet-engined)
Junkers Ju 287, built at Dessau, Germany.

Ground-to-air missile
Dr Wurster, *Germany – 'Enzian'.*

Marine wave energy absorber and diverter
Robert Lochner, *Royal Naval officer. Called 'Bombardon', it was used during the D-Day landings.*

Nerve gases
Nazi chemists, *but never used during conflict.*

Pyrex telescope lens (diameter 200 inches, weight 15 tons)
Corning Glass Works, to the design of George Ellery Hale, US *astronomer, for the Mount Palomar Observatory. Casting took 7 hours, cooling took 10 months, and polishing – using 30 tons of grinding paste – took 11 years. Although the lens was in position by 1944, the telescope was not in operation until 1949.*

1945
Artificial kidney
Dr Willem J Kolff, *Dutch surgeon. His early artificial kidneys were made of cellophane.*

Atomic bomb
Professors J R Oppenheimer, Arthur H Compton, Enrico Fermi *and* Léo Szilard *at the Los Alamos Laboratory near Santa Fé, New Mexico. Following a test explosion at*

Alamogordo, two bombs were dropped on Japan (Hiroshima and Nagasaki) with 130,000 victims.

Cephalosporium Acremonium (isolated)
G Brotzu, *Cagliari Institute of Hygiene, Italy.*

Contact lens (fenestrated solution-less)
Norman Bier, UK.

Flying ram aircraft
Northrop XP-79, US.

Microwave oven
Percy LeBaron Spencer; *manufactured by the Raytheon Co., Waltham, Massachusetts in 1947.*

Nylon sail.

Programming language (for electro-mechanical calculator)
Konrad Zuse, *Germany. He lost all but his Z4 machine in air-raids. He called his language 'Plankalkül'.*

Surface-to-air missile
US Western Electric Company's *Nike-Ajax.*

Torsion bar suspension
Douglas Motorcycle Co., *Bristol,* UK, *for their 350 cc Flat Twin model.*

Tupperware boxes (plastic)
Earl W Tupper, *former chemist with Du Pont de Nemours,* USA.

Vinyl floor covering.

Zoom lens
Dr Frank Back, *Zoomar Corporation, New York State. It was first used to cover the baseball world series as broadcast on* US TV *in 1947.*

1946
Artificially induced rainfall
Vincent J Schaefer *and* Irvin Langmuir *of General Electric Co., Schenectady, New York. Carbonic snow was dispersed into a cloud over Greylock, Massachusetts,* USA.

Artificial ski ramps
Major Tony Moore, *Staff Officer for Physical Training with the British Army at Klagenfurt, to 'break in' trainee soldiers to the disciplines of alpine and cross-country skiing.*

Bikini swimsuit
Louis Reard, *French couturier. It was first modelled by a dancer called Micheline Bernardi at a Paris fashion show, four days after the* USA *had detonated an atomic bomb at Bikini Atoll in the Pacific.*

Commercial radiotelephone service
between the public telephone system and motor vehicles, USA.

Electric window lift (automobile)
Cadillac and Lincoln, US, *and Daimler,* UK.

ENIAC (electronic numerical integrator and calculator)
Dr John W Mauchly *and* J Presper Eckert *of the University of Pennsylvania,* USA. *With 18,000 vacuum tubes, this was the first* US *vacuum-tube computer.*

Espresso coffee machine
Gaggia, *Milan, Italy.*

Kell blood group system
Race, R R Coombs *and* A E Mourant.

Lutheran blood group system
Race *and* Callender.

Photo-typesetting machine
René Higonnet *and* Louis Moyroud, *French telephone engineers, following two years' work in their own homes. Backed first by the* US *Lithomat Corp. then by the Graphic Arts Research Foundation, development of this machine continued throughout the 1950s and it was ultimately marketed as the Photon in the* USA *and the Lumitype in Europe.*

Radial-ply tyre (steel-reinforced)
Pierre-Marcel Bourdon *and* Marius Mignol, *Michelin et Cie, Clermont-Ferrand, France. Called the 'Michelin X', it was developed from the 'Pilote' and 'Metallic' tyres produced by this company 10 years earlier.*

Vinylite disc record
RCA-Victor, USA, *to replace records made of shellac.*

1947

Coenzyme A
Fritz Albert Lipmann, *Head of the Biochemical Research Department, Massachusetts General Hospital, Boston,* USA.

Flying-boat fighter (turbojet-engined)
Saunders Roe SR A/I, *Cowes, Isle of Wight,* UK.

Flying-wing jet bomber
Northrop YB-49, *powered by eight Allison* J-35 *turbojets,* US.

Gleep (graphite low energy experimental atomic pile)
Harwell, UK.

Holography (3-D)
Dennis Gabor, *Hungarian-born Professor of Electrical Engineering at Imperial College, London. He made his first experiments with mercury lamps, giving his system the name holography after the Greek 'holos' (whole).*

'Intelligent machinery'
Alan M Turing, *logician at Cambridge University,* UK. *His prophetic article remained unpublished for 30 years.*

Land-Rover
The Wilks brothers Spencer *and* Maurice, *had the idea of producing a go-anywhere vehicle at the latter's farm in Anglesey, Wales,* UK. *Since then, Land-Rovers have travelled across the globe.*

Motorboat (gas-turbine-engine)
Metropolitan-Vickers Electrical Co. Ltd, *who fitted a 2500 hp gas turbine into the British Admiralty triple-screw* MGB 2009.

Pernicious anaemia (Vitamin B12 as cure for)
Dr Karl August Folkers, *chemist working for Merck & Co., Rahway, New Jersey,* USA, *with assistance from* Dr Mary Shorb *of Maryland University.*

Pilotless transatlantic aircraft
Douglas Aircraft, US. *A* DC-4 *Skymaster flew some 2400 miles.*

Pion
Cecil Frank Powell, *Professor of Physics at Bristol University,* UK, *in conjunction with* Giuseppe Ochiatini, *Italy.*

Polaroid Land camera
Dr Edwin H Land, *who had originally formed his Polaroid Corporation in Cambridge, Massachusetts to manufacture polaroid sunglasses. The automatic developing and printing process was carried out in the camera itself. The polaroid colour photo, trade-named 'Polacolor', was introduced in 1963 – a direct positive being produced in one minute.*

Powerboat (converted to turbojet-propulsion)
Peter du Cane, *naval architect, Portsmouth,* UK *and* Major Frank Halford, *engineer, Hatfield,* UK, *who modified Sir Malcolm Campbell's 'Bluebird K4' hydroplane to take a de Havilland Goblin II jet engine.*

Promethium
J Marinsky *et al.*

Radiocarbon dating
Dr Willard Frank Libby, *Professor of Chemistry, Institute for Nuclear Studies, University of Chicago,* USA.

Supersonic aircraft
the Bell XS-1 rocket-plane, *'Glamorous' Glennis', launched from a Boeing B-29 Superfortress in flight and piloted by Captain 'Chuck' Yeager over Edwards Air Force Base, California. Its top speed was Mach 1.015 (670 mph).*

Synthesised Vitamin A
O Isler.

Tractor (power adjustments of rear wheel tread)
Allis Chalmers, US.

Tubeless tyres
B F Goodrich Co. *of Akron, Ohio,* USA.

Tungsten carbide dental drill.

1948

Atomic clock
Dr Willard Frank Libby, *Professor of Chemistry, Institute for Nuclear Studies, University of Chicago,* USA. *First model built at the National Bureau of Standards, Washington* DC.

Cybernetics
Norbert Wiener, US *mathematical logician, Massachusetts Institute of Technology.*

McDonald hamburgers
Maurice *and* Richard McDonald, *brothers in Pasadena, California,* USA.

Microgroove record (long-playing)
Dr Peter Goldmark *and* a fifteen-strong research team *at the Columbia Broadcasting System,* USA. *The* $33\frac{1}{3}$ *rpm speed was soon to oust the 78 rpm monopoly. Soon after, the Radio Corporation of America introduced a 45 rpm microgroove disc for 'pop' as opposed to 'classical' music.*

Passenger elevators (electronically controlled)
Otis Co. *at Universal Pictures Building, New York.*

Polyester-resin boatbuilding
Dr Irving E Muskrat *of Marco Resins Inc., New Jersey,* USA.

Rear-window heating (embedded tungsten wires)
Rolls-Royce Cars Ltd, UK.

Relascope (tree-measuring instrument)
Walter Bitterlich, *forestry officer of Salzburg, Austria, for angle-count sampling.*

'Scrabble' (board game)
James Brunot, *Connecticut,* USA.

Solid electric guitar (production model)
'Leo' (Clarence) Fender, 'Doc' Kauffman *and* George Fullerton, *of California. It was called the 'Fender Broadcaster', later renamed the 'Telecaster'.*

Soluble aspirin
Reckitt & Sons Ltd, *Hull,* UK – *called 'Disprin'.*

Transistor (point contact)
Dr William Shockley, John Bardeen *and* Walter H Brattain *at the Bell Telephone Laboratories,* USA.

Turboprop airliner
Rex K Pearson *and* George Edwards *at Vickers Aviation Ltd, Weybridge, Surrey,* UK. *The 'Vickers Viscount' was powered by four Rolls-Royce Dart turboprop engines.*

Universal digital stored program computer
Professors Sir Frederick Williams *and* T Kilburn *at Manchester University,* UK. *Called the 'Manchester Mark I', and using electrostatic tubes, it was manufactured by Ferranti Ltd in 1951.*

'Velcro' (fastening device)
Georges de Mestral, *Swiss engineer.*

1949

Aircraft (delta-wing)
the Avro 707, UK.

Berkelium
Glenn Theodore Seaborg *with* Stanley Gerard Thompson, *nuclear scientists at the Lawrence Radiation Laboratory, University of California, Berkeley,* USA.

Cable television (experimental)
Oregon, USA.

Colour television tube
David Sarnoff *and* a research team *at the Radio Corporation of America – called the 'Shadowmask'.*

EDSAC (electronic delay storage automatic calculator)
researchers at *Cambridge University,* UK.

EDVAC (electronic discrete variable automatic computer)
John von Neumann and a team *at Pennsylvania,* USA.

High-altitude liquid-fuel rocket
Viking, *fired to an altitude of 50 miles in 54.5 seconds, over White Sands, New Mexico.*

Jet-powered airliner
design team led by R E Bishop *of the*

de Havilland Aircraft Company, Hatfield, UK. The DH 106 Comet was powered by the de Havilland Ghost 50 engine.
Key starting (automobiles)
the Chrysler Corp., Detroit, Michigan, USA.
Magnetisable ferrite cores (for data storage)
Jay Wright Forrester of the Massachusetts Institute of Technology, Boston, for the 'Whirlwind' computer.
'Super amp' (rock music amplifier)
Leo Fender and Donald Randall, US.

1950
Californium
Glenn T Seaborg and Albert Ghiorso, nuclear scientists at the Lawrence Radiation Laboratory, University of California, Berkeley. USA.
Chlorpromazine (tranquilliser drug)
Paul Charpentier, French chemist working at the Rhône-Poulenc laboratories of the Specia drug firm. It took him only three months to synthesise a new phenothiazine derivative, which was to have remarkable effects on schizophrenic patients.
Credit card (for general purchase)
Ralph Scheider for the first 200 members of the New York Diners' Club.
Duffy blood group system
James S Cutbush, P L Mollison and D Parkin, chemists, US.
'Elmer and Elsie' (robots)
Dr Gray Walter of the Burden Neurological Institute, Bristol, UK. The robots, made of photo-electric cells, batteries, transistors, relays and electric motors, were capable of making simple decisions.
Ferroxdure (ferrite)
Philips Research Laboratories, Eindhoven, Netherlands.
Gas-turbine car
Rover Ltd, UK – the prototype two-seater Jet 1.
Oceanographic research ship
'Calypso', commanded by Jacques Yves Cousteau of the French Navy's Undersea Research Group.
Radio-immunoassay
Rosalyn Sussman Yalow, Veterans Administration Hospital, Bronx, New York.
Schottel-Rudderpropeller
Joseph Becker, Rhineland boatbuilder.
Xerographic-copying machine (production 'office model')
Haloid Co. of New York – called the 'Xerox 914'.

1951
Cinerama
Fred Waller, US, making use of three projectors and a wide curved screen to recreate a stereoscopic motion picture experience. Initial tests had been going on since 1936 at a former indoor tennis court near Oyster Bay, Long Island, New York. In 1952 'This Is Cinerama' opened in a Broadway cinema and ran for 2½ years. Its seven sound tracks were developed by Hazard Reeves.
Experimental breeder reactor No. 1
Walter H Zinn, one of Enrico Fermi's wartime assistants at the National Reactor Testing Station in central Idaho, USA. This was the first nuclear reactor to produce a significant amount of electric power.
Helicopter (gas-turbine-engined)
SO 1120 'Ariel III', France.
Isoniazid (drug)
simultaneously by a team of chemists at the Hoffmann-LaRoche pharmaceutical company, New Jersey, and by a team of scientists at E R Squibb & Son laboratories in New Brunswick, both USA. Isoniazid was

first used with streptomycin on tuberculosis patients by Doctors Selikoff and Robitzek at Sea View Hospital, Staten Island, New York, the world's largest TB hospital. Having successfully tested it on themselves, the doctors administered it to their worst 92 patients, nearly all of whom showed rapid signs of recovery.

Merchant ship (converted to gas-turbine engines)
'Auris' by John Lamb, *Head of Research and Development at Anglo-Saxon Petroleum Ltd,* UK.

Mobile genetic elements
Barbara McClintock, *cytogeneticist, Cold Spring Harbor Laboratory, New York,* USA, *while studying the maize plant.*

Motor-cruiser (glass-fibre)
M Rosenblatt & Son, US. *The 'Dreadnought 23' was built using the Marco vacuum process.*

Oil (Persian Gulf)
Aramco, *who discovered the great Safaniya field.*

Oral contraceptive pill
Gregory Goodwin Pincus *and* Min Chuch Chang, *biologists of the Worcester Foundation for Experimental Research, Shrewsbury, Massachusetts, in collaboration with* Dr John Rock, *obstetrician-gynaecologist of the Brookline Reproductive Clinic, Boston, and again in collaboration with* Dr Carl Djerassi *of the Syntax Corporation, Mexico, who discovered the progestogenic agent, 19-Norsteroids. The US Food and Drug Administration approved 'the pill' for public use in 1961, after extensive trials in Puerto Rico and Haiti.*

Power-steered automobile
Chrysler, USA.

Precision-base electric guitar
Leo Fender *and* 'Doc' Kauffmann, *California,* USA.

Transistor (junction)
Dr William Shockley, R L Wallace *and* Morgan Sparks *at the Bell*

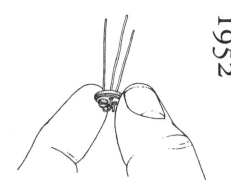

Transistor

Telegraph Laboratories, USA. *The original point-contact transistor was found to be noisy and unable to control high amounts of power, and had limited applicability. It was therefore superseded by the junction-transistor.*

1952

Artificial heart valve
Dr Charles A Hufnagel *at Georgetown University Hospital, Washington* DC, USA.

Automated automobile manufacture (partial)
the Ford Co. *at Cleveland, Ohio,* USA.

Bubble chamber
Donald Arthur Glaser, *physicist, at the University of Michigan,* USA.

Cosmotron (3 billion-volt proton synchroton)
Brookhaven National Laboratory, Long Island, New York.

Electric coffee pot (fully automatic)
Bill Russell *and* Peter Hobbs *of Croydon,* UK.

Electron (precise magnetic moment of)
Polykarp Kusch, *Professor of Physics at Columbia University,* USA.

Electronic calculator (germanium diodes as electrical components)
Compagnie Bull, *Paris – called the 'Gamma 3'.*

FM car radio
Blaupunkt, *Germany, – with push-button tuning.*

Hearing aid (transistorised)
Sonotone Corp. *of Elmsford, New York.*

Hydrofoil (turbojet-engined)
Frank Hanning-Lee, UK. *The 'White Hawk' was unsuccessful in trials on Lake Windermere.*

Hydrogen bomb
simultaneously by Edward Teller, US, and Igor Kurchatov, USSR. *The US test (6 megatons of TNT) was at Eniwetok Atoll, Pacific Ocean, and the Soviet test was in the Arctic in 1953.*

Integrated circuit (theoretical)
G W A Dummer, *Royal Radar Establishment, Malvern, UK.*

Pattern 5580 inflatable lifejacket
Dr E A Pask, RAF *Physiology Laboratory, RAF Farnborough, UK, after ten years' research and development.*

Powerboat (custom-built for turbojet power)
Peter du Cane *of Vospers Ltd*, Reid A Railton, *aerodynamicist, and* Major Frank Halford, *engineer. Powered by a de Havilland Ghost turbojet, 'Crusader' disintegrated at 240 mph on Loch Ness, Scotland, killing her pilot,* John Cobb.

Surface-to-surface missile
the 'Firestone Corporal', US.

Terylene yachting sail
Used on 'Sonda' of West Mersea, UK.

UNIVAC I (universal automatic computer)
Eckert and Mauchly Computer Co., *Philadelphia, as the first computer designed for mass production. Built by Remington Rand, it was delivered to the US Census Bureau.*

Videotape (experimental)
John Mullin *and* Wayne Johnson *at Bing Crosby Enterprises Laboratories, Beverly Hills, California, USA.*

'WD 40' (water-displacer and temporary rust preventative)
Norm Larson, *for the Rocket Chemical Co., San Diego, California, USA. It was first used on certain parts of the Atlas missile. The aerosol version of 'WD40' was available by 1955.*

1953

Automobile (glass-fibre bodywork)
Chevrolet Corvette, US.

'Cloudbuster'
Wilhelm Reich, US, *channelling his orgone energy through a bank of metal tubes to disperse hurricanes, UFOs and black clouds, and to make rain where there was drought – questionable.*

Collapsible polythene tube
Bradley Container Corp. *of Delaware, USA, for 'Sea and Ski', a skin-tanning lotion.*

Commercial passenger hydrofoil
Baron Hanns von Schertel, *Germany, of Supramar, the pre-war Schertel-Sachsenberg Research and Development Consortium. It first plied for service on Lake Maggiore between Switzerland and Italy.*

DNA (deoxyribonucleic acid)
Francis Harry Compton Crick, *molecular biologist, UK, and* James Dewey Watson, *biochemist, US, at the Medical Research Council Unit, Cavendish Laboratory, Cambridge, UK.*

Disc brakes (aircraft).

Einsteinium
Albert Ghiorso *et al., Heavy Ion Linear Accelerator, Lawrence Radiation Laboratory, University of California, USA.*

Fermium
Albert Ghiorso *and* Stanley Gerald Thompson, *Lawrence Radiation Laboratory, University of California, USA.*

Heart-lung machine
Dr John H Gibbon *at the Thomas Jefferson University Hospital, Philadelphia, Pennsylvania, USA, after some eighteen years' research.*

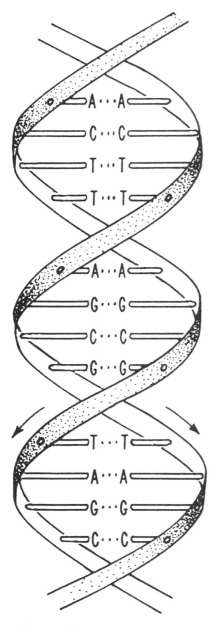

DNA molecule

Jet-prop helicopter
independently, Fairey Aviation Co.'s
Gyrodyne, UK, *and the* SO*1221* Djinn,
France.
Measles vaccine
Dr John Franklin Enders *and his
assistant,* Dr Thomas Peebles,
*paediatricians at the Children's
Hospital, Boston, Massachusetts,* USA.
*Peebles was able to isolate a measles
virus from the blood and throat
washings of a schoolboy.*
Nylon marine propeller
Ajax Marine Engines Ltd, UK.
Rapid computer print-out
Remington Rand, US, *for* UNIVAC. *It
was capable of 600 lines per minute.*
**Reserpine (anti-depressant
drug)**
Dr Nathan S Kline, *psychiatrist at
Rockland State Hospital,
Orangeburg, New York. Three years
later, he discovered iproniazid as a
second anti-depressant drug.*
Swing-wing aircraft
Grumman, XF 10 F *Jaguar jet-fighter*
US. *Only two prototypes were ever
built.*
Supercocotte (pressure cooker)
Lescure brothers (Frederic, Jean and
Henri), *for* SEB, *Boulogne, France.*

1954
**Atomic absorption
spectrophotometer**
Dr Alan Walsh, *Chemical Division of
the Commonwealth Scientific and
Industrial Research Organisation,
Melbourne, Australia.*
Ball-bearing handpiece (dental)
*enabling the drill to spin at 25,000
rpm.*
**Colour television (regular
broadcasts)**
the National Television System
Committee, USA.
**Computer memory magnetic
cores (commercial)**
IBM *at Poughkeepsie,* US, *for their 704
and 705 computers, developed in
conjunction with* Massachusetts

1954

Institute of Technology's Lincoln Laboratory, *based on the pioneering work of* A Wang *and* F W Viehe.

Electronic music synthesiser
Dr Harry Olsen *of the Radio Corporation of America; re-designed and installed at both Columbia and Princetown Universities as the* RCA *Mark II.*

Hatchback estate car (side opening rear door)
Hillman Husky, UK.

High-speed switching surface barrier transistor
Philco, US.

Input/Output computer channel
Bob Evans *on* IBM *704.*

Isotactic polypropylene
separately by Karl Ziegler, *German chemist, Director of the Max Planck Carbon Research Institute at Mülheim, and* Giulio Natta, *Italian chemist, at the Milan Institute of Technology. It was first marketed by Montecatini as 'Moplen' in 1957.*

Maser prototype
Charles Hard Townes, *physics professor, with* James P Gordon *and* Herbert J Zeiger *at Columbia University,* USA. *Also by* Aleksandre Mikhailovic Prochorov *with* Nikolai Gennadievic Bosov *at the Lebedev Physics Institute,* USSR. *The practical maser (three-level, solid state) was developed by* Nicholas Bloembergen *at Harvard University and built by Messrs Feher, Scovil and Seidel at Bell Telephone Laboratories,* USA.

Motor yacht (glass-fibre construction)
49ft 'Perpetua', moulded in 'Deborine' for Patrick David de Laszlo *at his own company, Halmatic Ltd of Portsmouth,* UK. *Given strength by de Laszlo's patent 'Top Hat' aluminium framework for her ribs and stringers. 'Perpetua' was fitted out with twin Gardner diesel engines.*

Non-stick pan
Marc Grégoire, *French research engineer at* ONERA, *by coating Téflon*

Patrick De Laszlo's 48ft glass-fibre 'Perpetua'

on to metal. The Téfan Co. was formed in 1956.

Passenger conveyor (permanent)
'Speedwalk' at Erie Station, Jersey City, USA.

Retractable ball-point pen (press-and-press-again)
Parker Pen Co., *Janesville, Wisconsin,* USA.

Rotary piston engine
Felix Wankel, *motor engineer from Lindau, Bavaria, after 30 years' research; production engine by* NSU *Motorenwerke in 1957.*

Solar battery (400 silicon cells)
G M Chapin, C S Fuller *and* G L Pearson *at Bell Telephone Laboratories,* USA.

Supersonic delta-wing seaplane
Convair's *jet-engined 'Sea Dart' for the* US *Navy.*

Transistor radio
Regency Electronics, *Indianapolis,* USA – *the 'TR-1'.*

Tubular-nibbed technical drawing pen
Rotring, *Hamburg, Germany – the 'Rapidograph'.*

Uranium, graphite-moderated, gas-cooled reactor
Marcoule Atomic Power Station, France – unsuccessful.

Xenon short-arc lamp
Dr Kurt Larché *and researchers at Osram GmbH, Germany – called the 'Wotan* XBO'.

1955

Air-to-air guided missile
USAF Hughes GAR-1 Falcon *for the*

Northrop Co.'s F-89 Scorpion fighterplane.

Anti-proton
Emilio Gino Segré, Owen Chamberlain, Clyde E Wiegand *and* Thomas Ypsilantis, *physicists at the University of California, Berkeley,* USA.

Automatic electric kettle (vapour-controlled)
Bill Russell *and* Peter Hobbs, *Dartford,* UK.

Bubble-oxygenator (cardiac)
Dr Clarence W Lillehei *and* Dr Richard A DeWall *at the University of Minnesota Hospital, Minneapolis,* USA.

Diego blood group system
M Layrisse, Arends *and* Dominguez, *Venezuela.*

Electric locomotive (remote radio-controlled)
SNCF, *France. It travelled from Paris to Le Mans at 80 mph.*

Felt-tip marker (refillable)
Esterbrook *of Birmingham,* UK – *called the 'Flo-Master'.*

Fibre optics
Dr Narinder S Kapany, *Imperial College, London, based on a discovery made 80 years before by* John Tyndall, UK *physicist.*

Graphic conversation terminal
Jay Wright Forrester *of Massachusetts Institute of Technology, Boston, with the 'Whirlwind' binary computer.*

Hovercraft
Christopher Sydney Cockerell, *electronics engineer with a small boatbuilding concern on the Norfolk Broads,* UK.

Hydropneumatic suspension (motor car)
Citroën, *France.*

Large radio telescope
Sir Bernard Lovell *at Jodrell Bank, Cheshire,* UK, *following five years of construction. Its 'dish' measured 250 ft in diameter.*

Lego® system of play

Godtfred Kirk Christiansen, *using interlocking plastic toy bricks as developed with his father,* Ole Kirk Christiansen, *carpenter and joiner of Billund, Denmark. 'Leg godt' (Danish) = play well. 'Lego' (Latin) = put together.*

Mendelevium
Albert Ghiorso, *University of California, Berkeley,* USA.

RNA (ribonucleic acid) synthesis
Severo Ochoa, *Spanish-*US *physician, Chairman of the Department of Biochemistry at New York University Medical College,* USA.

Spectacle lenses (unbreakable)
'Orma 1000' using CR-9 polymer.

Stereophonic tape-recordings
EMI *'Stereosonic' tapes.*

Submarine (nuclear-powered)
'Nautilus', under the direction of Captain Hyman G Rickover, US Navy. *On sea trials, 'Nautilus' cruised 60,000 miles on a single charge of fuel, using up no more than 8 lb of uranium.*

Super-aircraft carrier (100 jet aircraft)
'Forrestal', US Navy.

Tilting-rotor fixed-wing aircraft
Bell Aircraft Co.'s XV-3.

Tractor-mounted hydraulic digger-loader
J C Bamford Ltd, *Uttoxeter, Staffordshire,* UK.

Transoceanic telephone cable
American Telephone and Telegraph Company. *It was laid by the cableship 'Monarch' for a distance of 1950 nautical miles, from Clarenville, Newfoundland to Oban, Scotland.*

Ultrasound
Leskell, US, *to observe the heart.*

1956
Amniocentesis (routine procedure)
St Mary's Hospital, *Manchester,* UK.

Contact lens (plastic corneal lens)
Norman Bier, UK.

DNA biosynthesis
Arthur Kornberg, *Head of the Department of Microbiology at Washington University, St Louis,* USA.

'Dyform' (seven-wire and nineteen-wire pre-stressing strand)
the Steel Division of British Ropes Ltd, UK.

Flexible wing
Francis Melvin Rogallo, *aeronautical research scientist, and his wife* Gertrude, US.

Go-kart
Art Ingles, *Echo Park Road, Los Angeles, California.*

Human growth hormone (HGH)
Choh Hao Li, *Chinese biochemist working at the University of California, Berkeley,* USA. *Since used to wipe out dwarfism in children.*

Letraset
Fred McKenzie, *chemist, and* Dai Davies, *commercial artist,* UK, *who used the apparently simple principle of the child's waterslide transfer to develop a headline typesetting process. The Letraset business was founded in London in 1959. Two years later Letraset's Instant Lettering dry transfer process created a revolution in art-work and typographic layout.*

Merchant ship (gas-turbine/electric)
'John Sergeant,' US *freighter.*

Motorcycle (desmodromic valves)
Fabio Taglioni, *chief designer, Ducati Meccanica* SPA, *Bologna, Italy.*

Plutonium power reactor (large-scale)
Calder Hall Atomic Power Station, Cumberland, UK, *one year after an experimental pressurised-water system station had gone into operation at Obninsk, near Moscow,* USSR.

Record-players (in automobiles)
Chrysler, USA.

Video-tape (commercial)
Charles Ginsberg, Charles Anderson, Ray Dolby, Fred Pfost *and* Alex Maxey, *engineers at Ampex Co., Redwood City, California,* USA. *The* VTR *spool began to be replaced by cartridges from 1969.*

Vitamin B12 (cyanocobalamin)
Karl August Folkers, *Merck and Co. Inc., Rahway, New Jersey,* USA.

1957

Air-to-surface missile
the US *Bell GAM-63 Rascal, as carried by Boeing* B-47 *Stratojet bombers,* USAF.

Air-turbine dental drill
spinning at 300,000 rpm.

Animal-carrying spacecraft
USSR. *A bitch, Laika, orbited the earth, supplied with food and oxygen, whilst her respiration and heart behaviour, telemetred, were transmitted back to earth. The satellite completed 2370 orbits before it re-entered the denser atmosphere and burned up.*

BCS theory (superconductivity)
John Bardeen, Leon N Cooper *and* J Robert Schrieffer, US *physicists.*

Cosmodrome
launchpad complex for Soviet manned and unmanned spacecraft, believed to be at Tyuratam in Kazakhstan, USSR.

Fibre-optic tubes (medical)
Dr Basil Isaac Hirschowitz, *South African physician at the University of Michigan,* USA, *for exploring the interior of the human body.*

Fortran (computer programming language)
John Backus *and a* research team working for IBM Corporation, *New York City, for use with the IBM 704 computer. A Fortran manual accompanied each computer sold.*

Glass ceramics
marketed as 'Pyroceram' in the USA *and in the* UK *as 'Pyrosil'.*

Ground effect (vehicles)
Louis Duthion, *France.*

Intercontinental ballistic missile (ICBM)
USSR – *called the* SS6 *(Sapwood), with*

a 6000-mile range.

Interferon (protein)
Dr Alick Isaacs *and* Jean Lindemann *(Swiss), biochemists at the National Institute of Medical Research, Mill Hill, London.*

Mechanised sorting of mails
British Post Office engineers, *Southampton*, UK. *The system of post-coding was initiated experimentally two years later, but did not come into complete nationwide use until 1974.*

Mössbauer effect (gamma radiation)
Rudolph Ludwig Mössbauer, *German physicist at the Max Planck Institute, Heidelberg University.*

Pacemaker (internal)
Dr Clarence Walton Lillehie, *surgeon at the University of Minnesota Hospital, Minneapolis*, USA, *with technical assistance from* Earl Bakk, *electronics engineer.*

Plumbicon TV camera tube
Philips Co., *Eindhoven, Netherlands.*

Polio vaccine (live virus)
Dr Albert Bruce Sabin, *microbiologist, New York, following* Dr J F Enders' *isolation of the virus in a test tube, some 9 years before, at the Children's Hospital, Boston. This vaccine was first adopted in the* USSR, *because medical authorities in the* USA *were more interested in other types of vaccine which proved, in the long run, far less effective.*

Pressurised water-cooled reactor
Westinghouse Corporation *of Pittsburgh, Pennsylvania*, USA, *for their Shippingport atomic power station.*

Satellite
Dr Sergei Pavlovich Korolyov *and a Soviet research team. Called 'Sputnik' (fellow traveller) I this 23 inch diameter metal sphere, travelling at a speed of 17,000 mph, orbited the earth 1400 times. It measured, among other things, the concentration of electrons in the ionosphere, before returning to the lower atmosphere and burning up early in 1958.*

Semi-synthetic penicillins
Messrs Doyle, Naylor, Rolinson *and* Batchelor *of the Beecham Research Laboratories, Brockham Park*, UK. *The 10-year research and development programme may have cost Beechams £12 million, but the sales of 'Broxil', 'Celbenin', 'Penbritin', etc. were to repay that investment many times over.*

Silicon-grown junction transistor
Texas Instruments, USA.

Steel-rolling mills (automatic thickness control)
Davy-Ashmore Ltd *with* United Instruments Ltd, *Sheffield*, UK.

305 RAMAC (random access method of accounting and control)
Ray Clifford Johnson *and an* IBM research team, USA, *who developed a magnetic disc storage system.*

Tilt-wing aircraft
Vertol Model 76, the VZ-2A, USA.

1958

'Airstrip' (microporous plastic medical dressing)
Smith & Nephew Ltd, *Hull*, UK.

ALGOL (computer language)
Swiss computer scientists.

All-transistor FM radio
Sony Corporation, *Tokyo, Japan.*

Banking credit card
Bank of America.

Communications satellite
SCORE *(signal communication by orbital relay), built and launched from the* USA.

Float glass process
Alistair Pilkington, *St Helens, Lancashire*, UK. *From 1952 until 1958 Pilkington developed this process, ignorant that it had been investigated but never tested by* William Heal *in 1902 and* H K Hitchcock *in 1905, both in the* USA.

1958

Geodesic dome (large scale)
Richard Buckminster Fuller – *inventor-architect. The 384 ft diameter Union Tank car dome at Baton Rouge, Louisiana,* USA.

High-speed production of bread
Milling and Baking Industry Research Centre, *Chorleywood, Hertfordshire,* UK.

'Hula-hoop'
Richard P Knerr *and* Arthur K 'Spud' Melvin, *owners of Wham-O-Manufacturing Co., San Gabriel, California,* USA. *It was a tubular plastic hoop with a choice of colours.*

Intercontinental ballistic missiles (greater range)
USSR*'s* T-3 *followed in* 1959 *by the* USA*'s Atlas* D.

'Lexan', 'Makrolan', 'Merlon' (polycarbonates)
Farfabriken Bayer, *Germany, and* General Electrics, USA.

LISP (computer language)
John McCarthy, *computer scientist, Massachusetts Institute of Technology, Cambridge,* USA *for 'artificial intelligence' research.*

Nobelium
Albert Ghiorso, *Lawrence Radiation Laboratory, University of California, Berkeley,* USA. *It is called 'Joliotium' in the* USSR.

Optical-wavelength maser (laser)
Charles A Townes, *physicist, Columbia University, New York, with his brother-in-law,* Arthur Leonard Schawlow, *Bell Telephone Laboratories.*

Rotary blade electric lawn-mower
Wolf, *France.*

Stereophonic disc recordings
Audio Fidelity, US, *and* Pye, UK.

Transistorised echo sounder (marine)
Major Richard Gatehouse, *formerly of the Royal Radar Establishment,* UK. *It was tested on his Lymington one-design sloop 'Wavecrest II'. Two*

models of the 'Hecta Mark I' appeared at the 1959 *London Boat Show.*

Unitary fibreglass automobile
Colin Chapman's *Lotus Elite,* UK.

Van Allen radiation belts
James Alfred Van Allen, *Professor of Physics, University of Iowa,* USA. *Based on geiger counter readings made from the 'Explorer I' satellite.*

Wooden sailboard (prototype)
Peter Chilvers, *twelve-year-old schoolboy at Hayling Island,* UK.

1959

AG proton synchroton (high energy particle accelerator)
CERN *Laboratory for nuclear research in Geneva, Switzerland, designed and built with support from eleven European countries.*

Amphibious hydrofoil (gas-turbine-engined)
Lycoming Co., *under contract to the* US *Army Ordnance Corps.*

Animal radio-tracking (prototype system)
Messrs Le Munyan, White, Nybert *and* Christian, US, *to track the wild chipmunk.*

Anti-missile missile
Nike-Zeus, test-fired from White Sands, New Mexico, USA.

Artificial intelligence (AI) computer program (to play chess)
Claude E Shannon, *Bell Telephone Laboratories.*

BWR (heavy-water-moderated boiling water reactor)
Halden Atomic Power Station, Norway.

COBOL (computer language)
US *Conference on Data Systems Languages.*

Computer (performing eight-data-processing tasks simultaneously)
Honeywell, US – *the 'Model 800'.*

Computer-controlled chemical plant
Imperial Chemical Industries, UK.

Computer-controlled refinery
Texaco Oil Co., US – *for their Port Arthur plant.*

Full-sized hovercraft
Richard Stanton-Jones *and* research team at Saunders-Roe Ltd, *Cowes, Isle of Wight,* UK. *It was called the* SR.N1.

Fully simultaneous, large-scale computer system
Bull, *France – the 'Gamma 60'.*

Ice-breaker ship (nuclear-powered)
USSR. *The 440 ft 'Lenin' was capable of operation throughout the winter without refuelling.*

Identikit
Hugh C McDonald, *detective in the Los Angeles Identification Bureau. It was used by Sheriff Peter Pitchess of Los Angeles County Police to identify a criminal from McDonald's 500 master foils.*

Integrated circuit (phase-shift oscillator)
Jack S Kilby, *Texas Instruments, Dallas,* USA.

Ion engine (cesium contact)
Dr Alvin Theodore Forrester, *physicist, Rocketdyne,* USA.

Librium (drug)
Hoffmann-LaRoche Pharmaceutical Co., *Switzerland.*

Lunar observation orbiter
USSR. *Called 'Lunik III', it circled the moon and photographed the far side, radioing pictures back to earth.*

Microwave point-to-point radio system
Pacific Great Eastern Railway, between Vancouver and Dawson Creek-Fort St John, British Columbia, Canada.

Morris Minor 'Mini'
Arnold Constantine Issigonis, *British Motor Corporation,* UK.

Motor car safety belt (production model)
Volvo, *Sweden.*

Plastic fishing nets
Smith & Nephew Plastics Ltd, *Harlow and Hull,* UK.

PDP-1 (programmed data processor)
Ken *and* Stan Olsen *and* Harland Anderson *of the Digital Corp., Maynard, Massachusetts,* USA – *a small computer.*

Snow motor scooter
J Armand Bombardier, *Canadian mechanic – called the 'Ski-Doo'.*

Solid-state data-processing system
Olivetti, *Italy, for their 'Elea 9003'.*

Submarine (Polaris)
US *Navy's 'George Washington', capable of firing a Polaris missile, as developed two years previously, from beneath the water surface. It was built at and launched from Groton, Connecticut.*

Toughened float-glass car windscreens.

Transistorised computer
The IBM 1401.

Tunnel diode (commercial)
Sony, *based on work by* Leo Esaki, *Tokyo University, Japan.*

1960

Aircraft carrier (nuclear-powered)
USS *'Enterprise', 1101 ft long and capable of carrying 100 aircraft.*

Argon ion laser
D R Herriott, A Javan *and* W R Bennett *at Bell Telephone Laboratories, New Jersey,* USA.

Artificial ski slope
Len Godfrey *of Dendix Ltd,* UK, *manufacturers of industrial brushes in Vybak pvc monofilaments. The first 'Snowslope' was tested at Philbeach Hall, Earls Court, London.*

Balloon communications satellite
Echo 1, US.

Bathyscaphe (deep-sea)
Auguste Piccard, *Swiss physicist. His son* Jacques, *together with* US *Naval officer,* Lieutenant Don Walsh, *took the 'Trieste' to a depth of 35,800 ft*

into the Marianas Trench off the *Pacific Island of Guam. They observed the deepest ocean floor for half an hour through special plexiglass portholes.*

Bird radio-tracking system
Einar Eliassen, *Bergen, Norway, for mallard in normal flight.*

Closed-circuit TV camera as viewfinder for motion picture filming
David Samuelson, *cinematographer of Cricklewood, London.*

Computer (6000 additions or subtractions per second)
US Honeywell *'Model 400'.*

Computer-controlled production line
North Carolina Works of Western Electric, *Hickory, North Carolina,* USA.

Digital flight simulator
University of Pennsylvania, USA, *and manufactured by the Sylvania Corp. as* 'UDOFT'.

Electron bombardment engine
Harold R Kauffmann, *space propulsion engineer,* NASA, *Cleveland, Ohio,* USA.

Fibre-tip pen
Japan Stationery Co., *Tokyo – called the 'Pentel'.*

Intercontinental ballistic missile (8000-mile range)
US *Minuteman I.*

Ion source (implementation gun)
J Harry Freeman, *nuclear chemist,* UK *Atomic Energy Research Establishment, Harwell, Oxfordshire. Initially for isotope separation, it was developed to make materials tougher and more corrosion-resistant.*

Missile launch warning satellite
USAF.

Moulton bicycle
Alex Moulton *at the small research establishment of his family estate in Wiltshire,* UK. *By 1965 major cycle makers in seven countries had taken licences for the manufacture of the Moulton bicycle with its F-shaped*

frame and small wheels.

Nitinol (memory-metal)
William Buehler, *engineer at Silver Spring Research Laboratories,* USA.

Pulsed ruby laser
Theodore H Maiman, *physicist, at Hughes Research, Malibu, California,* USA.

Re-locatable building module (loda-strut-leg system)
Donald Shepherd, *building contractor, York,* UK. *To produce working facilities on building sites. Called the 'Portakabin'. 50–60,000 currently in versatile use worldwide.*

Turbojet-engined car
Dr Nathan Ostich, US. *His 'Flying Caduceus', when tested on the Bonneville Salt Flats, Utah, was found mechanically unsuccessful.*

Typesetting by computer
Imprimérie Nationale, *Paris.*

VTOL (vertical take-off and lift) aircraft
Frank Taylor *and a* design team *at Short Brothers & Harland at Sydenham, Northern Ireland. The* SC-1*'s test pilot was Tom Brooke-Smith and the powerplant was the Rolls-Royce* RB108 *engine.* VTOL *tests had been carried out some six years before with the Rolls-Royce-engined 'Flying Bedstead'.*

Weather satellite
NASA*'s* TIROS-1.

Back-pack aero-engine
Wendell F Moore, *engineer at the* US *Bell Aircraft Corporation. The* SRLD *(small rocket lift device), strapped on the back of Harold Graham, made its first untethered flight.*

Cargo and passenger ship (nuclear-powered)
'The Savannah', US, *21,800 tons with a passenger capacity of 60. It proved too costly and was laid up in 1967.*

Computer time-sharing experiment
Ferdinand José Corbató,

Computation Centre, Massachusetts Institute of Technology, Cambridge, USA, *for exploitation with the* IBM *'709' model.*

Computerised algorithmic music
Pierre Barbaud *and* colleagues *at Compagnie de Machines Bull, Paris.*

Electric toothbrush
Squibb & Co., *New York.*

Electronic calculator (commercial model)
Sunlock Comptometer Ltd, UK – *called the 'Anita'.*

Flexible skirt (hovercraft device)
C H Latimer-Needham, UK.

'Flying saucer'
Avro-Avrocar VZ-9V, *Canada. Powered by three Continental* J69 *turbojets, it was abandoned as uncontrollable.*

Gas-assisted gas-laser cutting tool
British Welding Institute, *marketed by the British Oxygen Co.*

Golf ball electric typewriter
'IBM 72', US.

Hatchback estate car (upward-opening rear door)
Renault 4, France.

'Kenyapithecus Wickeri' (upper jawbone)
Dr Louis Seymour Bazett Leakey, *prehistoric archaeologist. The jawbone, found at Fort Ternan in Western Kenya, was dated to 14 million years old.*

Lawrencium
Albert Ghiorso, *nuclear scientist, Lawrence Radiation Laboratory, University of California, Berkeley,* USA.

Magnetic ink character recognition
Honeywell, US, *to read banking computer print-outs.*

Man-carrying spacecraft
USSR. *Travelling in the 4.5 ton 'Vostok I', Flight Major Yurij Alexeyevich Gagarin, aged 27, completed a single*

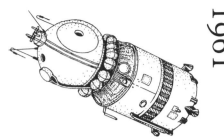

'Vostok I', cosmonaut-carrying spacecraft

orbit of the earth, making a safe landing in the USSR *one hour and 48 minutes later.*

Nuclear bomb (50 megatons +)
exploded by the USSR.

Observation/'spy' satellite
USAF.

Self-twist yarn spinner
David Henshaw, *physicist, Textile Division of the Commonwealth Scientific and Industrial Research Organisation, Geelong, Victoria, Australia.*

Space center
Johnson Space Center, Houston, Texas, USA. *Flight controllers at Mission Control were subsequently equipped with as many as 171 consoles including 133* TV *cameras and 557 receivers.*

Stereophonic radio broadcasts
Zenith *and* General Electric, USA.

'Stretch' (computer system)
IBM. *150,000 transistors enabled it to perform 100 billion computations per day, using an Octet (eight-bit) system.*

Transistorised video-tape recorder
Sony SV-201, *Tokyo, Japan.*

'US Patent No. 3,000,000'
Kenneth R Eldredge *for the General Electric Co., New York: automatic reading system.*

Valium (drug)
Hoffmann-LaRoche Laboratories, *Switzerland.*

Zone-toughened car windscreens
Triplex-Glass, UK.

1962

Aérotrain (tracked hovercraft)
J Bertin *and* P Guienne *of Bertin et Cie, France. In 1974, their vehicle achieved a speed record average of 425 kph.*

AG electron synchroton
Massachusetts Institute of Technology, *Cambridge, with* Harvard University, *supported by the* US *Atomic Energy Commission – known as the Cambridge Electron Accelerator.*

Air-bearing, air-turbine dental drill
to run at 800,000 rpm.

'Allopurinol' (drug)
Dr George H Hitchings *and the Burroughs Wellcome research staff, to give relief from gout. It was first administered to gout patients at Duke University Medical Center at Durham,* USA. *'Allopurinol' was Hitching's fourth great medical drug discovery – 'Daraprim', '6-*MP*' and 'Azathioprine' being the previous three.*

Amphicar
Hans Trippel, *Germany. It was developed from his 'Alligator' prototype of two years earlier. 3000 Amphicars were manufactured and sold by 1965.*

Autoflare (automatic landing equipment)
developed with BOAC, UK, *for the* Vickers VC10 *airliner.*

Ballistic-missile submarine
US Navy's *'Ethan Allen'.*

Computer simulation languages
US Rand Corp.'s SIMSCRIPT and IBM's GPSS.

Computerised 'real time' flight reservation system
IBM *in conjunction with American Airlines – called 'Sabre'.*

Experimental atomic power station (fast breeder reactor type)
UK *Atomic Energy Authority, Dounreay, Caithness, Scotland.*

Gallium-arsenide semiconductor laser
R N Hall, *General Electric Co., Schenectady, New York,* USA.

Gamma-ray sterilisation (medical and surgical products)
UK *Atomic Energy Authority, as suggested by* G S Murray.

Laser microbeam
Marcel Bessis *and* a team of Parisian biologists, *producing a 2.5 micron diameter spot for work in cell biology.*

LED (light-emitting diode)
Nick Holonyak Jr, *Advanced Semiconductor Laboratory, General Electric, Syracuse, New York,* USA.

Lunar-reflected optical maser beam
Bell Telephone Laboratories, USA.

Magnox (magnesium non-oxidising) atomic power stations
Berkeley, Gloucestershire and Bradwell, Essex, both UK.

Micro-television (5 inch screen)
Sony, *Tokyo, Japan – called the* TV5-303.

Minicomputer
Digital Corporation *of Maynard, Massachusetts,* USA *– called the* PDP-5.

Muon neutrino (elementary particle)
Leon Max Lederman *et al. at Columbia University, New York,* USA.

Passenger hovercraft
Saunders-Roe Ltd, *Cowes, Isle of Wight,* UK. *The* SR.N2 *had a carrying capacity of 70 passengers.*

Primitive industrial robot
Unimation Inc., *Dambury, Massachusetts,* USA.

Silicon integrated circuits
Texas Instruments, USA *– complex circuit functions in a single chip.*

Transatlantic satellite transmission
American Telephone and Telegraph Co., *using 'Telstar' to transmit an image from Maine,* USA *to Cornwall,* UK.

X-ray astronomy
using rocket-borne equipment to identify the 'Crab Nebula'.

1963
Anti-Xi-Zero (atomic particle).
Cassette audio-tapes
Philips, *Eindhoven, Netherlands.*
Computerised stylometric analysis
Rev. Andrew Q Morton *at the University of Edinburgh, UK, to discover and authenticate the authorship of the Pauline Epistles.*
Etorphine (M-99)
Dr Kenneth W Bentley *and* D G Hardy *of Reckitt & Sons Ltd, Hull, UK.*
Geosynchronous satellite
NASA – *called 'SYNCOM 2' and 'SYNCOM 3'.*
'Liberator' concept
automatic translation of computer programs from IBM's Model 1401 to Honeywell's Model 200.
Motor car (Wankel engine)
NSU *'Wankel Spyder', Germany, although large-scale production came in 1968 with the 'Mazda 110S', Japan.*
Navigation satellite
NASA's *'Transit'.*
Quarks
Murray Gell-Mann, *theoretical physicist, California Institute of Technology, Pasadena, USA.*
Quasars (quasi-stellar radio sources)
Marten Schmidt, *Dutch astronomer, Hale Observatory, Palomar, California, USA.*
Rotary lawn-mower with hover
Flymo, UK.

1964
'A–Z System' (new knitting yarns from man-made fibres)
Linen Industry Research Association, *Lisburn, County Antrim, Northern Ireland, with* McCleery and L'Amie Ltd, *Belfast.*

Acrylic paints
Reeves Ltd, UK – *called 'Acrylic Vinyl Co-Polymer Paints'.*
Airflow ventilation (via automobile dashboard)
Ford Motor Co, USA.
BASIC (beginner's all-purpose symbolic computer instruction code)
Thomas E Kurtz, *teacher, and* John G Kemeny, *mathematician, Dartmouth College, New Hampshire, USA.*
Carbon fibre
RAF Farnborough, UK, *following 28 years' research and development. It was manufactured in the UK by Courtaulds Ltd as 'Grafil' and in the USA by Hercules Inc., ISD, Utah.*
Computer-aided graphic design system
IBM 2250
Crew-carrying spacecraft
USSR. *The three-cosmonaut crew remained in 'Voshkod I' for sixteen orbits (or 24 hours) to test collaboration on board.*
Fibreglass sailboard
S Newman Darby *and his wife* Naomi, *at Falls, Pennsylvania, USA. The plans were published in 'Popular Science' magazine soon after.*
Fibrillated polypropylene
British Ropes Ltd *with* Plasticisers Ltd *for rope and cordage manufacture.*
Home-use, all-transistor, video-tape-recorder
Sony, *Tokyo, Japan – called the* 'CV-2000'.
Hovercraft (production model)
British Hovercraft Corporation, *Cowes, Isle of Wight, UK. The* SR.N5 *45 were built and in service by 1968.*
'IBM System 360'
Gene M Amdahl, M Blaun *and an* IBM research team *at New York – the high-volume, automatic, microminiature production of Solid Logic Technology (SLT), semiconductor, ceramic module circuits.*

Laser eye surgery
H Vernon Ingram, *ophthalmic surgeon*, USA, *using a* 'NELAS'.
Laser gun (prototype)
USA.
Mellotron
the Bradley brothers (Norman, Frank *and* Leslie) *of Birmingham*, UK. *The* BBC'*s Mellotron Mark II of the following year could produce 1260 special effects.*
Rhesus haemolytic disease (treatment for)
simultaneously by Doctors Clarke, Sheppard *and* Finn *at Liverpool University Medical School*, UK, *and* Doctors Gorman, Freda *and* Pollock, *sponsored among others by the Ortho Pharmaceutical Foundation, New York. The first trial of the anti-Rh gamma globulin vaccine was on Dr Gorman's own sister-in-law. By 1967 it was being administered to women en masse worldwide.*
3-D laser-holography
Emmett Norman Leith *and* Juris Upatnieks *at the Willow Run Laboratory, University of Michigan*, USA.
Satellite navigation
US Government *for use by their 'Polaris' submarines.*
Unnilquadium
Albert Ghiorso *et al., Heavy Ion Linear Accelerator, Lawrence Radiation Laboratory, University of California*, USA. *The American name is Rutherfordium; the Soviet name is Kurchatovium.*
Voltage-control music synthesiser
Dr Robert Moog, US *engineer*.

1965
Airliner (automatic landing)
Paris-London BEA *de Havilland 'Trident' at Heathrow Airport, London.*
Artificial tooth
researchers at the University of Michigan School of Dentistry, USA,
to record and broadcast dental pressures. It contained six radio transmitters, 28 electronic components and two rechargeable batteries.
Carbon fibres as prosthetic material
Henry G. de Laszlo, *physicist, Englefield Green*, UK.
Cartridge audio tapes
Learjet, *Netherlands.*
Computer-carrying spacecraft
NASA's *'Gemini 3', as piloted by astronaut Commander Virgil Grissom*, US.
Cosmic microwave background radiation
Arno Allen Penzias, *astrophysicist,* and Robert Woodrow Wilson, *radio astronomer, Bell Laboratories, Holmdel, New Jersey*, USA.
Global communications satellite
NASA's *'Early Bird', the first of the five 'Intelsats'.*
Hydraulic mobile and truck cranes (synchronised telescopic jibs)
Coles Cranes Ltd, *Sunderland*, UK.
'Kevlar'
Stephanie Kwolek *and* researchers *at Du Pont de Nemours*, USA. *By 1969, as* PRD-49, *this aramid was available to select* US *government departments. By 1973 it had been marketed under the computer-chosen names of 'Kevlar 49' and 'Kevlar 29'.*
Medium density fibreboard
USA – *dry-formed panel manufactured from lignocellulosic fibres bonded with a synthetic resin.*
Minicomputer (mass-produced)
Digital Corporation *of Maynard, Massachusetts*, USA – *called the* 'PDP-8'.
Mini-skirt (woman's dress)
Mary Quant, *fashion designer, King's Road, Chelsea, London.*
'Multics' (computer time-sharing system)
General Electric *for project 'Mac' at both the Massachusetts Institute of Technology and Bell Telephone*

Laboratories, USA.
Natural gas (North Sea)
British Petroleum, UK. *Natural gas was piped ashore from the West Sole Field for the* UK's *domestic gas system by 1967.*
Radial tyre (asymmetric tread)
Michelin et Cie, Clermont-Ferrand, France. Called the 'XAS'.
Rubella vaccine
Dr Paul D Parkman *and* Dr Harry M Meyer Jr *at the Division of Biologic Standards of the National Institute of Health in Bethesda, Maryland,* USA.
Scanning electron microscope
Engineering Department of the University of Cambridge *with* Cambridge Instruments Ltd, UK. *Called the 'Microscan* SEM', *it gave a depth of field 300 times greater than any previous optical or electron microscope.*
'UK Patent No. 1,000,000'
Ernest Benson, *Garforth, near Leeds,* UK: *'improvements relating to extensible loft ladders'.*
Word processor
IBM – *magnetic tape storage system linked to an electronic typewriter.*

1966
Integrated circuit radio
Sony, *Tokyo, Japan – called the '*ICR-100'.
International chess match by computer
Professor John McCarthy's *computer, 'Stanford University', versus Abram Alikhanov's computer, Russian 'Institute of Theoretical and Experimental Physics'. The* USSR *won 3–1.*
Josephson effect device
Brian David Josephson, *physicist, University of Illinois,* USA.
Laser automobile design system
Hitachi Ltd *and* Mitsubishi Electric Corp., *Japan.*
Linotron
Mergenthaler-Linotype Co., US, *in collaboration with* CBS *Laboratories to develop an electronic machine using a cathode ray tube to form letters at rates over 1000 characters per second. The initial prototype for a cathode-ray-tube machine had been developed 20 years before by E Dinga.*
Noise reduction system (audio recording and transmission)
Dr Ray M Dolby, US-*born electronics engineer and physicist in South-west London,* UK. *Called 'Dolby A', it was first used professionally by the Decca Record Co.*
Racing car with aerofoils
the Chaparral 2E, US, *as driven by Jim Hall at the Bridgehampton Can-Am event.*
Skateboard
California, USA.
Supertanker
the 205,000 ton 'Idemitsu Maru', built in Japan.
Telephonic optical-fibre link (experimental)
Dr Charles Kao *and* Dr George Hockham *at Standard Telecommunications Laboratory, Harlow, Essex,* UK.

1967
Bubble memory (prototype)
A H Bobeck *and* researchers *at Bell Telephone Laboratories, Holmdel, New Jersey,* USA.
'Cache'/Buffer memory circuit card
IBM *System/360 for their Model 85 computer.*
Computerised electronic fuel-injection system (motor car)
Robert Bosch GmbH, *Stuttgart, Germany.*
Dynamic memory cell
IBM.
International computer time-sharing
General Electric, US, *between Toronto, Canada and London,* UK. *It was later extended to Sydney, Paris, Milan, the Hague, Brussels, Cologne, Copenhagen and Stockholm.*

Laser-lancet (medical)
D R Herriott, E I Gordon, H A S Hale *and* W Gromnos *of Bell Telephone Laboratories, New Jersey,* USA. *Called the 'Light Knife', it was used to cut or cauterise a wound with minute precision.*

Levodopa (drug)
George C Cotzias, *Greek physician, to control the horrifying symptoms of Parkinson's disease.*

Louma crane (cinematography)
Jean Marie Lavelou *and* Alain Masseron, *Paris. It was improved and marketed by Samuelson's of Cricklewood, London.*

Magnetic-motor train
Pueblo, Colorado, USA.

Precision-surveying laser
used during the construction of tunnels for the Bay Area Rapid Transit Railway, San Francisco, USA.

'Raudive voices' (communications from the dead)
Konstantin Raudive, *Latvian psychologist, living in Germany. He used a tape recorder to register 'white noise' from either a radio or crystal diode set with a very short aerial. Playback appeared to reveal 'voices'.*

Rocket car (liquid-fuelled)
Reaction dynamics 'X-1' dragster, designed and built at Milwaukee, Wisconsin, USA.

Satellite docking
USSR, *with automatic docking of the unmanned 'Cosmos 186' and 'Cosmos 188' satellites.*

Spinal meningitis vaccine
Doctors Artenstein, Gotschlich *and* Goldschneider, *working for the* US *Army at the Walter Reed Institute.*

Tidal-electric power project
the Rance Plant at the Gulf of St Malo, Brittany, France.

Tobiscope
Mikhail Gaikin, *Soviet surgeon in Leningrad, together with* Semyon Kirlian, *for the precise location of acupuncture points.*

1968
Carbon dioxide gas laser.
Electronic signature verification
a team of psychologists and electronic engineers *working for three years at* the National Physical Laboratories, *Teddington, Middlesex,* UK. *The prototype machine, to read human handwriting, was called 'Cyclops 3'. Commercial versions included 'Securisign' from the Transaction Security division of Analytical Instruments Ltd,* UK.

Grass-skis
Kurt Kaiser, *knitting-machine manufacturer, Stuttgart, Germany, who pioneered the sport of running down grassy hillsides on 1 ft long 'caterpillar-track' skis.*

Haemoglobin molecule structure (complete)
Dr Max Ferdinand Perutz, *Director, Medical Research Council's Molecular Biology Laboratory, Cambridge,* UK, *following 30 years' research – with assistance from his colleague* Dr John Kendrew.

Jumbo helicopter
Mikhail Mil, USSR: *the four turboshaft-engined Mil-V12 'Homer'.*

Lithium niobate (holographic storage material)
Bell Telephone Laboratories, *Holmdel, New Jersey,* USA.

Long-life, high-output argon and krypton laser
Shibaura Electric Co. Ltd, *with the Japan Central Research Laboratory, Tokyo.*

Pulsars
Dr Jocelyn Bell Burnell *at the Mullard Radio Astronomy Observatory, Cambridgeshire,* UK.

SGHWR (steam-generating heavy water reactor)
100 MW *Power Station at Winfrith, Dorset,* UK.

Submersible (glass-fibre construction)
Vickers Oceanics Ltd, *Barrow,* UK.

Supersonic airliner
*the Soviet-built Tupolev Tu-144, with
a maximum speed of Mach 2.37 and a
passenger capacity of 130. The Anglo-
French 'Concorde', designed and built
by British Aerospace with
Aerospatiale, first flew in 1969 and
soon reached Mach 2.0. 'Concorde' has
proved more viable commercially.*
Tractor (hydrostatic steering)
David Brown Tractors Ltd, UK.
**Trinitron (colour television
picture tube)**
Sony, *Tokyo, Japan – called the* 'KV-
1310'.

**1969
ALSEP (Apollo lunar scientific
experiment package)**
NASA's *'Apollo 12' astronauts, Conrad
and Bean, deployed six scientific
experiments each with a nuclear-
powered battery giving operational
equipment life of one year on the lunar
surface.*
**Antibodies (chemical molecular
structure of)**
Rodney Robert Porter, *Professor of
Biochemistry at Oxford University,*
UK *and* Gerald Maurice Edelman,
*Professor of Biochemistry at
Rockefeller University, New York,*
USA.
**Command and service lunar
modules**
NASA *with aerospace companies led by*
North American Rockwell *and*
Grumman Aircraft. *They were used
by the 'Apollo 11' astronaut team for
their landing on the moon, and for five
subsequent landings.*
Digital avionics flight computer
Honeywell Inc., *for the McDonnell-
Douglas* DC-10 *airliner.*
Gamma-ray astronomy.
**Gear wheels (carbon-fibre-
reinforced plastic)**
British Rail Engineering Ltd, UK.
Jumbo-jet airliner
Joe Sutherland *and a* design team *at
Boeing, Seattle, Washington. Called*

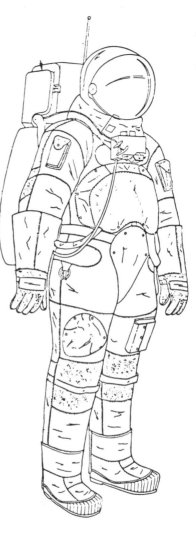

'Apollo' astronauts' spacesuit

*the Boeing 747, it made its first test
flight from Paine Field, watched by
PanAm Airways officials who had
already ordered 25 such 'Jumbo's.
Passenger capacity was 385-500.*
**Laser-ranging retro-reflector
(part of** ALSEP**)**
*Able to measure the distance of the
earth to the moon to an accuracy of 1.5
miles.*

Manned spaceflight docking machinery
Soviet spacecraft 'Soyuz 4' and 'Soyuz 5', followed by the 37-minute spacewalk transfer of two space-suited cosmonauts from one craft to the other.

PASCAL (computer language)
Professor Niklaus Wirth, *computer scientist, ETH, Zurich, Switzerland.*

Twin-engined offshore catamaran
James Beard, Clive Curtis *and* Chris Hodges *of Twickenham, Middlesex, UK – called 'Volare II'.*

VTOL fighter-plane (vectored lift)
Hawker Siddeley Aviation Ltd.
Called the 'Harrier', it was developed during the 1960s with the Hawker 'P1127' and the 'Kestrel'.

Video-cassette system (plastic-boxed)
Sony, *Tokyo, Japan – called the 'U-Matic'.*

1970
Airfoilcraft
VFW-Fokker GmbH, *Germany.*

'AI' computer programs (for understanding language)
Roger C Shank *of Yale University, USA – called* 'MTRANS', 'ATRANS' *and* 'ATTEND'.

Bar code (computer-scanned binary signal code)
simultaneously by Monarch Marking *of Dayton, Ohio, USA with 'Codebar' for retail trading, and* Plessey Telecommunications Ltd *of Poole, Dorset, UK for industrial use.*

Floppy disc (magnetic storage device for computer systems)
IBM.

Laser/hologram art exhibition
Finch College Museum of Art, New York. It was called 'N-Dimensional Space'.

Lunar vehicle (remote-controlled)
USSR's *'Lunokhod I', which travelled 6½ miles on gradients of up to 30 degrees*

in the Mare Imbrium and remained functional for almost one year.

Noise reduction system (cassette tapes)
Dolby B-type Encoder 320, UK.

Oil (North Sea 'Forties' field)
discovered by British Petroleum's drilling barge, 'Sea Quest'.

Overhead travelling crane (linear-induction-powered)
Linear Motors Ltd, *Loughborough, Leicestershire, UK, based on 15 years' research and development by* Professor Eric Roberts Laithwaite, *Heavy Electrical Engineering Department, Imperial College of Science and Technology, London.*

Relational data base
Dr Edgar Codd *and an* IBM *research team.*

SHRDLU (AI computer program)
Terry A Winograd *of the Massachusetts Institute of Technology, Cambridge, USA, to simulate actions/responses of a person in a 'Blocks World'.*

Sialon
Professor K H Jack *at the University of Newcastle-upon-Tyne and* Y Oyama *of Tokyo, Japan. It was developed in subsequent years as an engineering ceramic by* Lucas Research Centre, *Solihull, UK.*

Sixteen bit minicomputer
Digital Equipment Corp. – *called the* 'PDP-11'.

Synthetic lubricant for supersonic aircraft engines
British Petroleum Ltd *with* Geigy (UK) Ltd.

SYSTRAN (computerised translation system)
Dr P Thomas. *It was adopted by the European Economic Community in 1981.*

Unnilpentium
Albert Ghiorso *et al., Lawrence Radiation Laboratory, University of California, US. The American name is Hahnium; the Soviet name is*

Nilsbohrium.
Videodisc (black and white)
Decca, UK, *and* AEG Telefunken,
Germany – called 'Teldec'.
Windsurfer (commercial sailboard)
James Drake *of Santa Monica, California, and* Hoyle Schweitzer *of Pacific Palisades, California.*

1971
'Anorthosite' (crystal moon rock)
Pilot-Astronaut David R Scott, *commander of* NASA's *'Apollo 15,' during a 28 km drive in the Lunar Rover Vehicle, when he collected 28 kg of rock samples.*
Earth-orbiting space station (USSR)
'Salyut', visited by cosmonauts from the 'Soyuz 11' spacecraft.
LCD (liquid-crystal display)
Hoffmann-LaRoche, *Switzerland.*
Micro-processor
Doctors Robert Norton Noyce *and* Gordon Moore, US. *Launched by the Intel Corporation, Santa Clara, California, USA, it was called the 'Intel 4004'.*
Orbital engine
Ralph Sarich, *engineer, Australia.*
PROLOG (computer language)
Alain Comerauer *and* Philippe Roussel, *French researchers at Strasbourg University.*
Quadraphonic disc recordings.
Quartz digital watch (LED)
George Theiss *and* Willy Crabtree, US *engineers, Dallas, Texas – called the 'Pulsar'.*
Shirt-pocket radio pager (tone and voice)
Motorola, US.
Sony memory transistors
Sony, *Tokyo, Japan.*

1972
Advanced Passenger Train
British Rail's Research Centre *at Derby, UK. Its top speed was 152 mph.*

Anti-matter particles (anti-helium 3)
Professor Yuri Dmitriyevich Prokoshkin *and a research team at the atom-smashing particle accelerator at Serpukhov, Armenia,* USSR.
Artificial hip
John Charnley, *British orthopaedic surgeon at the Wrightington Hospital, Wigan, Lancashire,* UK, *using high-density polyethylene as the constructional material for the ball and socket.*
Auto-focusing still camera (commercial)
Polaroid SX 70, *using an ultra-sonic signal.*
Black holes (proven)
Professor Robert L F Boyd *and his team at University College, London, using X-ray telescopes on board the 2 ton* US *orbiting astronomical observatory No. 4. They were discovered in the binary-star X-ray source, Cygnus X-1.*
Earth Resources Technology Satellite
NASA'S ERTS I, *later renamed* LANDSAT I *and followed by* LANDSAT 2 (1975) *and* LANDSAT 3 (1978). *They were used to make a world atlas, taking photographs from a height of 560 miles.*
Electronic pocket calculator
simultaneously by Jack St Clair Kilby, *with* J D Marriman *and* J H Van Tassel *of Texas Instruments, Dallas,* USA, *and* Clive Marles Sinclair *of Sinclair Radionics Ltd, Cambridge,* UK.
Enkephalin (brain chemical)
Dr John Hughes *and* researchers *at the Drug Research Institute of Aberdeen University,* UK. *It was isolated, purified and analysed and finally synthesised during the subsequent two years.*
Fly-by-wire (air navigation system)
McDonald-Douglas' F4 *'Phantom II' was the first to use this system.*

1972

Minehunter ship (glass-fibre)
150 ft HMS 'Wilton,' *built by Vosper-Thornycroft, Portsmouth,* UK.

Outer solar system probe
NASA's *'Pioneer 10', which was still transmitting information about spatial conditions 13 years later, having left the solar system in 1983.*

Polypropylene motor car bumpers
Renault 5, France.

Programmable hand-held calculator
Hewlett-Packard, *Palo Alto, California – called the* 'HP 35'.

Satellite-tracking of wildlife
Frank *and* John Craighead, US, *using a 'Nimbus 3' weather satellite to track an elk for 28 days.*

'SLURP' (self-levelling unit for the removal of pollution)
Esso Oil Corporation, USA.

Telepen
George Sims, *physicist, of Friern Barnet, Middlesex,* UK. *A bar-code light pen reader, it was manufactured two years later, for use particularly in libraries and factories.*

Videodisc (laser-read)
Philips, *Eindhoven, Netherlands. It was marketed in 1980 (*USA*) and 1982 (Europe).*

Video game
Noland Bushnel, *28-year-old* US *student. Called 'Pong', it enabled him to form his own company – Atari.*

1973

BMX (bicycle-motor-cross)
Marvin Church Senior *of California for his son,* Marvin Jr. *The* BMX *frame was marketed the following year by Webco, motorcycle accessory manufacturers of Southern California.*

Computerised axial tomography X-ray scanner (CAT)
Allan Macleod Cormack, *South African physicist, Tufts University, Medford, Massachusetts,* USA, *and* Godfrey Newbold Hounsfield, *Central Research Laboratories,* EMI,

Hayes, Middlesex, UK. Professor I Isherwood *of Manchester University developed its medical implications during initial clinical trials at Atkinson Morley's Hospital, Wimbledon. Two years later* EMI *produced a fast whole-body* CAT *scanner.*

DNA (recombinant)
Paul Berg, Stanley N Cohen *and* Annie C Y Chang *of Stanford University, California, and* Herbert W Boyer *with* Robert B Helling *of the University of California, San Francisco,* USA.

Earth-orbiting space station (US)
Nearly 26,000 scientists *and other workers were deployed on preparations for* NASA's *'Skylab' mission. It was subsequently visited by three separate Apollo astronaut crews.*

Extra-corporeal shockwave lithotripter
Professors Eisenberger *and* Chaussy *in the Department of Doctor Egbert Schmidt, Munich, West Germany. Built by Dornier GmbH of Friedrichshafen. First used in Munich in 1980 for the disintegration of kidney stones.*

Fast-reacting photochromatic ophthalmic glass
Optical Division of Chance-Pilkington, *St Asaph, Clwyd,* UK.

Flying saucer (nuclear-powered)
British Rail – *patented but never built.*

Gamma-ray hunting satellite
NASA *Special Astronomical Satellite –* SAS-2.

Microcomputer
Trong Truong, *Director of* RE2 Co., *France – called the 'Micral'.*

Motor car headlamp wipers
Saab 99, *as developed by* Robert Bosch GmbH, *Stuttgart, Germany, in 1967.*

Sailing dinghy (injection-moulded polypropylene)
Ian Douglas Ben Proctor, *yacht*

designer, UK, *with* Rolinx Ltd *of Manchester*, UK *and* J C Dunhill Boats *of Basingstoke, Hampshire, using* ICI *propathene. Called the 'Topper', 21,000 dinghies were produced by this system in a very short time.*

Sears Tower (skyscraper)
the headquarters of Sears, Roebuck & Co., Chicago, Illinois, USA. *It is 1,454 ft high with 110 storeys. The building's population of 16,700 is served by 103 elevators, eighteen escalators and 16,000 windows.*

Skateboard (urethane plastic wheels)
Frank Nasworthy, *Californian skateboard enthusiast.*

Sydney Opera House
Jørn Ψtzon *et al., following 16 years' design and construction by many various companies.*

Thermographic duplicating machine
Rapid Data Ltd, UK.

Turbocharged production motor car
BMW, *Munich, Germany – called the 2002 Turbo.*

1974

ARECIBO message
Professor Frank Donald Drake, *astronomer*, US. *Using a radio telescope in Puerto Rico, he sent out a powerful series of radio pulses to communicate with extra-terrestrial intelligences elsewhere in the universe. So far, no replies . . .*

18,000-lines-per-minute, non-impact printing system
Honeywell Inc., USA.

Electronic memory credit card
Roland Moreno, *self-taught engineer, France.*

Laser-engraving of printing plates
using the Pagitron System of Optronics Inc., with the Laserplate of Lasergraphics Inc., *for Gannett Newspapers Inc.*, USA.

'Mailgram'
NASA's *'Westar' I and II satellites for Western Union, the* US *postal and telegraph service.*

Motorboat (Kevlar construction)
Fabio Buzzi, *Italy.*

Motorcycle (Wankel-engined)
Suzuki, *Japan.*

NRDC 143 and 161 (pesticides)
research team at Rothamsted Experimental Research station *at Harpenden, Hertfordshire*, UK. *These highly active, safe insecticides are 1000 times more potent than* DDT.

'Nodding boom' (wave-power device)
Stephen Salter, *Edinburgh*, UK.

Passenger hydrofoil (waterjet-propelled)
Marine Systems Division of Boeing Aircraft Corporation, *Seattle, Washington*, US. *The Jetfoil had a carrying capacity of 250–400 passengers. The first Jetfoil, 'Madeira', went into operation between Hong Kong and Macao in 1975.*

Psi (atomic particle)
Professor Burton Richter, *physicist, at the Stanford Linear Accelerator complex, California, and* Professor Samuel Chao Chung Ting, US *particle physicist of Brookhaven National Laboratory, New York – using a colliding beam machine called* SPEAR.

'Sensurround' (cinema)
Waldo O Watson *and* Dick Stimf *of Universal Pictures*, US.

Sony Mavica system
Sony, *Tokyo, Japan. It uses magnetic video card for recording and playback.*

Teletext (computer-based television information service)
the engineering departments *of the* BBC *and the Independent Broadcasting Authority, using redundant space in* TV *waveform.*

Unnilhexium
Glenn Theodore Seaborg *and* Albert

Ghiorso *at the Lawrence Berkeley Laboratory, University of California,* USA.

1975

Carbon brakes (commercial aircraft)
Dunlop Ltd, British Aerospace *and* British Airways, UK.

Computerised astrology
Esther Blumenthal's *'Astrocan 2000'.*

Condeep
Aker *and* other Norwegian contractors *at Arendal, Stavangar, Norway – a concrete production platform for the* UK *sector of Mobil Oil's 'Beryl A' offshore oilfield. A similar version was built the following year by* McAlpine Seatank, UK.

Docking module assembly (Apollo-Soyuz test project)
both NASA *and* Soviet engineering teams *and* aerospace companies.

Electronic sewing machine
Singer, USA.

High-speed train
British Rail's Research Centre *at Derby,* UK. *Its top speed was 125 mph.*

Home videotape systems
Sony's *'Betamax' and* Matsushita/ JVC's *'Video Home System' (*VHS*), both of Japan.*

Hybrid cells (half animal/half plant)
Professor Jack Lucy, *Royal Free Hospital, London, and* Ted Cocking *of Nottingham University,* UK.

Laser cell-sorting machine
Biophysics and Instrumentation Group, Los Alamos Scientific Laboratory, *New Mexico – to detect cervical and vaginal cancer.*

Mass storage system (magnetic tape computer cartridges)
IBM 3850, *using a honeycomb shell structure.*

Monoclonal antibodies (practical production)
Dr César Milstein, *Head of the Protein and Nucleic Acid Chemistry*

Dr César Milstein

division at the Medical Research Council's Molecular Biology Laboratory, Cambridge, UK, *in conjunction with* Dr Georges Köhler, *German scientist working at the Institute of Immunology in Basle, Switzerland, and* Neils Jerne, *Danish professor.*

Printing ink (ultra-violet)
a research and development team at Coates Brothers Inks *in Kent,* UK.

Submarine pipeline
Ekofisk-Emden gas pipeline, stretching 260 miles under the North Sea.

'Supral' (superplastically-deformable copper-aluminium alloy)
British Aluminium Co., Tube Investments *and* Superform Metals Ltd.

Video games (mass-produced for consumers)
Atari Corp., USA.

1976

AGR (advanced gas-cooled reactor) atomic power station
Hinkley Point, Bridgwater, Somerset, with an output of 1320 MW, based on

a decade of research with the prototype AGR at *Windscale, Cumbria*, UK.

Automatic still camera (microprocessor)
Canon, *Japan – called the* 'AE-1'.

'Charm' (sub-atomic particle)
Stanford Linear Accelerator Center, California, then at the Fermi Laboratory, near Chicago, USA.

Covert Rebus (theory of)
Jack Leslau, *London, who discovered hidden messages in the paintings of Hans Holbein the Younger (fl. 1497–1543).*

D° Meson
Gerson Goldhaber *and* Françoise Pierre *at Stanford University*, USA.

Male contraceptive pill
Professor Salat-Baroux, *Paris, France.*

Marine propeller (tip-vortex-free, TVF)
Gonzalo Perez Gomez and a research team *at Spanish Shipbuilding* SA, *Madrid.*

Mars space probes
NASA's *'Viking 1' and 'Viking 2'.*

'Rapid bursters' (astronomy)
several X-ray satellites, including the 'Copernicus' orbiting observatory and the US *Special Astronomical Satellite,* SAS 3.

Solar-powered boat
Alan T Freeman, *65-year-old former* UK *consultant on railway electrification. The 2.4 metre catamaran 'Solar Craft 1' was powered by two panels containing 50 photovoltaic cells, which drove an electric motor. It proved capable of speeds of 3 mph for two hours on the British Waterways canal at Rugby,* UK.

Super-computer
Seymour R Cray *of Cray Research Inc., Minneapolis, Minnesota*, USA. *With 200,000 integrated circuits to perform 200 million operations per second, it was called the 'Cray 1'.*

Talking card system
Sony, *Tokyo, Japan, to teach tiny*

children language and other kinds of knowledge.

Tandem-aerofoil boat (practical)
Gunther W Jörg, *German engineer. The* TAB VII-5 *'Jörg II' was capable of 135 kmph.*

Unnilseptium
Organesyan *et al.*, USSR.

'US Patent No. 4,000,000'
Robert L Mendenhall, *Las Vegas, Nevada*, USA *– a process for recycling asphalt-aggregate compositions.*

1977

Alkyd paints (fast-drying oil colours)
Messrs Winsor & Newton Ltd, *Whealdstone, North London.*

Artificial pancreas
Dr Wan Jun Tze, *Vancouver, Canada.*

Drug biosynthesis (recombinant DNA techniques)
Genentech, *San Francisco*, USA.

Human-powered aircraft
Dr Paul Macready, US. *Pilot-pedalled by Bryan Allen, the 'Gossamer Condor' flew a figure-of-eight course totalling 1.35 miles (7½ minutes airborne) at Shafter Airport, California to win a £50,000 Kremer Prize.*

Laser-based glass-fibre manufacturing system
Bell Telephone Laboratories, *Murray Hill, New Jersey*, USA, *for harder-than-steel fibre-optic communication lines.*

Neutron bomb
US Military, *based on the work of* Samuel Cohen *in 1958.*

Planet-to-planet unmanned space probes
NASA's *'Voyager I' and 'Voyager II'.*

Pocket television receiver (2 inch screen)
Sinclair Radionics Ltd, UK.

Programmable, polyphonic voice-assignable synthesiser
Sequential Circuits Co.'s *'Prophet 5'.*

Solar power plant
Solar Thermal Test Facility, Sandia
Laboratories, *Albuquerque, New
Mexico, Sunlight from 222 helipostats
is concentrated on a target 114 ft up in
the power tower.*

Space shuttle orbiter
NASA *and allied aerospace companies
for a re-usable spacecraft programme
initiated in 1972. Orbiter No. 101,
'Enterprise,' made a series of approach
and landing tests, captive and free, on
the back of a specially converted
Boeing 747 Jumbo Jet. By 1981,
Orbiter No. 102, 'Columbia', made a
spaceflight and landed on a runway
with astronauts John Young and Bob
Crippen on board during a 54½ hour
flight.*

Telephonic optical-fibre cables
General Telephone Co. *of
California, between Long Beach and a
local exchange 5½ miles away in
Artesia.*

Visitronic auto-focus (cameras)
Honeywell Inc., US.

1978
**ABS anti-skid braking system
(motor cars)**
Robert Bosch GmbH, *Germany.*

Argon laser (medical use)
Coscas, *France.*

**Arneson drive (fixed-shaft,
surface-piercing propeller)**
Howard Arneson, *Californian
inventor, and* Dan Arena, *hydroplane
driver, with their prototype offshore
powerboat 'Sea Sweep'. Two years
later, an adjustable drive was
developed and sold to the Borg-
Warner Corporation, USA.*

Automatic pressure cooker
Prestige Ltd, *Derby, UK.*

Cyclosporin A (transplant drug)
Dr Tony Allison *in the Clinical
Research Centre at Northwick Park
Hospital, London, and* Roy Calne,
*surgeon at Addenbrooke's Hospital,
Cambridge, UK, to assist with combined
heart and lung transplant operations
and combined kidney and pancreas
transplants.*

Electronic typewriter
Olivetti *of Italy and* Casio *of Japan.*

**Human insulin (recombinant
DNA process)**
Genentech *of San Francisco, US,
biotechnology firm. It was produced
and marketed by Eli Lilly & Co. as
'Humulin', first in Europe, then in the
USA.*

**Integrated bodywork bumpers
(motor car)**
Porsche 928 and 928S, Germany.

**'Intelligent' typewriter (word
processor)**
Qyx, USA.

**International ultra-violet
explorer satellite**
*to study ultra-violet radiation in outer
space.*

**Laser annealing (silicon chip
manufacture)**
Stanford University, *California, US.*

**Laser for metal removal in
aircraft construction**
British Aerospace Aircraft Group,
Filton Division, Bristol, UK.

Magnetic-lift linear-motor car
*Japan. It did 310 mph on a test track
near Miyazaki.*

Plant growth regulators (PGR)
ICI Research Station, *Jealott's Hill,
Berkshire, UK. It was marketed in
1984 as 'Clipper', 'Cultar', 'Bonzi'
etc.*

Pluto's satellite (Charon)
James Walter Christy *at the US Naval
Observatory at Flagstaff, Arizona.*

Scanning acoustic microscope
Cambridge Instrument Co. Ltd, UK.

**64,000-bit memory chip (64K
RAM)**
IBM, *using their innovation, SAMOS
(silicon and aluminium metal oxide
semi-conductor).*

**'Speak and Spell' (instruction
machine)**
Texas Instruments Inc., *Dallas, USA.*

**Store and press telephone call
maker**
British Telecom.

Test-tube baby
Patrick C Steptoe, *gynaecologist, and* Dr Robert G Edwards, *physiologist,* UK. *Louise Brown was conceived in a laboratory dish and born two weeks prematurely by Caesarian section, at Oldham General Hospital, Lancashire,* UK.

1979

Anti-dive motorcycle suspension (variable geometry)
simultaneously: Honda's *'Pro Link',* Kawasaki's *'Uni Track' and* Suzuki's *'Full Floater', all in Japan.*

Artificial blood
Ryochi Naito, *doctor at the Fukushima Medical Centre, Japan.*

'AI' computer program (to play backgammon)
Hans Jack Berliner *of Carnegie-Mellon University. Called 'Mighty Bee', it defeated the world backgammon champion, Luigi Villa, at Monte Carlo.*

Ceramic diesel engine (prototype)
Cummins Technical Center, *Columbus, Indiana, with the* US *Army Tank Command.*

Computer information-processing system (Japanese characters)
IBM. *A 254-key, multiple-shift keyboard enabled its operator to call up more than 7000 Japanese Kanji ideographs.*

Far infra-red laser
Columbia University, *New York City, with the* Naval Research Laboratory, *Washington* DC, USA.

Human-powered aircraft
Dr Paul Macready, US. *Pilot-pedalled by Bryan Allen, the 'Gossamer Albatross' crossed the English Channel – 22 statute miles – in 2 hours 49 minutes.*

Human speech machine
IBM *System 370. It is programmed to understand a vocabulary of 1000 English words and print out what it hears.*

'Intelligent Pig' (on-line inspection vehicle for high pressure natural gas trunk distribution pipelines)
British Gas.

Masuda wave power barge
Japan Marine Science and Technology Centre's *'Kaimei,' based on* Yoshio Masuda's *self-regenerating navigational buoys.*

Mega-submarine (nuclear-powered)
USS *'Ohio', 560 ft long and carrying 24 Trident I missiles with a 4600 mile range before the need to refuel.*

Ricardo comet catalytic unit
Ricardo Consulting Engineers, *Shoreham, Sussex,* UK, *with* Johnson Matthey, *refiners and distributors of platinum metals from the Rustenburg Platinum Mines, South Africa: an anti-pollution device.*

Rubik's cube
Ernö Rubik, *Professor at the University of Budapest.*

Satellite navigator (miniature)
Thomas Walker & Son, UK, *for motor-cruisers and yachts.*

Single cell protein process
ICI Agricultural Division, *Billingham, Cleveland,* UK; *marketed as 'Pruteen', a high protein animal feedstuff.*

Two-megawatt wind turbine
US Department of Energy. *The Boeing* MOD *1, with blades 68 metres in diameter, was installed at Boone, North Carolina. It was mechanically unsuccessful.*

Ultrasonic scanners (cross-section views of unborn babies)
based on the pioneer work of Professor Ian Donald, *Regius Professor of Midwifery, University of Glasgow,* UK.

Viewdata systems
Videotex Section of British Telecom Research Department, *Martlesham Heath, Ipswich,* UK – *called 'Datel' and 'Prestel'.*

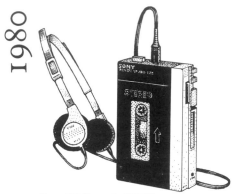

Sony 'Walkman' – TPS L2

'Walkman' (personal stereo tape-replayer)
Sony, *Tokyo, Japan – called the* 'TPS L2'.

1980
'Balans' chair
Hans Christian Mengshoel, *helicopter pilot, Norway.*
Carbon-fibre tennis racket (injection-moulded)
Dunlop Sports Co. Ltd, *Wakefield, West Yorkshire,* UK.
'Cloned' interferon
Biogen *of Zurich, Switzerland, followed by several other companies. It was produced from genetically engineered bacteria.*
CMOS (complementary metal oxide semiconductor)
Compact disc digital audio system (laser-optically recorded)
Sony, *Tokyo, Japan, with* Philips, *Eindhoven, Netherlands.*
'Dolby C' noise reduction system
Dolby Laboratories, *London,* UK, *and San Francisco,* USA.
Foil propeller (boat propulsion by wave power)
Einer Jakobsen, *Wave Control Co., Norway.*
High quality underwater video camera
Sony Broadcast Ltd, *with the* British

Admiralty's Research Laboratories *at Teddington, Middlesex,* UK.
Holographic cinema (experimental)
N Aebisher *and* C Bainier, *researchers at Franche-Comté University, France, showing the flight of a gull.*
Holographic supermarket terminal
IBM.
Laser cutting of cloth
Hughes Aircraft Corp., *Malibu, California,* USA.
Laser-driven optical system for phototypesetting
Monotype International – *called the 'Lasercomp'.*
Laser repair system (for high-density computer memory chips)
Western Electric Co., USA.
Load-monitored fasteners
Rotabolt Ltd, *Dudley, West Midlands,* UK.
Merchant ship (auto-controlled, rigid, vertically folding sails)
Nippon Kokan, *Japanese shipyard, with the 1600 ton Shin-Aitoku-Maru.*
Microwave thermography
Professor Y Leroy, *University of Science and Technology, Lille, France.*
Personal stereo (weatherproof)
Sony *Tokyo, Japan – called the* 'WM F5'.
Plastic-tipped pen
Berol Ltd *of King's Lynn, Norfolk,* UK, *to improve children's handwriting.*
Quantity-processing computer
Mikroelektronik, *East Germany.*
Rocket boat (liquid-fuelled)
Lee Allen Taylor Jr *of Bellflower, California,* USA. *He was killed when* 'US *Discovery II' disintegrated at high speed.*
Solar maximum mission (satellite)
carrying spectrometers, photometers and imaging devices to monitor the sun

in U/V, x-*ray and Gamma radiation. It
was retrieved and repaired by the space
shuttle, 'Challenger', in 1984.*

Solar-powered aircraft (for sustained flight)

Dr Paul Macready, *scientist and
aviator, California. Piloted by Janice
Brown, 'Gossamer Penguin' made a
1.95 mile flight at Edwards Air Force
Base in the Mojave Desert, sustaining
an altitude of 12 ft for 14 minutes. The
following year Macready's 'Solar
Challenger' made a 230 mile flight
from France to the* UK, *including the
Channel crossing. The 'Challenger',
powered by 16,000 photovoltaic cells,
flew for 5½ hours at a height of 11,000
ft.*

Streamlined cabin tricycle (high-speed)

Allan A Voigt, John Speicher, Steve
Wojcik *and* Doug Unkrey *of
Versatron Research Inc., Anaheim,
California,* USA. *Their Vector tandem
reached almost 63 mph through the
timed 200 metre traps, whilst their
Vector solo reached 59 mph. This was
without either stored energy or motor-
pacing.*

Triple quasar

Steward Observatory *of Arizona
University together with the*
Smithsonian Astrophysical
Laboratory *and the* Bell Telephone
Laboratory, *New Jersey,* USA.

Turbocharged motorcycle (production model)

Honda *of Japan – the* 'CX 500 *Turbo'.*

Waterworks sludge treatment plant

Water Research Centre, *Stevenage,
Hertfordshire,* UK, *using a continuous
polymer-thickening process. The first
full-scale plant went into operation in
1982 for the Anglian Water
Authority.*

1981

Air pollution detector (twin laser beam system)

ICI, UK – *the 'Eagle' system.*

Anti-interferon

Medical Research Council's
Molecular Biology Laboratory *at
Cambridge,* UK; *marketed by Celltech,
the government-backed* UK
Biotechnology Group.

First full 32-bit single-chip CPU

Hewlett-Packard, USA – *450,000
individual transistors on a silicon chip
measuring ¼ inch square.*

Fully aerodynamic saloon car

*The five-cylinder Audi 100. It had a
drag factor of 0.30, obtained from such
features as a flush windscreen and
windows.*

Infra-red explorer balloon (Montgolfière type)

*launched from Pretoria, South Africa,
with four French scientists on board.*

Laser-enhanced etching and plating

IBM.

Lightweight plastic engine manifold

Ford Europe *with* Fiberglass Ltd *and*
BTR Permali RP, *moulding specialist.*

Micro-chip electric toaster

Russell Hobbs Ltd, *Staffordshire,*
UK.

Multi-function bubblejet printing

Canon Inc., *Ota-ku, Tokyo, Japan.*

Nuclear magnetic resonance (NMR) scanner

Thorn-EMI Research Laboratories,
Hayes, and Nottingham University,
UK. *It was installed at the
Hammersmith Hospital, West
London.*

Simulated listening typewriter (prototype)

Messrs Gould, Conti *and* Hovanyecz
at IBM, *Yorktown, Virginia,* USA. *It
was capable of recognising a limited
vocabulary of 5000 words.*

Super-computerised musical instrument

Giuseppe di Giugno, *Italian
engineer, for Pierre Boulez, French
composer and his 'keyboard' assistant
Gerzo, at the Institute of Musical and*

Acoustic Research and Co-ordination, Paris. With the '4X', capable of 200 million operations per second, Boulez composed 'Répons'.

Tandem accelerator mass spectrometer
University of Arizona, *Tucson*, USA, *to carbon-date very small samples of material. The first experiments were carried out by* Paul E Damon, *Professor of Geosciences at the university.*

1982

Abnormal genes (cancer-tumour-causing)
Dr Robert Weinberg *and scientists at the Massachusetts Institute of Technology and* Dr Mariano Barbacid *in the National Cancer Institute, Bethesda, Maryland,* USA.

Artificial gill (sub-aqua breathing device)
Celia *and* Joseph Bonaventura, *Duke University, North Carolina,* USA – *called the 'Hemosponge'.*

Artificial heart (compressed-air-driven)
Dr Robert Jarvik. *The 'Jarvik Mark 7' was implanted in Dr Barney Clark of Wisconsin by William De Vries,* US *surgeon at the Utah Medical Center, Salt Lake City. Clark lived on with the device for 112 days.*

Astronomical observatory (airborne)
NASA, *using a modified* C-141 *jet airplane.*

Automated moulding machines and plating robots (computer-controlled production line)
Oita-Canon Inc., *Ota-ku, Tokyo, Japan, for producing advanced autofocus cameras such as the* 'AF 35 MII'.

Automatic still camera (disc-cassette film)
Kodak, *Rochester*, NY, USA.

Chip for echo cancellation in TV
Philips Research Laboratories, *Eindhoven, Netherlands.*

Colour TV pictures of Venus
taken consecutively through red, blue and green filters, by the descent module of Soviet spacecraft 'Venera 13' (unmanned) which had landed on the planet after a two-year voyage.

Computer-controlled electrical stimulation of paralysed limbs
Dr Jerrold Petrovsky *at the biomedical laboratory of Wright State University, Dayton, Ohio,* USA – *called 'Band Aid'.*

Computer floppy-disc pen
Berol Ltd, *King's Lynn, Norfolk,* UK.

Computer robotic system (RSI)
IBM, *using* AML, *a programming language, for manufacture.*

Denloc tyres
Rothmans-Porsche '956 C' racing car.

Domestic robot
Heathkit, *part of the* Zenith Corp., USA – *called 'Hero 1'.*

Electronic still camera
Sony, *Tokyo, Japan, with their Mavica System, using a disc-shaped film which was an electronic video-tape allowing colour pictures to be phoned through to whichever news media required them. While the Sony 'Mavica' had 280,000 picture elements, the* Canon 'ESC', *developed jointly with* Texas Instruments Inc., *had 400,000 picture elements and a charge-coupled device (*CCD*).* Canon *put their camera on trial at the 1984 Los Angeles Olympic Games.*

'The Fastest Act of Man'
C V Shank *and co-workers at Bell Telephone Laboratories, New Jersey,* USA – *the shortest ever flash of laser light, its pulse lasting only 30 femtoseconds or 0.03 picoseconds.*

Laser gyro inertial navigation system
Honeywell Avionics Division *for the* Boeing 757 *and* 767 *airliners.*

Magnetic bubble binary memory devices (production model)
Texas Instruments Inc., *Dallas*, USA.

Megawatt wind turbine farms
US Department of Energy, *siting three machines with 90 metre-diameter blades at Goldendale Columbia River Gorge, Washington State. They were capable of producing 2.5 MW. A five-turbine farm was also in operation at Byron, California, generating 1.2 MW.*

Monoclonal antibodies (to combat leukemia)
Doctors John Kersey *and* Norma Ramsay *at the University of Minnesota with assistance from immunologist* Tucker LeBien.

Motor car (solar-powered for long distance)
Hans Tholstrup *and* Larry Perkins. *Their 'Solar Trek 1', with a solar-panel roof powering an electric motor through batteries, travelled 4084 km across Southern Australia from Perth to Sydney at an average 23 kmph. Total running time was 172 hours.*

Solar-powered trans-Australian car

Nuclear structure facility (NSF)
UK Government Physics Laboratory, *Daresbury, Cheshire. It included an ion accelerator designed and built for smashing atomic nuclei.*

Organic waste earthworm exploitation process
Dr Clive Edwards *and* researchers *at Rothamsted Experimental Station, Harpenden, Hertfordshire, UK, for use in agricultural and horticultural industries.*

Portable electronic typewriter
Brother Corp., *Japan. Called the 'EP 20', it weighed 2 kg and used a 35-dot matrix system.*

Quasar PKS 2000–330
A Wright *and* D Jauncey *at the Parkes Radio Astronomy Observatory, New South Wales, Australia. It is estimated at 13,000 million light years away.*

Rising sector gate (Thames Flood Barrier)
Rendel, Palmer & Tritton, UK *consulting engineers, for the Greater London Council. After two decades of planning and design, the 10 year construction of the barrier was carried out by 26 main building contractors, including two major consortia and several hundred subcontractors.*

Solar power plant
Mojave Desert, California, USA. *A 300 ft tower surrounded by 1800 giant movable mirrors generated 10,000 Kw.*

Space station (intermediate term)
USSR's *'Salyut 7', with cosmonauts Anatoly Berezovoi and Valentin Lebedev on board, remaining in orbit for 211 days, during which time 87.6 million miles were computed.*

Thermophilic anaerobic effluent treatment plant
researchers at H P Bulmer Ltd, *cider manufacturers, Hereford,* UK.

Unnilenium
G Munzenberg *et al., West Germany.*

'Voice advice' motor car
Peugeot 505 Turbo Injection, France, soon followed by the Renault 11, France and the Austin Maestro, UK.

Word-teaching game (using optical pencil and bar codes)
Texas Instruments Inc. – *called 'Magic Wand'.*

1983
Anti-sound device
University of Leeds *with* British Gas, *using a computer-controlled sound emitter to cancel out noise.*

'Aquacolor' (true underwater colour ciné)
W Tuckerman Biays *and* Alice de T P Biays, *Islamorada, Florida,* USA.

Aspirated-cylinder-sail catamaran
Commander Jacques Yves Cousteau, *France – the 'Moulin-à-Vent II'*.
Atomic resolution microscope
National Optical Centre, *Japan*.
Biopol (biodegradable plastic)
ICI Agricultural Division, *Cleveland, UK. After 14 years' research and development, this thermoplastic polymer was produced through a bacterial fermentation process.*
Biosensors
Cambridge Life Sciences *and other companies. Enzymes from living bacteria were linked with silicon chips.*
Carbon-fibre aircraft wing
developed for the UK *government-backed 'Lear Fan 2100' project.*
Colour negative film (extreme speed)
Kodak ASA 1000.
Computer-regulated automatic route-setting and railway-signalling
British Rail, Westinghouse *and* GEC.
Digital scene simulation
John Whitney Jr *and* Gary Demos, *Los Angeles, California,* USA. *Using a Cray X-MP super-computer, the process was first successfully used for 'The Last Starfighter', Lorimar Productions.*
Electronic laser engraving of printing cylinders
Crosfield Electronics Ltd, *Hemel Hempstead,* UK *– called 'Lasergravure'.*
512 K-bit 120 nanosecond (Dynamic Access Memory chip)
IBM, *Essex Junction, Vermont,* USA. *It measured ⅔ inch square.*
Infra-red astronomical satellite (IRAS)
USA *with the* UK *and the* Netherlands. *Discovered or suspected over half a million sources of heat in the universe – including, perhaps, 'Planet 10'.*
Integrated electronic filing system
Canon Inc., *combining microfilm with*

electronic data bases.
Intermediate vector bosons (W and Z particles)
Dr Carlo Rubbia, *Italian scientist,* and Dr Simon van der Meer, *Dutch engineer, at* CERN, *the European Centre for Nuclear Research, near Geneva, Switzerland, using a machine that involved the collision of matter and anti-matter beams.*
Joint European Torus (JET)
Experimental Fusion reactor started up at Culham, Oxfordshire, UK.
Laser ranging
Crustal Dynamics *Project at* NASA's *Goddard Space Flight Center, using a reflector-bearing Laser Geodynamics satellite (* LAGEOS *) and a chain of Neodymium –* YAG-*yttrium-aluminium-garnet – lasers.*
MAGLEV (magnetic-levitated linear-propulsion passenger car)
People Mover Group, GEC, Brush Electrical Machines Ltd *and* Balfour Beaty Power Construction – *from Birmingham Airport to the National Exhibition Centre,* UK.
Motor cycle chassis (carbon-fibre)
Armstrong Motor Cycles Ltd, UK.
Newspaper printing plate (double-width plastic)
Mitsubishi, *Japan*.
'Quiteron' (superconducting transistor)
Sadeg M Faris *of* IBM.
RMS (remote manipulator system)
NASA. *A 50 ft long mechanical arm in the payload bay of space shuttle orbiters, for deploying satellites into orbit and collecting them, it was controlled from the flight cabin by astronauts using both a console and TV cameras.*
Robotic manufacture of robots
Yamazaki, *Japan*.
Spacelab
ESA (European Space Agency) *with* NASA.

Tactel
ICI Fibres Division research team –
an attempt to bring the appearance of
synthetic fibres closer to that of cotton.
Tension leg oil production
platform
Conoco *for the Hutton Field, North*
Sea.
Tracking and data-relay
satellite (TDRS)
deployed by NASA's *space shuttle,*
'Challenger', the first of four such
satellites.
288,000-bit computer-memory
chips (mass-produced)
IBM.
Videodisc arcade game
'Dragon's Lair'.
'Voice-advice' still camera
(multilingual options)
Minolta – *called the* 'AF SV'.
Winged keel (racing yacht
design)
Benjamin Lexcen, *marine architect,*
Norport Pty Ltd, Australia. It was
used on 'Australia II', which won the
America's Cup.

1984
All-plastic aircraft
Avtek 400, built in Amarillo, Texas.
Arc-welding robots (real-time
3-D vision guidance system)
Oxford University Robotics Group,
Fairey Automation Ltd, GEC *and* BL –
called the 'Fairey MetaTorch'.
Artificial comet
NASA.
Artificially-induced rainfall
Professor A Gagin *and a* team of
Israeli scientists, *by bombarding*
clouds with a fine powder of silver
iodine filings.
Asbestos safe waste recycling
process
Dr David Roberts *of King, Taudevin*
& Gregson – the 'Vitrifix' process.
Automatic still camera (built-in
8-bit microprocessor)
Canon Inc. – *called the* 'T-70'.

Car compact disc player
simultaneously by Sony *and by*
Fujitsu *in Japan and by* Philips,
Eindhoven, the Netherlands.
Car engine (90 per cent plastic)
Polimotor Research, *New Jersey,*
USA : *tested in a Lola T616 racing car.*
Cloned antitrypsin gene (body
enzyme)
National Institute of Health, USA,
Transgene, *France, and* Otago
University, NZ.
Computerised proof-reader
Bert de Pamephilis, US *typographer,*
using a standard ASC II *word-*
processing code and voice-synthesiser
board loaded into an IBM *personal*
computer.
Computerised translation
machine
Fujitsu, *Japan. The 'Atlas II' was*
able to translate Japanese into English
at 60,000 words an hour.
Contact X-ray electron
microscopy
Thomas J Watson Research Center
of IBM, *New York, in conjunction with*
the National Institute of Health *in*
Bethesda, Maryland, the New
England Deaconess Hospital *and the*
Harvard Medical School, *Boston,*
USA, *by adapting photolithography*
normally used for etching micro-
circuits on silicon chips to create
permanent 3-D images of the inside of
living human cells and blood platelets.
Genetically-engineered 'Factor
VIII' (haemophiliac blood-
clotting factor)
21 researchers at Genetech *of San*
Francisco and the Genetics Institute,
Boston, Massachusetts, USA, *with*
Edward Tuddenham *of the*
Haemophilia Centre of the Royal Free
Medical School, London.
High-speed direct-injection
diesel engine
Ford Motor Co., *Europe, as used in*
their Transit van.
Holocamera
Newport Corporation, *Fountain*

Valley, California, USA.
Holocopier (rapid hologram copying machine)
Applied Holographics, *Essex*, UK. *It produces over 400 copies per hour.*
Hologram-illustrated magazine cover
'The National Geographic', Washington DC, USA.
Holographic credit-card
VISA, *soon followed by banks throughout the world.*
Hubot (completely integrated robot)
Hubotics Co., *Carlsbad, California,* USA – *a TV Screen/computer/FM stereo receiver.*
Human-powered hydrofoil (catapult launch)
Alec Brookes *and* Allan Abbott *of Pasadena, California*, USA. *The 'Flying Fish' was successfully pedalled for 2 km in 6½ minutes by Steve Hegg.*
Kodavision (amateur video film)
Kodak, *Rochester, New York*, USA.
Laser gravity-wave detector
Ronald Drever *and* researchers *at the California Institute of Technology, Pasadena*, USA.
Laser light compressor
Jean-Marc Halbout, Daniel Grischkowsky, Anne C Balant *and* Hiroki Nakatsuka, *scientists at* IBM's *Thomas J Watson Research Center, Yorktown Heights, New York, to generate 12-femtosecond, ultra-short light pulses – at the rate of 800 per second.*
'LIDAR' (transportable laser radar)
researchers at Lincoln Laboratory, *Massachusetts Institute of Technology*, USA.
'LIF' (laser-induced fluorescence)
Richard Neil Zare, *laser chemist at Stanford University, California*, USA, *to detect chemicals in gases and liquids.*
MMU (manned manoeuvring units)
NASA, *enabling space shuttle*

astronauts, McCandles and Stewart, to make the first-ever untethered spacewalk 110 yards from 'Challenger'.
1024K-bit integrated circuit
simultaneously by Hitachi, NEC, NTT Atsugi Electrical Communications *and* Toshiba, *all of Japan.*
1,048,576-bit computer memory chip
IBM.
Parking meter (credit-card operated)
GEC Traffic Automation Ltd, *Borehamwood*, UK.
Pedalectric tri-cyclecar
Sinclair Vehicles Ltd *of Coventry,* UK, *in conjunction with* Lotus Cars *of Norfolk*, UK – *called the 'Sinclair C5'.*
Pocket television (flat screen)
simultaneously by Sony, Seiko *and* Casio *of Japan. A wristwatch television followed soon after.*
Portable compact disc player
Sony, *Tokyo, Japan* – *called the 'Sony* D50'.
Robotic electronic component assembler (high-speed)
ASEA, *Sweden* — *the* 'IRB 1000'.
RTM III (rock-trenching machine)
remotely controlled submarine robot to cut two trenches across the sea-bed of the English Channel for the 2000 MW *interconnection between both the Central Electricity Generating Board and Electricité de France power station networks.*
64K-bit 20-nanosecond dynamic access memory chip
Stanley Schuster *and* researchers *at* IBM's *Thomas J Watson Research Center, Yorktown Heights, New York*, USA. *It is capable of discharging sixteen bits of information at a time, using a four-device memory 'cell'.*
Spring and damper (plastic and rubber)
National Engineering Laboratory, *Glasgow*, UK.

Supersonic forward swept-wing aircraft
US *Grumman* X-*29, a single-seater fighter-plane with wing-joints using Mylar and carbon-fibre. First tests were held at Edwards Air Force Base, California,* USA.

'Top Quark' (sub-atomic particle)
Carlo Rubbia *et al.,* CERN, *Geneva, Switzerland.*

Unniloktium
G Munzenburg *et al., West Germany and the* Dubna Research Institute, *Moscow.*

1985
Jewish Year 5746

1985
ALERT (automatic logging of events and ruptures in tyres)
Lartrack, UK, *with* Honeywell, US. *A sensor is fitted between each tyre and its rim.*

Anti-ICBM/Spy satellite high-energy ground-baser laser
by USA *and* USSR *separately.*

'Anxiety chemical' (human brain)
Alessandro Guidotti *and* Erminio Costa, *National Institute of Mental Health, Maryland,* USA.

Artificial tooth root
Furukawa Electric Company *with the* Dentistry Department of Nihon University, *Japan. Made of 'shape memory' nickel-titanium alloy.*

Atmosphere trace molecule spectroscope (ATMOS)
Honeywell *for use by* NASA's *Jet Propulsion Laboratory in Pasadena, California,* USA. *An infrared interferometer installed on Spacelab 3 and pointing towards the sun to investigate the chemicals which affect the ozone layer.*

Captopril (orally active angiotensin-converting enzyme-inhibitor)
E R Squibb & Sons, *Princeton, New Jersey,* USA. *Based on over a decade of research by* Messrs Cushman *and* Ondetti *at the Squibb Institute for Medical Research. This* ACE *inhibitor was marketed as 'Capoten'. It was followed by 'Enalapril' as developed by researchers at Merck, Sharp and Dohme, West Point, Pennsylvania,* USA.

Car body (self-coloured, injection-moulded polypropylene)
Austin Rover, *Coventry,* UK. *As incorporated in the* 'MG EX-E'.

Catamaran powerboat (transatlantic)
Cougar Marine, *Hampshire* UK. *The 65 ft Virgin Atlantic Challenger. After completing 2973 nautical miles at an average 41.5 knots, the Challenger hit submerged debris and sank less than 200 miles from the* UK *coast.*

CD-ROM (compact-disc read-only memory)
Hitachi, *Japan under licence to Philips, Eindhoven, Netherlands. Their '1502s', with a parallel interface for the* IBM-PC. *By storing 522 Megabytes, 270,000 A4 pages can be placed on a single CD.*

Cellular radiophone (nationwide network)
Racal-Vodaphone Ltd, *Newbury, Berkshire,* UK. *Also 'Cellnet' by* British Telecom *and* Securicor.

Cloned leprosy genes (for vaccines)
Ron Davis *et al., Stanford University,* USA.

Cockpit safety capsule (circuit catamaran powerboats)
Chris Hodges, *boatbuilder, Norfolk,* UK, *following the death of his partner Tom Percival in a race. First saved a life when Bob Spalding,* UK, *somersaulted his catamaran at the Sacramento Grand Prix,* USA.

Comet-intercepting spacecraft
Space & Communications Division of British Aerospace Dynamics

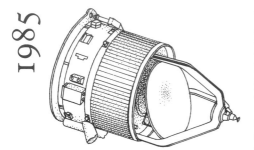

'Giotto': comet-intercepting space probe

Group *leading a component consortium from 9 different countries for the European Space Agency. Called 'Giotto', it was designed to chemically analyse and also take 'close-up' photographs of Halley's Comet (from 300 miles) before being destroyed by it.*

Compact disc automobile-navigation system
Philips, *Eindhoven, Netherlands – called 'Carin', a compact-disc-personal-computer for maps, information and music.*

Computer data wristwatch terminal (using microsoft BASIC)
Seiko, *Japan. Called the* 'RC 1000'; *80 'pages' of text transmittable from any computer via an* RS-232C *interface. Includes the time in up to 80 cities around the world.*

Computer-generated, synthetic text-to-speech system (intonation and voice-range options)
Dr Dennis H Klatt, *computer scientist at Massachusetts Institute of Technology,* USA, *in conjunction with* Ed Bruckert *and* Walt Tetschner *of the Digital Equipment Corporation, Maynard, Massachusetts,* USA. *Called 'DECtalk' it is able to generate exact pronunciations for over 20,000 words.*

Computerised submarine-hunting detection device
Gresham-CAP, *New Malden, Surrey,* UK *for the Royal Navy. Called*

'SEPADS' = *Sonar Environmental Prediction and Display System.*

Cryostatically-immersed traction magnets
Bangor *and* Oxford Universities, UK, *using liquid helium.*

Domesday project
Teachers, schoolchildren et al., throughout the UK. *Two videodiscs storing both digital data and moving and still pictures concerning the* UK *during this year. 'Re-read' by combining a Philips LaserVision player,* BBC *micro-computer with* VDU *and cursor.*

Driverless six-wheeled automobile (experimental)
Researchers at the Robotics Institute, *Carnegie-Mellon University, Pittsburgh,* USA. *Using stereo* TV *cameras, acoustic soundings and computer memory bank.*

Electricity-conducting polymer
Terje Skotheim *and* researchers *at Brookhaven National Laboratory, New York. Called the* 'PP-CoPc'.

Filmless X-ray system
Koyo *of Japan.*

Gallium-arsenide, 6000-transistor gate array chips (mass-produced)
Honeywell *at their factory in Richardson, Texas,* USA, *under contract to the* US Rockwell *Corporation for use in aerospace and defence.*

Geodesic-domed house (triple-skin ventilation roof system)
Nectar Domes Ltd, *Purley, Surrey,* UK.

Image digitiser
Optronics, *Cambridge,* UK. *It converts an image on a page to a digital electronic signal that can be printed out or transmitted over vast distances via computers, fibre-optic data transmission lines and satellites.*

Integrated circuits (half-micron line width manufacture)
IBM scientists *at Thomas J Watson Research Center, Yorktown Heights, New York* USA.

Largest high-definition television screen in the world
Sony, *Tokyo, Japan. Called the 'Jumbotron' and measuring 40 metres wide by 25 metres high, it was demonstrated at Expo '85, Tsukuba, Japan.*

Laser coronary angioplasty
James Forrester, Warren Grundfest *and* Frank Litvak, *physicians at the Cedars-Sinai Medical Center, Los Angeles, California,* USA, *using a 'cool' Xenon chloride laser.*

Laser-disc electronic filing system
Philips, *Eindhoven, Netherlands – called 'Megadoc', with screen or facsimile-printout retrieval options.*

Laser-optical storage system
Jerome Drexler, *Drexler Technology Corp.,* USA – *called the 'Lasercard'.*

Lean-burn automobile engine (experimental)
research engineers, Ford Motor Company, *Dunton, Essex,* UK – *using variable inlet geometry.*

Linear-motor car
Japan Airlines.

Miniature radio (3 mm thick)
Sony, *Tokyo, Japan – called the 'ICR-101', nicknamed the 'Credit Card Radio'.*

Mission-adaptive wing
Boeing F1-11 *as modified in conjunction with* NASA *and the* USAF. *Tested at Edwards Air Force Base, California,* USA.

Motor yacht (entirely moulded in cord carbon-fibre)
147 ft 'Sterling Lady'. Jon Bannenberg, *London yacht designer,* Sterling Yacht and Shipbuilders *of Miami, Florida,* USA, *and* Nishii Shipyards, *Japan.*

Oil spill solidification process
BP Research Centre, *Sunbury-on-Thames,* UK, *after 5 years' field research.*

Omnigripper (robotic hand device)
Centre for Robotics, *Imperial College, London,* UK.

Positron emission tomography
Michael Phelps *et al, University of California Medical School, Los Angeles,* USA. *To predict, by scanning, likely sufferers of Huntington's Chorea.*

Proton-antiproton collider (synchroton energy equivalence of 400,000 GeV)
CERN, *Switzerland.*

Scanning tunnelling microscope
simultaneously by Calvin F Quate *and* researchers *at Stanford University,* USA *and* Gerd Binnig, Heinrich Rohrer *and researchers at* IBM *Research Laboratory, Zurich, Switzerland. Able to resolve features one-hundredth the size of an atom.*

Soft bifocal contact lenses
Softsite Contact Lens Laboratory, *Coral Gables, Florida,* USA, *following 7 years' research and 20,000 experimental lenses.*

Solar-powered pay-phone
Plessey Telecommunications Ltd, UK, *for use in mountain chalets and developing countries.*

Spaceflight-proof maser atomic clock
Hughes Aircraft Co., *Malibu, California, under a 3-year contract to the* US *Naval Research Laboratory, Washington* DC. *The clock is accurate to within 1 second in 30 million years.*

'Spectra 900' (synthetic ultra-strong fibre)
Allied Corp., *New Jersey,* USA. *It is ten times stronger than steel, but light enough to float.*

Super-scanning electron microscope (computer-controlled)
Professor Albert Crewe, *Enrico Fermi Institute, Chicago University,* USA. *8 ft tall, weighing 1 ton and able to resolve detail down to half of one angstrom.*

Telescope (for millimetre and sub-millimetre wavelengths)
Rutherford Appleton Laboratories, UK, *in conjunction with* Genius

Company, *Ijmuiden, Netherlands. Installed near the summit of Maunea Kea, Hawaii – remote computer-controlled at the Royal Observatory, Edinburgh,* UK.

Thermal electronic typewriter
IBM, *Lexington, Kentucky,* USA. *Called the 'Thermotronic I'.*

3-D television
Matsushita Electric Industrial Co., *Japan, using a 14 inch screen and requiring no special glasses.*

Three-rotor rotary car engine (experimental)
Mazda, *Japan. As installed in the* 'MX-03', *also fitted with four-wheel steering.*

Ultra-fast Gallium-arsenide micro-chip
Honeywell Physical Sciences Center, *Bloomingdale, Minnesota,* USA. *Enabling the electrical signal to travel from the point of input to output point in just over 11 trillionths of a second.*

Ultrasonic bubble-cleansing domestic washing machine
Japan Ace, based on research at Nihon University – *detergentless.*

Unducted fan aero-engine
General Electric Aero-Engine Co., *Cincinatti, Ohio,* USA. *Principle also under evaluation by engine manufacturers in Europe and the* USSR.

Video 8 camera
Sony, *Tokyo, Japan. It weighs less than 5 lb.*

Proposed for 1986

Airbus A320 (civil airliner).
Electra 225
British Rail; *Inter-City streamlined locomotive, electrically-powered, capable of 225 kmph (140 mph).*

'Hipparchus' (spaceprobe)
European Space Agency – *to plot precisely the accurate position of 100,000 stars.*

International solar polar mission
NASA – *'Ulysses'.*

Jupiter space probe
'Galileo'.

Oil-production vessel (dynamically positioned)
BP Petroleum Development Ltd, UK – *directly connected to sub-sea well-heads.*

Shipbuilding arc-welding robot
Kemppi, Rosenew *and* Wartsila Shipyards, *Japan.*

Space telescope (called 'The Hubble')
NASA, *fitted into orbit by the space shuttle and used outside the earth's atmosphere to observe planetary systems as charted by 'Hipparchus'.*

Variable-geometry vertical-axis wind turbine
Dr Peter Musgrove *of Reading University,* UK, *and built by a consortium headed by McAlpine & Sons at Carmarthen Bay, Wales.*

Notwithstanding a 'nuclear winter', the 'greenhouse effect', an alien invasion and other, unforeseen, cataclysms, the following innovations might be expected during the next 50 years:

Adiabatic engine (production model).
Air-phone satellite system.
Artificially intelligent, fully robotic domestic servants and/or companions.

Biological computer.
Biological pesticides (mass-produced).

Cellular holophone.
Circular and revolving floating restaurant.
Civil airliner (vtol, composite plastic construction and hydrogen-fuelled).
Cloned superhumans, superlivestock and supervegetables.
Computer holographics.
Constellation of laser battle stations in space.

Discovery of the cellular causes of certain types of cancer.
Discovery of new elements and new sub-atomic particles.
Discovery of Planet 10 in our solar system.
Cryogenic deep-freezer (domestic production model).

Drugs, administered via nasal and specifically sited transdermal routes.

Earth-orbiting hotel and planetary cruise-liner (laser-propelled).
Electric-engined, plastic-bodied saloon car (long-range and solar/light rechargeable).
'Electronic secretary' (production model).
Electronically-stored and displayed sea charts.
Energy ultra-efficient, pollution-free, 'domed-in' conurbations.
England-France Channel Tunnel.
Eradication of acquired immune deficiency syndrome (AIDS).
Exchangeable cassette-array traction batteries.

Fibre-optic world communication network.
Fifth-generation artificially intelligent computerised personal adviser (expert in medical/legal/accounting/estate agency/do-it-yourself/educational matters).
Fleets of 'commercial sail' merchant ships (steam/electric/magnetic/propulsion options).
Fleets of freight-carrying airships and chains of airship mooring masts.
Fleets of OTECS (oceanic thermal-energy collectors).

The shape of things to come . . .

Fleets of 'space tugs' (orbital manoeuvring vehicles), to journey between shuttles, spacelabs and lunar/planetary bases for retrieval and delivery work.

400-mph powerboat.

400-passenger, computer-controlled, five-masted motor-sailing barque.

Genetically engineered disease-and-pest-resistant crops.

Geothermal-energy power stations (practical).

Giant artificial oasis (irrigation technology).

Hologramophone (compact discs).

Holovideo-polyphonic domestic games.

HOTOL re-usable spaceship (horizontal take-off and landing).

Hydrogen factory (using chloroplast technology).

Iceberg-irrigation systems.

Interstellar precursor.

Ion-drive-engined projectile (100,000 mph capacity).

Inference engine.

Knowledge-based processors.

Laser-equipped weather satellites.

Long-range manned spacecraft.

Lunar survey and mining station (including hotel).

Manned planetary landing (other than lunar).

Motor car (production model): to include constantly variable transmission, computer-controlled 'voice-advice' navigation systems, four-wheel steering, ID-car operation and adjustment, plastic-injected tyres in matching colours and voice-command of dashboard instrumentation.

Nationwide chain of lead-free petrol stations.

Nationwide chain of nuclear-fusion power stations.

Natural tooth enamel (synthetically produced).

Nuclear-pumped, X-ray laser gun (star wars).

Offshore metropolis.

Offshore wind-turbine platforms.

Permanently manned sea-bed laboratories.

Omni-purpose international credit/ identification micro-chip (implanted subcutaneously).

'Particle beam' weapon (star wars).

Permanently manned space station factory: to produce drugs, super-glass, ultra-strong cast-iron, fibre-optic materials and gallium-arsenide crystals; powered by solar energy.

Public lending library stocking computer programs, videodiscs, compact discs, videocassettes and holographically illustrated books.

Self-service banks (rapid deposit and encashment).

SETI probes (search for extra-terrestrial intelligences).

Ships and submarines (propeller-less and rudderless): using superconducting electro-magnetic thrusters, on both hull and sea-bed, to attain 100-knot speeds.

700-passenger airliner.

'Smart pills' (total control of the precise site and rate of release).

Solar heliospheric observational spacecraft.

Supersonic land speed record car.

'Talking typewriter' (production model): able to read back what is typed into it.

Three-masted cutter-rigged yacht (112 ft length overall): requiring only one man to operate the computer-controlled automated

rigging and electronic navigational
equipment.
Tidal power stations.
Time machine (long-range).
200-miles-per-gallon motor car.

Ultra-violet explorer satellite.
Unducted-fan aero-engine (contra-
rotating, multiple-layer
propellers).

Video-discs and compact discs
(erasable).
Voice-operated, personal-typefaced,
proportional-spaced word-
processor.

Wind-rain-sun-sensitive energy-
versatile roof tiles.

*(Notification to the publisher when
these, or other, ideas are realised would
be most welcome.)*

Bibliography

*Annual Register of World Events,
The,* (from 1945), London,
published annually

BRYDSON, JA, *Plastics Materials,*
London, 1982

HAYDN, JOSEPH and VINCENT,
BENJAMIN, *Haydn's Dictionary of
Dates and Universal Information
relating to all Ages and Nations,*
London, 1881

*Innovation and Achievement:
Scotland's Place on the World Map*
(Scottish Development Agency),
Edinburgh, 1984

FRANZBERG, MELVIN and PURSELL,
CARROLL W, JR, (editors)
Technology in Western Civilisation,
Vols I and II, Oxford, 1966/1967

LARSEN, EGON, *A History of
Invention,* London, revised
edition 1971

NEEDHAM, JOSEPH with WANG LING,
*Science and Civilisation in Ancient
China,* Vols I–V, Cambridge,
1956–1980.

*Queen's Annual Award to Industry,
The,* published in the London
Gazette, London, 1966–1984

READERS' DIGEST ASSOCIATION, *The
Inventions that Changed the World,*
London, 1982

ROBERTSON, PATRICK, *The Shell Book
of Firsts,* London, 1974

Smithsonian Book of Invention, The,
Washington DC, 1978

Entries in this index are followed by the year in which they appear, not by page numbers. Entries falling in the same year are listed alphabetically.

A

abecedarium 1874
aberration of light 1729
accelerator, foot 1897
accordion 1822
accounting machine 1910
accounts statement, pictographic 3500 BC
accumulator 1859
acetylene 1836
acetylene headlamp 1896
'Acousticon' (hearing-aid) 1901
acre, standardised 1305
acrylic paints 1964
actinium 1899
actinometer 1825
activated sludge process 1912
adding machine, recording 1888, 1889
advanced gas-cooled reactor (AGR) 1976
Advanced Passenger Train (APT) 1972
aeoline (musical instrument) 1829
aerated bread 1856
aerial (radio) 1895
aerial archaeology 1924
aerial crop-dusting 1921
aero-advertising 1935
aerocycle 1921
aero-engine *see* aircraft engine
aerofoil 1884
aerograph 1912
aerophore 1865
aeroplane *see* aircraft
aerosol can 1926
    insect spray 1941
aero-steam engine 1869
aerotherapy 1878
aérotrain 1962
aetherophon 1924
Aga kitchen range 1924
Agfacolor (photographic film) 1935
AGR (nuclear reactor) 1976
agricultural motor, Ivel 1902
agrimotor, petrol-engined 1897
air-brush, aerograph 1893

air-conditioning 1902, 1906
    of a railway carriage 1930
air filters, charcoal-based 1854
air navigation system 1972
air pollution detector 1981
air-raid
    airship 1915
    monoplane 1911
air vacuum pump 1654
airbed 1813
airbus A320 1986
aircraft
    all-metal cantilever wing 1915
    all-plastic 1984
    controllable wing-warping 1903
    delta-wing 1949
    flying ram 1945
    flying wing tailless 1940
    geodesic construction 1935
    human-powered 1977, 1979
    jet-engined 1939
    jet-engined forward-swept-wing 1944
    monoplane 1907
    passenger-carrying 1890, 1914
    pilotless flight 1924
    pilotless transatlantic 1947
    solar-powered 1980
    Spitfire fighter 1936
    steam-powered 1847, 1884, 1894
    supersonic 1947
    supersonic forward-swept-wing 1984
    swing-wing 1953
    tailless 1931
    tilting-rotor fixed-wing 1955
    tilt-wing 1957
    transatlantic 1919
    vertical take-off (VTOL) 1960
        fighter-plane 1969
    see also airliner; airship; helicopter
aircraft brakes
    carbon 1975
    disc 1953

aircraft carrier
  adapted ship 1911
  custom-built 1922
  nuclear-powered 1960
  super 1955
aircraft detection, electromagnetic 1934
aircraft engine
  back-pack 1961
  rotary 1887, 1907
  unducted fan 1985
aircraft radio-location 1935
aircraft refuelling, in-flight 1923
aircraft undercarriage
  hydraulically operated retractable 1929
  pneumatic-tyred 1906
  retractable 1911
  wheeled 1906
aircraft wing
  carbon-fibre 1983
  mission-adaptive 1985
  slotted 1921
airflow ventilation 1964
airfoilcraft 1970
airliner
  Airbus 320 1986
  automatic landing 1965
  jet-powered 1949
  jumbo jet 1969
  pressurised passenger cabin 1938
  supersonic 1968
  turboprop 1948
airscrew, variable-pitch 1924
airship
  diesel-engined 1929
  non-rigid/helium-filled 1918
  rigid 1900
  rigid helium-filled 1925
  steam-powered 1852
airship hangars 1916
airship-tethering mast 1911
'airstrip' (medical dressing) 1957
alarm-clock tea-maker 1902
alcohol still 1832
ALERT (tyre sensor) 1985
ALGOL (computer language) 1958
alizarine red 1831
alkalimeter 1815
Alka-Seltzer 1931
alkyd paints 1927
allopurinol 1962
alpaca 1848
alphabet
  Arabic 512
  Aramaic/Assyrian 720 BC
  Aryan 300 BC
  Chinese 1880
  complete 1300 BC
  Ionian-Greek 400 BC
  Phoenician (22-letter) 1100 BC (30-letter)
    1500 BC
  Sinaitic 1600 BC
  Slavonic 862
alphabetic script, phonetic 775 BC
Alphonsine tables 1253
ALSEP (experiment package) 1969
alternation of generations 1819
alternator
  electric 1878
  high frequency (Alexander) 1906
aluminium 1827
  chemically produced 1854
  electrolytically produced 1886

ambulance 1792
americium 1944
amines 1849
ammonia 1774
  synthesis process 1909
ammonia-soda reaction 1810
ammonium picrate 1885
amniocentesis 1956
amphibious vehicle 1805
  DUKW 1942
  petrol-engined 1905
amphicar 1962
amplifier, rock music 1949
amplitude modulation 1915
amputation 1679
amyl cinnamic aldehyde 1926
anchor
  C.Q.R. 1933
  stockless 1821
  Trotman's 1853
anchor escapement 1680
anchor fluke 592 BC
androsterone 1931
anaesthesia, ether 1842
anaesthetic
  chloroform 1847
  ether 1846
  intravenous 1902
  local 1884
  nitrous oxide 1844
  novocaine 1904
  spinal 1855
Anémocorde 1789
anemometer 1709
  cup 1846
  pressure 1775
aniline 1826
aniline dye 1856
aniline yellow dye 1858
animal cells 1839
animal husbandry 11000 BC
animalcules (bacteria) 1683
Anno Domini dating system 532
anorthosite 1971
anthropoglossus 1864
antibiotic drug 1943
antibodies
  chemical molecular structure of 1969
  monoclonal 1975, 1902
anti-cyclone 1863
anti-depressant drug 1953
anti-freeze 1927
anti-helium 3 1972
anti-interferon 1981
anti-knock petrol compound 1921
anti-matter particles 1972
antimony 1490
anti-proton 1955
anti-sound device 1983
anti-tank grenade projector 1941
anti-tank gun 1917
antitrypsin gene, cloned 1984
anti-Xi-zero (particle) 1963
'anxiety chemical' (human brain) 1985
Apollo experiment package 1969
Apollonicon (barrel-organ) 1817
apple, Granny Smith 1868
aquacolor (cine) 1983
aqueduct 700 BC, 312 BC
aqualung 1943
arc furnace, electric 1894
arc lighting 1879

arc-welding, electric 1865
ARECIBO massage 1974
arch, true 400 BC
archaeology 1924
Archimedean screw 260 BC
areometer 1768
argand lamp 1780
argon 1894
argon ion laser 1960, 1978
arithmetic, binary 1703
arithmographe (calculator) 1889
arithometer 1820
ark (ship) 1568 BC
armillary sphere 225 BC
armoured vehicle 1896
  Ivel 1904
  steam-powered 1855
Arneson drive 1978
arsenic 1733
  standard test 1836
artery 200
artificial blood 1979
artificial comet 1984
artificial gill 1982
artificial heart 1982
artificial heart valve 1952
artificial hip 1972
artificial insemination 1780
  human 1785
artificial intelligence (AI) 1959
  computer program: backgammon 1979
    language understanding 1970
    SHRDLU 1970
artificial kidney 1945
artificial limbs 1540
artificial pancreas 1977
artificial silk 1889
artificial ski slope 1980
artificial tooth (pressure recorder) 1965
artificial tooth root 1985
artificially induced rainfall 1953, 1984
artillery 1343
asbestos 1868
  waste recycling process 1984
aspirin 1899
  soluble 1948
assembly line, moving 1913
astatine 1931
asteroids 1801
astigmatic lens 1827
astigmatism 1793
astrolabe 130 BC, 1613
astrology, computerised 1975
astronomy
  airborne observatory 1982
  gamma-ray 1969
  radio 1931
  X-ray 1962
astro-photography 1840
atlas, modern 1570
atmospheric railway 1840
atomic absorption spectrophotometer 1954
atomic bomb 1945
atomic clock 1948
atomic nuclei, bombardment 1919
atomic pile
  experimental 1947
  reactor 1942
atomic power station
  advanced gas-cooled 1976
  fast breeder experimental 1962
  Magnox-type 1962

atomic weights, true 1858
atoms
  theory of 1803
  theory of structure of 1911
audiometer 1879
audio-tapes, cartridge 1965
audiphone 1880
auriscope 1862
Australopithecus 1936
Austria (asteroid) 1874
autoflare 1962
autogyro 1923
automatic doors 1922
automaton
  chess-player 1769
  coach and horses 1649
  dancing men 1200
  doll 1760
  draughtsman 1773
  duck 1738
  eagle 1474
  flying dove 400 BC
  lady and Cavalier 1632
  musician 1772
automobile *see* motor car
autopilot 1913
  electronic 1941
  equipment 1920
aviation, basis of 1889
Avogadro's Law 1811
axes, copper 4000 BC
axles, swinging 1903
'A–Z system' (knitting yarns) 1964
azathioprine 1962

B

Babitt metal 1839
Babcock test 1890
baby incubator 1891
back-pack aero-engine 1961
back-staff 1594
bacteria 1683
  tests for 1884
bacteriophage 1915
Bailey bridge 1943
baked beans
  canned 1875
  canned in tomato sauce 1895
bakelite 1908
balance
  Roberval 1670
  torsion 1776
balance spring, spiral 1675
balance spring regulator 1658
'Balans' chair 1980
ball bearings 1794
ball-valve 1845
ballistic pendulum 1742
ballistite 1888
balloon
  gas-filled 1783
  hot-air 1783
  infra-red explorer 1981
  vulcanised toy 1847
ballpoint pen 1938
  retractable 1954
bamboo gun 1259
'Band Aid' (medical system) 1982
bandage, adhesive 1928

band-saw 1808
bank 808
bank notes
    Bank of England 1695
    Bank of Stockholm Sweden 1661
banknote numbering machine 1806
bar code 1970
Barany indication test 1909
barbed wire 1867, 1873
barbiturate drugs 1903
barge
    motor 1894
    transportation 1421
    wave-powered 1979
barium 1774, 1808
barometer 1644
    aneroid prototype 1758
    aneroid spring-type 1843
    floating man 1672
    syphon wheel 1665
barrel-organ 1502, 1817
BASIC (computer language) 1964
bassoon 1500
    double 1620
bat bombs 1941
bath tub
    gas-heated 1882
    royal 1700 BC
bathyscaphe 1960
bathysphere 1930
battering ram 850 BC
battery 1770
    bichromate 1855
    galvanic 1842
    Leclanché 1868
    rechargeable 1881, 1908
    silver-zinc 1941
    solar 1954
    solid depolariser 1877
    storage 1859, 1908
battleship, steam turbine 1906
bauxite 1822
bayonet 1647
bazooka gun 1942
BCG vaccine 1908
BSC theory (superconductivity) 1957
Beaufort wind scale 1806
bed, hydrostatic 1832
bee-keeping system 1890
beer 6000 BC
beer-pump handle 1785
'Beetle' plastic 1924
Bélinographe 1925
bell 1100 BC
    electric 1831
Benday process 1879
benzene 1849
    cyclic structure 1865
benzole 1849
Bereguardo Canal 1458
berkelium 1949
Berlin (carriage) 1660
beryllium 1828
Bessemer converter 1856
betatron 1939
Biblia Pauperum 1260
bibliographic system 260 BC
bicycle
    amphibious 1931
    BMX 1973
    boneshaker 1868
    decumtyplet 1897

derailleur gear 1905
    electric 1890
    farthing-penny 1881
    front-wheel pedal 1865
    kangaroo 1884
    Moulton 1960
    Ordinary (penny-farthing) 1871
    pedal prototype 1839
    safety 1885
    semi-bodyshelled 1913
    supine-recumbent 1933
bicycle tyre, pneumatic 1888
Biela's comet 1826
bifocal lenses 1884
'Big Bertha' (gun) 1914
bikini (swimsuit) 1946
binary arithmetic 1703
binary calculator 1939
Binet intelligence scales 1911
binoculars 1859
biodegradable plastic 1982
biopol (plastic) 1983
biosensors 1983
biotin 1940
biplane
    amphibious 1911
    cabin 1912
    hydrofoil 1911
Bird Flight (book) 1889
bird radio-tracking system 1960
Birdseye frozen food 1923
bireme (galley) 850 BC
Biro pen 1938
Bismarck brown 1867
bismuth 1529
bismuth effect 1897
bitumen 2500 BC
    for damp-proofing 850 BC
'black holes' 1972
blade manufacture 1636
blanket, electric 1883
blind, raised letters for 1786
blood, artificial 1979
blood bank 1931
blood circulation 1610
blood-clotting factor for haemophiliacs 1984
blood group systems
    Diego 1955
    Duffy 1950
    Kell 1946
    Lutheran 1946
    rhesus 1940
blood groups
    A1 and A2 1911
    four primary 1901
    MNS 1927
blood test, pre-transfusion 1908
blood transfusion 1818
    animal to human 1667
bloomers 1849
blowing engines 1310
blowlamp 1880
blue sugar paper 1665
BMX bicycle 1973
boat
    aluminium 1891
    box-kite 1903
    concrete 1848
    internal combustion engine 1886
    iron 1777
    plank-built 1500 BC
    polyester resin 1948

remote-radio-controlled 1898
solar-powered 1976
steam-powered, *see* steamboat; steamship
steam-powered model 1769
timber-built 3000 BC
umbrella 1897
boat-cloak 1844
bobbin-net machinery 1808
bogie truck 1831
boiler
    solar 1860
    tubular 1830
bolometer 1881
bolting machine 1764
bomb
    atomic 1945
    bat 1941
    bouncing 1943
    hydrogen 1952
    neutron 1977
    nuclear 1961
    rocket-powered glide 1940
bomb sight, equal-distance 1915
bombard (cannon) 1586
bomber
    flying-wing jet 1947
    multi-engined heavy 1914
book dust-jacket 1833
book pagination 1470
bookcase, revolving 1588
books
    disposable 1841
    paperback 1935
    printed 868
    vellum 198 BC
Boolean algebra 1854
boot-sewing machine 1790
borax 1702
Bordeaux mixture 1879
boron 1702, 1808
bosons, intermediate vector 1983
bottle caps, aluminium foil 1914
Boulder Dam 1936
bounce-back system (ship detection) 1930
bouncing bomb 1943
bourbon (whisky) 1789
bow and pike weapon 1634
bowl, bevel-rimmed open 4000 BC
bowler hat 1849
box-girder construction 1840
boxing gloves 1747
Boyle's Law 1662
braille 1837
brain chemical 1972
brain nerves 1807
brakes
    compressed-air 1868
    hydraulic four-wheel 1920
    servo-assisted four-wheel 1919
brassière 1913
'Brasso' (polish) 1905
bread, aerated 1855
bread-making
    high-speed process 1958
    machine 1862
breakfast cereal 1893
    flaked 1895
    Shredded Wheat 1893
breeder reactor, experimental 1951
Bren gun 1938
brewing saccharometer 1784
brewing thermometer 1762

brick-making machine 1839
bridge
    cast-iron 1779
    floating 480 BC
    iron-chain suspension 580
    iron-girder 1755
    long-span suspension 1883
    segmented arch 610
    stone 142 BC
    temporary 1943
    tubular steel 1874
    tubular suspension 1850
brine-pumping machine 1682
bromine 1826
bronze-working 2800 BC
Brooklands motor-racing circuit 1907
brougham (carriage) 1838
bubble chamber 1952
bubble memory, prototype 1967
bubble-oxygenator 1955
buffer memory circuit card 1967
building
    cast-iron construction 1849
    concrete 121 BC
    mobile 1960
building blocks, patterned concrete 1922
bulldozer 1926
bullet, concoidal cup rifle 1836
Buna S (synthetic rubber) 1926
bunsen burner 1855
Burbank potato 1873
burglar alarm 1858
bus *see* omnibus
butadiene, synthetic 1912
BWR (nuclear reactor) 1959

C

cable car, aerial 1866
cable television 1949
cabriolet, hackney 1823
cacodyl 1760
cadmium 1818
Caesarian section 1500
caesium 1860
caffeine 1821
caisson, compressed air 1852
calcium 1808
calculating machine 1642, 1843
    full automatic 1911
    key-driven 1886
    mechanical 1623
    pocket: Arithmograph 1889
        mechanical 1868
calculator
    automatic sequence controlled 1944
    binary 1939
    desk 1935
    electro-mechanical 1936
    electronic 1952
        commercial model 1961
    delay storage automatic (EDSAC) 1949
    pocket 1972
    programmable hand-held 1972
calculus, integral 1666
calendar, thirteen-month 1850
    Jewish 385
calendar reform 1582
californium 1950
calliope 1859

camera
    35 mm Leica 1914
    auto-focussing still 1972
    automatic still 1976, 1984
    automatic still disc-cassette 1982
    electronic still 1982
    Kodak 'Brownie' 1900
    microfilm 1939
    miniature 1859
    panoramic 1845
    Polaroid Land 1947
    roll-film 1888
    stereoscopic still 1851
    still 1858
    vest-pocket Kodak 1912
    video 8 1985
    visitronic autofocus 1977
    'voice advice' 1983
    wide-angle lens 1859
    see also cine camera
camera film, colour 1935
camera lucida 1807
camera obscura, portable 1560
camera tube, electronic 1923
camshaft, overhead 1902
canal
    Bereguardo 1458
    Languedoc 1692
    Suez 1869
    Panama 1914
canal-building 870 BC
canal locks 983, 1458
canal rays 1886
    Döppler effect 1909
canal tunnel 1777
canals and rivers, control of 300 BC
cancer-tumour-causing genes 1982
candle
    electric 1876
    stearine 1850
candle measurement system 886
candle wick, plaited 1824
canned food 1812
    key opening 1866
cannon 1260, 1326
    giant 1586
    wooden toy 1383
canoe 1865
canvas proofing 1878
capillary motor-regulating mechanism 1905
captopril 1985
car see motor car
car-park, multi-storey 1901
caravan, holiday 1885
carbolic acid 1847
carbon, quadrivalent nature 1858
carbon dioxide 1756
carbon disulphide 1796
carbon fibre 1964
    as prosthetic material 1965
    tennis racket 1980
carbon-filament lamp 1879
carbon paper 1805
carbon suboxide 1906
carborundum 1891
carburettor 1884
    surface 1876
cardiac catheter 1941
carding machine 1743
cargo and passenger ship, nuclear-powered
        1961
caricatures 1330

carillon 1487
carpet loom 1851
carpet sweeper 1876
Carrel-Dakin treatment 1915
carriage
    Berlin 1660
    brougham 1838
    Dennett Gig 1815
    Moray car 1885
    horse-drawn 1815
    sailing 552
    steam 1824
    Tilbury gig 1820
    'Volontas' 1822
    windmill-powered 1784
carriage spring, elliptical 1804
carronade 1779
cart, wheeled 3500 BC
cartes de visite, photographic 1854
cartoon film
    animated 1906
    technicolor 1933
cartridge 1813
casein plastics 1897
cash register 1879
cassette audio tapes 1963
cast iron
    cog wheels 1760
    pipes 1455
    water-wheel shaft, 1769
catalytic cracking (petroleum) 1928
catalytic unit, anti-pollution 1979
catamaran
    aspirated-cylinder-sail 1983
    twin-engined offshore 1969
    yacht 1662
Caterpillar/crawler tractor 1904
catgut (surgical thread) 1920
cathode ray tube 1878, 1897
cathode rays 1859
cat's-eye road-stud 1934
cavalry sword 1800
cave murals 13000 BC
cave painting 15000 BC
cavity magnetron 1940
celesta 1886
'cell-from-cell' theory 1858
cellophane 1908
    moisture-proof 1926
cells
    animal 1839
    hybrid 1975
    plant 1837
celluloid 1869
cellulose film 1888, 1892
cellulose paint finish 1924
Celsius scale 1742
cement 1824
    artificial 1824
    coloured-fabric 1634
centrifugal clutch, hydraulic 1905
Cephalosporium acremonium 1945
Cerenkov effect 1934
cerium 1803
cerium-iron alloy 1902
chain drive 1779
chain-pump 189
chair
    electric 1890
    on springs 1828
    rocking 1853
    sedan 1634

chairlift 1937
'chaise volanté' 1743
champagne 1688
Chanel No. 5 (perfume) 1924
charcoal filter 1811
chariot
    horse-drawn 1400 BC
    lightweight 2350 BC
    sailing 1600
chariot wheels, copper-tyred 2000 BC
Charles's Law 1798
charm (subatomic particle) 1976
charm, block-printed Buddhist 770
Charon 1978
charts, geographical/celestial 570 BC
cheese, Gorgonzola 879
chemical plant, computer-controlled 1959
chenille-weaving process 1839
cheque 1659
    printed 1763
Chernikeef ship's log 1917
'Cherry Blossom' (shoe polish) 1906
chess (game) 1150 BC
chess match, international, by computer 1966
chess match, international, by radio 1924
chewing gum 1848
    chicle-based 1870
chimney cowl 1934
chimney-sweeping machine 1805
Chinese white pigment 1834
chiropractic 1895
chlorine 1774
    as war gas 1915
chloroform anaesthetic 1847
chlorophyll, fluorescence in 1833
chlorpromazine 1950
chocolate
    blocks 1819
    milk 1875
cholera bacillus 1882
Christmas card 1843
chromatography 1906
chromium 1797
chronograph 1855
chronometer
    marine 1735
        pocket-sized 1761
chronoscope 1840
cigarette-making machine 1844, 1880
cigarettes, commercially produced 1843, 1853
cinchonine 1803
cine camera
    mirror reflex 1936
    pneumatic drive 1912
    true underwater colour 1983
cinema
    drive-in 1934
    experimental holographic 1980
    projection system 1907
    'Sensurround' 1974
    see also motion pictures
Cinemascope 1925
cinematography see motion pictures
'Cineorama' 1900
Cinerama 1951
cipher, Caesarian 50 BC
cipher device, cylindrical 1891
cipher-writing machine 1838
circular saw 1777, 1816
circumferentor 1612
Circus Maximus 605 BC
citric acid cycle 1937

city plan 1400 BC
    circular 224–241
    engraving 2200 BC
    open-plan 1666
Clarence carriage 1840
clarinet 1690
classified advertisements 1631
clavecin electrique (harpsichord) 1761
clavicylindre 1799
clavilux (colour-organ) 1922
clay objects, fired 11000 BC
clementine (fruit) 1900
clock
    astronomical 1500 BC, 1092, 1364
    atomic 1948
    automatic striking 1335
    compressed-air 1881
    diapason 1886
    mechanical 1360
    polar 1849
    practical electric 1918
    quartz crystal 1929
    repeating 1676
    striking 1367
    synchronised 1878
    travelling 1451
clock dial, illuminated 1826
clock movement manufacturing machine 1800
clock pendulum 1657
clock tower 1790
clockwork, eternally running 1598
clod-breaking roller 1841
cloth, woven 5000 BC
cloth cutting, laser 1980
cloud chamber 1911
cosmic ray magnetic 1930
'cloudbuster' (artificial rain) 1953
cloverleaf intersection 1928
CMOS (complementary metal oxide
        semiconductor) 1980
clutch, multiplate 1863
coal gas lighting 1792
coaxial cable 1929
COBOL (computer language) 1959
Coca Cola 1886
cocaine (anaesthetic) 1884
cocchio (wagon) 1288
cockpit safety capsule 1985
cocoa 1828
coding machine 1923
coenzyme A 1947
coffee 1000
    instant 1938
coffee machine, filter 1908
coffee pot 1806
    automatic electric 1952
cog wheels, cast-iron 1760
coherer 1890
coin-milling machine 1617
    London Mint 1811
coin-weighing machine, Napier's 1844
coinage
    copper 400 BC
    device-struck 600 BC
    disc 300 BC
    gold 770 BC
    regal portrait on 420 BC
    silver and gold 550 BC
coke 1611
coke smelting 1709
collapsible tube
    metal 1892
    polythene 1952

colliery winding engine 1784
collodion process 1851
Colossae of Memnon 1417–1379 BC
Colossus of Rhodes 280 BC
colour-blindness 1793
colour printing 1836
   machine 1851
Colt chimney cowl 1934
Colt revolver 1835
combine harvester 1836
command and service lunar modules 1969
communications satellites, *see* satellites
compact disc
   automobile navigation system 1985
   digital audio system 1980
   read-only memory 1985
compact disc player
   for cars 1984
   portable 1984
compass 1269
   distant-reading 1932
   gyroscopic 1908
   hanging 1608
   liquid-filled 1813
   magnetic 1115
   mariner's direct-reading 1766
   measuring 1602
component assembler, robotic electronic 1984
compressed-air brake 1868
compression-ignition engine 1890
comptograph 1889
comptometer 1886
Compton effect 1923
computer
   data-processing 1889
   digital avionics flight 1969
   electronic discrete variable automatic (EDVAC)
      1949
   first vacuum-tube 1946
   Honeywell 400 1960
   information-processing, Japanese characters
      1979
   programmable electronic 1943
   quantity-processing 1980
   simultaneous data-processing 1959
   transistorised 1959
   universal automatic 1952
   universal digital stored program 1948
computer-aided graphic design system 1964
computer channel, input/output 1954
computer control
   chemical plant 1959
   production line 1960
   refinery 1959
computer floppy-disc pen 1982
computer languages
   ALGOL 1958
   BASIC 1964
   COBOL 1959
   FORTRAN 1957
   LIST 1958
   PASCAL 1969
   PROLOG 1971
   simulation 1962
computer memory magnetic cores 1954
computer print-out, rapid 1953
computer proof-reader 1984
computer robotic system (RSI) 1982
computer system
   fully simultaneous large-scale 1959
   'stretch' 1961

computer terminal, wristwatch 1985
computer time-sharing
   experiment 1961
   international 1967
   'Multics' system 1965
computer translation system 1970
   automatic 1963
   by machine 1984
concertina 1825
Concorde airliner 1968
concrete
   boats 1848
   pre-stressed 1880, 1965
   reinforced 1865
   shell structure 1916
concrete mixer 1857
'Condeep' (platform) 1975
condenser, electrical 1746
conic sections 350 BC
conservatory 1545
contact lenses 1887, 1944
   plastic corneal 1956
   plexiglass 1936
   soft bifocal 1985
   solution-less 1945
contour lines 1737, 1791
contraceptive 1500 BC
   diaphragm 1880
   intra-uterine 1909
   male pill 1976
   oral pill 1951
controller, magnetic 1892
conveyor, passenger 1954
copper-aluminium alloy (Supral) 1975
copper coinage 400 BC
copper-working 4500 BC
copper-zinc alloy 1732
copying machine 1778
   hectographic 1923
   thermographic 1973
cordelier 1792
cordite 1889
Corliss valve 1848
corn cultivator 1820
'Corn Flakes' 1895
cornflour 1854
corrugated iron 1828
cortisone 1938
cosmic radiation 1911
   microwave background 1965
cosmodrome 1957
cosmotron 1952
cotton 2500 BC
cotton-dressing machine 1803
cotton gin (machine) 1794
cotton-lace machine 1808
cotton mill, gas-lit 1805
cotton-picker 1889
cotton printing 1783
cotton spinning and roving machine 1790
cotton wool 1874
coumarin 1879
Covert Rebus, theory of 1976
crane
   hydraulic 1845
   overhead travelling 1970
crankshaft compression phase 1891
crankshaft torsion dampers 1908
crash helmet, motor 1904
cream separator 1876
crèche 1844

credit card
  banking 1958
  electronic memory 1974
  general purpose 1950
  holographic 1984
crematoria 1876
creosote 1833
Cristallo glass 1530
croissant (bread) 1683
crop domestication 9000–8000 BC
crop dusting service, aerial 1925
crossbar (telephone exchange) 1926
crossbow 300 BC
crossbow gun 1488
crossword puzzle 1913
crown top bottle opener 1892
crucifix, commemorative metal 410
Crum Brown's Rule 1892
cryptograph 1868
crystal diffraction 1912, 1926
crystal glass, Venetian system 1549
crystal radio detectors 1906
crystal structure, X-ray analysis 1920
crythrosin 1876
cube sugar manufacturing process 1875
cuprammonium rayon 1890
curare drug 1740
curium 1944
cutting-out machines 1853
cutting tool, gas-assisted gas-laser 1961
cybernetics 1948
cyclometer 1877
  sound signal 1886
cyclosporin A 1978
cyclostyle pen 1881
cyclotron 1930
cylinder seal 3500 BC
cystoscope, gas-lit 1853

D

2, 4-D (herbicide) 1941
Da Vigevano's treatise 1335
'Daddy Longlegs' pier-boat 1895
Daedalum 1832
daguerreotype 1837
  disc 1873
Daltonism 1793
dam
  horizontal-arch 1859
  masonry 2950–2750 BC
damp-proofing 850 BC
Daniell cell 1836
daraprim 1962
dark lines (solar spectrum) 1821
data-processing system, solid-state 1959
data storage cores 1949
date-stamp, postal 1661
'Datel' viewdata system 1979
Davis's Quadrant 1594
'Dayglo' pigments 1933
Davyum 1877
DDT (insecticide) 1939
De Rebus Bellicis 370
De Triangulis Omnimodis 1464
Debusscope 1860
Decca navigator-receiver 1944
decimal system 520
decipium 1879
'DECtalk' system 1985

deep-sea bathyscaphe 1960
defibrillator 1932
defence tower, round 1803
Dellinger Effect 1935
democracy 505 BC
Denloc tyres 1982
Dennett gig 1815
dental amalgam 1819
dental drill 1863
  air-bearing air-turbine 1962
  air-turbine 1957
  electric 1875
  tungsten carbide 1947
dental forceps 1525
dental handpiece, ball-bearing 1954
dental plate 1817
dental pressure measurement 1965
dentures, vulcanite 1851
department store 1848
'dephlogisticated air' 1774
derailleur gear 1905
derrick 1850
Derringer pistol 1843
de-sulphurisation process 1933
detector lock 1818
detergent
  household 1907
  synthetic 1913
Dettol (antiseptic) 1933
deuterium 1932, 1940
Dewey decimal system 1876
diapason clock 1886
diastase 1833
dictionary 1100 BC
didot points 1737
didymium 1841
Diego blood group system 1955
Diels-Alder reaction 1928
diesel-electric traction 1934
diesel engine 1890, 1892
  ceramic 1979
  high-speed direct-injection 1984
  reversible four-stroke 1908
  reversible gear 1901
  reversible two-stroke 1905
diet deficiency problem 1747
dietheroscope 1876
differential analyser 1930
differential calculating machine 1833
differential gear 1828, 1877
digestion, physiology of 1897
digger-loader, tractor-mounted hydraulic 1955
digging stick 10000 BC
digital scene simulation 1983
digitalis 1775
dinner jacket 1886
diode, tunnel 1959
dioptric system 1819
diorama 1822
diphtheria 1820
diphtheria antitoxin 1890
diphtheria bacillus 1883
dis-assembly line 1869
disc brakes
  aircraft 1953
  electric vehicle 1888
  motor car 1902
disc-cassette camera 1982
disc recording
  quadraphonic 1971
  stereophonic 1933, 1958
disposable book 1841

Disprin 1948
dish-washing machine 1889
dividing engine 1740, 1768
dividing machine, circular 1872
diving bell 1531, 1717
diving dress 1715
  closed 1837
  open 1819
DNA 1953
  biosynthesis 1956
DNA, recombinant 1973
Döbereiner lamp 1823
Dolby noise reduction systems 1966, 1970, 1980
doll
  talking 1823
  toy 1413
doll's house 1558
dome
  design 1432
  iron 1811
  single-shell 532
Domesday project 1985
domesticated animals 11000 BC
  dogs 7700 BC
  goats 8050 BC
  pigs and cattle 7000 BC
  sheep 7200 BC
domesticated crops 9000–8000 BC
  peas and beans 7000 BC
Donati's comet 1858
door sockets, stone 4500 BC
doors, Baptistry 1452
Döppler effect 1851
  in canal rays 1909
double-decker bus, enclosed 1909
drainage engine 1623
draw-loom 100 BC
drawing pen, tubular-nibbed 1954
dredger 1561
drill
  dental see dental drill
  electric hand 1895
  stone-working 2500 BC
drilling, compressed-air 1857
drive-in bank 1937
driving belt, rubber 1825
driving chains 1864
drosophila 1910
drug biosynthesis 1977
Drummond light 1826
dry cleaning, clothes 1855
Duffy blood group system 1950
DUKW (amphibian vehicle) 1942
dumet 1913
Dunlopillo (foam rubber) 1929
duplex burner 1865
duplicating machine
  hectographic 1923
  thermographic 1973
duralmin 1910
dust filter 1906
dustbin 1883
dust-jacket, book 1833
dye-coupler colour process 1914
'Dyform' strand (prestressing) 1956, 1965
dynamic memory cell 1967
dynamite 1867
dynamo, self-acting 1866
dynamometer 1750
  rolling-road testbed 1904
dynastic divisions 250 BC
dysprosium 1886

E

Earth Resources Technology Satellite 1972
earthquake alarm 132
earthquake measurement scale
  Mercalli 1902
  Richter 1935
eagle, mechanical 1245
ear, tube structure of 520 BC
eau de Javel 1787
eau Grison (fungicide) 1852
ebonite 1851
echo-sounder 1917
  transistorised 1958
EDSAC (calculator) 1949
EDVAC (computer) 1949
EEG 1929
effluent treatment plant, thermophilic
  anaerobic 1982
eidograph 1821
eidophusikon 1781
Eiffel Tower 1889
einsteinium 1953
ejection seat 1938
elastic web 1830
Elastoplast (bandage) 1928
Electra 225 (locomotive) 1986
electrets 1919
electric-arc lamp 1808
'electric aura' 1939
electric battery 1770
electric bell 1831
electric candle 1876
electric car 1874
electric chair 1890
electric currents, secondary 1838
electric-generator windmill 1890, 1931
electric light 1834
  for towns 1880
electric machine
  frictional 1660
  Leyden-jar-powered 1855
electric motor 1829, 1837
  direct-current 1873
  induction 1888
  linear 1985
  miniature 1880
  synchronous 1902
electric telegraph 1837
electric telegraph key 1833
electrical properties (amber) 600 BC
electrical resistance 1827
electricity, galvanic 1790
electrocardiogram 1887
electrocardiograph 1903
electro-convulsive therapy 1937
electro-encephalogram (EEG) 1929
Electrolux refrigeration system 1922
electromagnet 1824
electromagnetic balance 1831
electromagnetic induction 1831
electromagnetic manufacture 1854
electromagnetic motor 1829
electromagnetic radiation 1873
electromagnetic wave transmission 1864
electromagnetic waves 1887
electron 1896, 1952
electron accelerator, Cambridge 1962
electron bombardment engine 1960
electron diffraction by crystals 1926
electron microscope 1929
electronic filing systems, integrated 1983

electrophone 1880
electrophorus 1775
   reciprocal 1862
electroplating 1840
electrotype 1839
elektromote 1881
elevator 1743
   electric 1887
   passenger 1857
   passenger electronically controlled 1948
ellipse 220 BC
'Elmer and Elsie' (robots) 1950
'emanium X' 1904
embalmed person 2950
embroidery machine 1804
Empire State Building 1931
'Empyreal air' 1772
enamel 1557
enamel-mosaic 1861
Encke's comet 1818
encyclopaedia 1270 BC
engine
   alcohol 1797
   compound marine 1854
   compression-ignition 1890
   condensed wind 1799
   diesel see diesel engine
   solar-electric 1893
   Wankel 1963
engine bearings, tapered oil film 1905
engine manifold, lightweight plastic 1981
engine mountings, rubber 1904
ENIAC (calculator) 1946
Enigma coding machine 1923
enkephalin 1972
envelope, adhesive 1844
enzymes
   crystalline 1926, 1930
   oxidation 1934
eosin scarlet dye 1875
epicyclic gearbox 1895
Epsom salts 1695
equals sign 1557
Equation of the Third Degree 1545
eraser 1770
erbium 1843
erinoid 1897
escalator 1894
escapement
   anchor 1680
   dead beat 1715
esparto pulp 1861
Esperanto 1887
espresso coffee machine 1946
esterone 1929
ether
   anaesthetic 1842, 1846
   inhaler 1847
ethylene 1785
ethylene glycol 1856, 1927
etorphine (M-99) 1963
Euclid's elements 320 BC
eugenics 1869
europium 1869
exposure meter, photoelectric 1931
extra-sensory perception 1940
eye surgery, laser 1964

F

'Factor VIII' 1984

factory gaslighting 1798
Fahrenheit scale 1714
false teeth 1770, 1788, 1845 see also dentures
fan
   commercial electric 1882
   ventilation 180
farthing-penny bicycle 1881
   steam-powered 1884
fasteners, load-monitored 1980
'Fastest Act of Man' 1982
felt-tip marker 1955
Fermat's Last Theorem 1631
fermentation
   cell-free 1897
   yeast growth during 1836
fermium 1953
ferrite cores, magnetisable 1949
ferroxdure 1950
ferryboat, air-cushioned 1865
fertiliser, artificial 1839, 1842
fibre, ultra-strong synthetic 1985
fibre optics 1955
   medical tubes 1957
fibre-tip pen 1960
fibreboard, medium-density 1965
fighter, flying-boat 1947
fighter-planes, aerially launched 1941
filaria 1897
filing/shaping machine 1836
filing system, laser-disc 1985
film (photographic)
   colour negative 1983
   slow-motion 1906
filter coffee machine 1908
fire
   domestic electric 1892
   fire-clay frame 1912
   man-made 12000 BC
fire alarm 1873
fire annihilator 1849
fire-engine 250 BC, 1518, 1632
   foam-equipped 1920
   motor 1898
   motor-chemical 1903
   petrol-propelled 1901
   steam-powered 1829, 1853
fire escape 1766
fire-extinguisher
   bomb 1734
   carbon dioxide gas 1866
   chemical foam 1905
   compressed air 1816
   sal ammoniac 1762
   water-chemical 1792
fire-fighting vessel 1855
fire-lances 1100
fire pump 1699
fire-ship (explosion vessel) 1585
   catamaran 1804
   steel-hulled/diesel-engined 1935
firearm silencer 1908
'fish joint' 1847
fish paste 1895
fishing nets, plastic 1959
fishnet-making machine 1804
Fiskeoscope 1919
'Five and Ten Cent Store' 1879
fives (ball game) 1692
fizzy mineral water 1741
flageolet (musical instrument) 1599
flashing apparatus, automatic 1906
flax-spinning machinery 1787, 1810

fléchettes 1914
Flettner rotorship 1922
flight reservation system, computerised 1962
'Flight Ship' (yacht) 1595
flight simulator, digital 1960
   electrical-mechanical 1929
   electronic 1941
Flinders bar 1804
flint arrowheads 9000–8000 BC
flint glass 1675
flintlock gun 1610
float glass process 1958
floatplane, ellipsoidal wing 1906
floorboard planing machine 1827
floppy disc 1970
floppy-disc pen 1982
flour, self-raising 1871
flour mill, transportable 1593
flue-gas desulphurisation process 1933
fluid coupling 1905
fluorescence 1852
   in chlorophyll 1833
   in fluorspar 1838
fluorescent lighting 1938
   hot-cathode 1935
fluorescent pigment 1933
fluorine 1771
flute 1847
fluxions (integral calculus) 1666
fly-by-wire system 1972
'flying bedstead' 1885
flying-boat
   fighter 1947
   giant 1921
   giant twelve-engined 1929
   transatlantic 1919
'flying saucer' 1961
   nuclear-powered 1973
flying shuttle 1733
'Flying Wing' 1940
foam rubber 1929
folacin 1938
food, heat-bottled 1809
food cans, tinplated 1875
food mixer, electric 1918
food preservation
   by canning 1875
   by freezing 1923
   by ion irradiation 1930
foot-and-mouth disease vaccine 1899
football pools 1922
foot-rule 1675
Ford Model T car 1913
formaldehyde 1867
FORTRAN (computer language) 1957
fossil fishes, glacial 1833
fossils, Palaeozoic 1854
fotosetter, intertype 1936
Foucault's pendulum 1850
fountain pen 1884
   tubular-nibbed 1928
francium 1930
franking machine, steam-powered 1857
Fraunhofer (dark) Lines 1821
frequency modulation (FM) 1933
Fresnel's rhomb 1819
friction gear, continuously variable 1925
friction match 1826
Friedel-Crafts reaction 1877
frigate 1649, 1825
frogman flippers 1926
fruit machine 1889

frozen food 1923
frying pan, electric 1911
fuchsine dye 1856
fungicide, 1852, 1878
fusée 1540

G

'Gabardine' 1879
gadolinium 1880
galalith 1897
Galleas (boat) 1515
gallium 1875
gallium arsenide chips 1985
Gallup Poll 1932
galvanic battery 1841
galvanic electricity 1790
galvanised iron 1837, 1840
galvanometer 1824
   multiplying 1821
   string 1903
gaming boards 2500 BC
gamma-ray astronomy 1969
gamma-ray sterilisation 1962
garden cities 1898
gas chamber 1924
gas cooker 1839
gas engine, internal-combustion 1853
gas fire, practical 1856
gas lighting
   city 1820
   cottage 1792
   cotton mill 1805
   factory 1798
   incandescent mantle 1885
   room 1784
   street 1807
gas mask/respirator 1915
gas meter 1813
   coin-in-the-slot 1887
gas stove 1826
   regulator 1915
gas turbine 1905
   constant-volume 1906
   early design 1791
gasogen 1883
gastrophetes 400 BC
Gatling machine-gun 1862
Gay-Lussac's Law 1804
gearbox
   automatic 1940
   pre-selector 1901
gearing system
   compound epicyclic 1919
   sun-and-planet 1781
gearwheels
   carbon-fibre-reinforced plastic 1969
Geiger counter 1928
Geiger-Müller tube 1928
gelatine, blasting 1879
generator
   high-frequency 1890
   ring armature 1867
genes
   abnormal 1982
   mobile elements 1951
   theory 1910
genetics 1865
Genoa jib 1927
geodesic dome 1958

geodesic dome house 1985
geothermal energy 1818
germanium 1886
geyser, gas 1868
gimbals 180
Giotto (spacecraft) 1985
girdle, inflated 1698
glaciarum (ice-skating rink) 1876
glass
 cast plate 1688
 fast-reacting photochromic 1973
 flint 1675
 float 1958
 optical 1798
 photosensitive 1937
 safety 1905
 sheet 1918
 Triplex laminated 1910
 unbreakable 1888
glass-blowing 200 BC
 machine 1895
glass ceramics 1957
glass fibre 1931
glass-making 1500 BC
 float process 1958
 texts 650 BC
 Venetian system 1000
glass vessels, polychrome 1470 BC
glassis (plaster) 1677
glasspaper 1833
Gleep (atomic pile) 1947
glider
 float-mounted 1905
 man-carrying 1852
 sea-going 1855
globe, terrestrial 1492
gluthathione 1921
glycerine 1779
glycogen 1856
glycol 1856
gnomon, astronomical 1276
go-kart 1956
gold coinage 770 BC, 550 BC
gold extraction, cyanide process 1887
gold mining 3500 BC
golden syrup 1885
goniometer, reflecting 1806
goods elevator 1852
goods-wagon, steam-powered 1870
gout treatment 1962
governor, centrifugal 1783
graham cracker 1829
grain binder 1850
grain grinder, rotary 3500 BC
gramophone, all-electric 1925
gramophone record 1887
 double-sided 1904
 microgroove 1948
Grand Prix engine 1914
graphic conversation terminal 1955
graphite-clay pencil 1792
graphite pencil 1565
graphometer 1597
graphophone 1886
grass-skis 1968
gravitational theory 1684
Great Arch of Ctesiphon 550
Great Galley, The (boat) 1515
'Greek fire' 672
Greenwich Mean Time 1884
grenade discharger and musket 1681
Grignard reagents 1901

grill, infra-red 1922
grinding machine
 heavy-production 1900
grist mill factory 1785
ground effect (vehicles) 1957
GR-S (synthetic rubber) 1942
guillotine 1791
guitar
 commercial electric 1931
 precision-base electric 1951
 semi-acoustic electric 1935
 solid electric 1948
guitar pick-up 1931
gun 1260
 anti-tank 1917
 bazooka 1942
 breech-loading 1859
 compressed steel 1869
 leather 1626
 muzzle-loading 1537
 naval 1779
 revolving cylinder 1835
 steam 1824
 steel 1847
gunboat, ironclad 1866
gun-cotton 1846
gunpowder 221 BC
 uninflammable 1865
gutta-percha 1843
gyroscope, ship-steadying 1908
gyroscopic compass 1908

H

hackney cabriolet 1823
hackney coach 1625
Hadley's quadrant 1731
haemin, synthesis 1929
haemophiliac blood-clotting factor 1984
haemoglobin, molecular structure 1968
hafnium 1923
hair brush, electric 1883
hairdressing, permanent waving 1906
hairpin, Kirby grip 1893
halftone process 1871
Halley's comet 1682, 1985
hall-marking 1300
hammerklavier 1709
hand-drying apparatus, hot-air 1928
handkerchiefs, disposable paper 1924
hanging, drawing and quartering 1241
Hanging Gardens of Babylon 570 BC
hansom cab 1834
 'Parlour' 1887
hardboard, prototype 1924
Harecastle tunnel 1777
harmonic analyser 1878
harmonic balancers 1911
harmonic strings 540 BC
harmonica, glass 1766
harp 3875 BC
 chromatic 1894
 pedal 1720
'Harpic' (cleaning agent) 1924
harpoon 12000 BC
harpoon gun 1863
harpoon log 1861
harpsichord, optical 1720
harquebus 1525
harrow, iron 1839

harvester 1858
hat
  Bowler 1849
  high beaver 1797
  mechanically-tipping 1896
hatchback estate car 1954, 1961
hearing aid
  electric 1901
  electronic 1935
  transistorised 1952
  valve system 1923
heart-lung machine 1953
heart pacemaker internal 1957
heart valve, artificial 1952
heat-bottled food 1809
heat theorem 1906
heater, electric 1887
heavy water 1940
heckelphone 1904
hectographic duplicating 1923
helicopter
  contra-rotating rotors 1936
  gas-turbine-engined 1951
  jet-prop 1953
  jumbo 1968
  main and small tail rotors 1939
  petrol-engined 1907
  production model 1940
  steam-engined 1870
helicopter model 1784
helicopter top 320
'heliographie' 1816
heliotrope (surveying instrument) 1821
heliotropin 1879
helium 1894
  in sun's chromosphere 1868
  liquid 1908
herbicide
  2, 4-D 1941
  MCPA 1941
hemp-spinning machine 1810
hexameter 190 BC
hexameter machine 1846
hieroglyphic texts 2600 BC, 2400 BC
High-Speed Train (HST) 1975
'Hipparchus' (space probe) 1986
histidine 1896
'History of the World' 1311
hobby horse 1790, 1817
hodometer 1528
hodometer-compass 1596
hoe blade, stone 5000 BC
hoist, hydraulic 1912
'Holland Circle' (circumferentor) 1612
holmium 1879
holocamera 1984
holocopier 1984
holograms 1947
  uses of 1968, 1980, 1984
holographic storage material 1968
homeopathy 1810
hormones 1902
  human growth (HGH) 1956
  sex 1927
  see also insulin; thyroxine
horn
  bulb 1871
  electric motor 1905
horse, artificial 1673
horse-bit 1500 BC, 1200 BC
horse-box 1836
horse-shoes 890
hose-pipe, fire-fighting 1672

hosiery factory, children's 1838
hosiery-knitting machine 1589
hospital, public 372
'hot bulb' engine 1890
'hounds and jackals' (game) 1700 BC
hourglass set 1475
hovercraft 1955
  flexible skirt 1961
  full-sized 1959
  passenger 1962
  production model 1964
  tracked 1962
'Hubble' (telescope) 1986
Hubot (robot) 1984
'hula-hoop' (game) 1958
human anatomy book 1543
human speech machine 1979
humidity control 1906
Huntington's Chorea 1985
hydrapsis 1690
hydraulic crane 1965
hydraulic press 1648, 1795
hydraulis (organ) 250 BC
hydrazine 1887
hydro-aeronaut 1810
hydro-aeroplane 1910
hydrochloric acid 1648
hydrocycle
  aerial propeller 1913
  triplet 1895
hydro-electric power 1867, 1881
hydrofoil
  air-cushion swing-wing 1904
  amphibious gas-turbine-engined 1959
  boat 1861
  commercial passenger 1953
  high-speed 1919
  human-powered 1984
  passenger waterjet-propelled 1974
  rotary foil 1876
  steam-powered 1897
  turbojet-engined 1952
hydrofoil ladder 1905
hydrogen 1766
  heavy isotope 1940
  see also deuterium
hydrogen bomb 1952
hydrogen peroxide 1818
hydrogenation of oils, catalytic 1899
hydrokimeter 1874
hydroplane 1872
  aerially propelled 1907
  multi-step/petrol-engined 1908
hydropneumatic suspension 1955
hydroponics 1929
hydrotherapy, revival 1797
hydrostatic bed 1832
hydrostatics 220 BC
hydroxy citronellal 1905
Hygeiopolis 1876
hygrometer 1664
  ether 1820
  hair 1781
hymnbook, Czech 1501
hyperbola 220 BC
hypodermic syringe 1869
hyposulphite of soda 1842

I

'I Speak Your Weight' machine 1928
'IBM System 360' 1964

ICBM (missile) 1957
ice-breaker ship, nuclear-powered 1959
ice-cream
    cones 1896
    dairy 1851
    sundae 1890
ice-skating rink 1876
ice yacht 1790
iconoscope 1923
Identikit 1959
ilmerium 1877
image digitiser 1985
immersion heater, tubular unit 1922
incompleteness theorem 1931
incubator, thermostatic 1609
indanthrene 1902
indium 1863
induction coil system 1882
induction furnace, high-frequency 1922
induction motor, a.c. electric 1888
Indulgences 1313
in-flight refuelling 1923
infra-red photography, medical 1932
infra-red region 1800
    map 1880
infra-red sensitising agent 1904
injection moulding press, automatic 1934
ink cartridge 1935
inks
    coloured 1772
    invisible 1653
    ordinary 2600 BC
    quick-setting 1953
    ultra-violet printing 1975
inoculation 1880
insulating board 1914
insulin
    animal 1921
    human 1978
integrated circuit
    1024k-bit 1984
    half-micron line width 1985
    silicon 1962
    theoretical 1952
integrated circuit radio 1966
    phase-shift oscillator 1959
'Intelligent Machinery' (article) 1947
interferometer 1881
interferon 1957
    'cloned' 1980
Intertype fotosetter 1936
intertype process 1911
intra-uterine contraceptive (IUD) 1909
intravenous anaesthetic 1902
invar alloy 1895
invisible ink 1653
iodine 1811
iodoso-compounds 1892
ion engine 1959
ion-irradiated food 1896, 1921
ion source 1960
ionium 1906
ionone 1893
ionosphere 1902
    measurement 1925
iproniazid 1953
iridium 1803
Iris (asteroid) 1847
iron
    manufacture, puddling furnace 1784
    smelting 1500 BC
    hot blast 1828
    vertical casting process 1846
    wrought 1851
iron, electric
    domestic 1880, 1882
    domestic steam 1938
    domestic thermostat-control 1936
'iron fire bombs' 1221
iron lung 1927
iron ship 1787
ironclad ship 1862, 1866
ironware manufacturing machinery 1683
irrigation, large-scale 3100 BC
    Nile 3150 BC
isoniazid 1951
isotopes 1911
isotopic labelling technique 1913
Ivel agricultural motor 1902
Ivel armoured car 1904

J

Jacquard loom 1805
Jasperware 1763
jaunting car (Irish) 1815
jeans (trousers) 1850
Jeep (car) 1940
jet engine 1936
Jetfoil 1974
Jewish calendar 385
jigsaw puzzle 1763
Joint European Torus (JET) 1983
Josephson effect device 1966
joule 1843
judo 1882
juke-box 1905
    all-electric 1927
Julian calendar 46 BC
jumbo-jet airliner 1969
Jupiter space probe 1986

K

kala-azar parasite 1900
kaleidoscope 1816
kaolin 1755
kapok lifebelts 1904
Karpol (cleaner) 1927
Kell blood group system 1946
Kelvinator refrigerator 1918
Kennelly-Heaviside layer 1902
'Kenyapithecus Wickeri' (upper jawbone) 1961
kettle, electric
    automatic 1955
    resettable 1931
kettledrum, pedal-operated machine 1855
'Kevlar' (aramid) 1965, 1974
key-starting (automobile) 1949
kidney, artificial 1945
kindergarten 1832
kinedrome projection system 1907
kinemacolour 1906
kinematograph 1890
kinematoscope 1861
Kinescope 1923
kinetographic theatre 1893
kinetoscope 1889
Kirby grip (hair pin) 1893
kitchen, model electric 1893
kitchen range, free-standing 1834
kitchen utensils, enamel 1799

kite 400 BC
  meteorograph-carrying 1894
  meteorological 1823
  reconnaissance 206 BC
kite-carriage 1826
'knives and hoes' coinage 400 BC
knotting-bill 1876
Kodachrome (film) 1935
Kodak camera 1888
  'Brownie' 1900
  vest pocket 1912
Kodavision (video film) 1984
Kroll process 1937
kryptographic pen 1833
krypton 1898
kymograph 1847

L

lace-making machine 1776
lacteals 1622
Lamarckism 1809
lamp
  carbon-filament electric 1879
  Döbereiner 1823
  earthen 100
  fluorescent 1859
  paraffin 1854
  quinquet 1804
  sodium vapour 1933
  stone 11000 BC
Land-Rover 1947
landing equipment (aircraft), automatic 1962
LANDSAT (satellites) 1972
language laboratory
  gramophone-based 1924
  wire-recorder based 1938
language-teaching course 1893
language-teaching system 1908
Languedoc Canal 1692
lanthanum 1839
laryngoscope 1857
laser 1958
  annealing 1978
  anti-ICBM 1985
  argon 1960
    medical use 1978
  argon and krypton 1968
  automatic design system 1966
  carbon dioxide gas 1968
  cell-sorting machine 1975
  cloth-cutting 1980
  coronary angioplasty 1985
  eye surgery 1964
  far-infra-red 1979
  fluorescence induction 1984
  gallium arsenide semiconductor 1962
  gravity-wave detector 1984
  metal removal 1978
  microbeam 1962
  phototypesetting 1980
  precision-surveying 1967
  pulsed ruby 1960
  repair system 1980
  transportable radar 1984
laser-based glass-fibre manufacturing system
  1977
laser-disc electronic filing system 1985
laser engraving (printing plates) 1974
laser-enhanced etching and plating 1981

laser gun, prototype 1964
laser gyro inertial navigation system 1982
laser/hologram art exhibition 1970
laser holography, 3-D 1964
laser-lancet 1967
laser light compressor 1984
laser-optical storage system 1985
laser ranging 1983
  retro-reflector 1969
latent heat 1762
'Latham Loop' (camera device) 1895
lathe
  automatic turret 1865
  chuck 1853
  horizontal turret 1845
  lens-grinding 1664
  ornamental turning 1569
  oval turning 1615
  profile 1818
  Ramsden's 1770
  screw-cutting 1569, 1800
  slide-rest 1798
launderette 1934
lavatory cleaning agent 1924
lawn-mower 1830
  electric-mains-powered 1926
  pedal 1898
  petrol-engined 1902
  rotary-blade electric 1958
  rotary hover 1963
  steam-powered 1893
  tricycle-powered 1881
lawrencium 1961
LCD (liquid-crystal display) 1971
leather-cloth 1848
Leclanché battery 1868
LED (light-emitting diode) 1962
LEGO play system 1955
lens
  apochromatic objective 1878
  astigmatic 1827
  bifocal 1784, 1985
  combination 1758
  composite glass 1733
  contact see contact lenses
  photographic zoom 1932, 1945
  plastic 1936
  Pyrex telescope 1944
  reading 66
  telephoto 1891
  trifocal 1827
  unbreakable 1955
lens-polishing machine 1818
leprosy genes, cloned 1985
lethane 1932
Letraset (dry-transfer process) 1956
letter-box 1653
lettering, instant 1956
leverage 215 BC
levodopa 1967
Lexan 1953
ley lines 1921
Leyden jar 1746
'Liberator' concept (computer programs) 1963
library 540 BC
  private 334 BC
library cataloguing 1876
library of Cordoba 970
Librium 1959
LIDAR (radar) 1984
lie detector 1921
LIF (laser-induced fluorescence) 1984

lifebelt
  cork 1852
  kapok 1904
lifeboat, steam-powered 1889
lifebuoy, navigator's 1810
lifejacket 1769
  cork 1763
  inflatable 1940, 1952
life-preserving dress 1874
ligature 1550
light
  aberration of 1729
  finite velocity 1690
  velocity 1849
  wave theory 1801
light bulb, electric 1886, 1898
light bulb and vacuum tube mass-production
    machine 1920
light-emitting diode (LED) 1962
light filament (incandescent) see Tantalum
'light-knife' (laser-lancet) 1967
lighthouse
  automatic flashing 1906
  Eddystone 1759
  glass reflectors 1780
  magneto-electric 1858
  'Pharos' 285 BC
  talking beacon 1928
lighting
  coal-gas 1792
  electric 1834, 1857, 1880
  fluorescent 1935, 1938
lightning-conducting umbrella 1786
lightning conductors 1752, 1754, 1766
lime scrubbing process 1933
limelight 1826
linear motor 1985
liner, ocean 1881
Linnean system (botany) 1737
linoleum 1860
linotron 1966
linotype machine 1886
lip-reading 1866
liquid-crystal display (LCD) 1971
LISP (computer language) 1958
lithium 1817
lithium disintegration 1932
lithium niobate 1968
lithofracteur 1869
lithography 1796
lithotomy 240 BC, 17, 1475, 1680
lithotripter 1973
lock
  detector 1818
  safety combination 1784
lock-gates 825
locomotive 1804
  compound 1875
  diesel-engined 1912
  Electra 225 1986
  electric remote-controlled 1955
  gas-turbine-engined 1941
  rack and pinion 1812
  Rocket 1829
  regular route 1825
lodestone 83
log line (for navigation) 1570
logarithmic calculator, helical 1878
logarithms 1614
logograph 1874
long-wave M-lines 1916
loom
  cam-operated 1794

electric 1854
Northrop 1889
shuttle change 1678
warp-weighted 3000 BC
lorry
  diesel-engined 1923
  petrol-engined 1894
loudspeaker 1898
  electrodynamic 1925
  portable 1934
louma crane 1967
LSD (drug) 1943
lubricants, synthetic 1970
lucifers 1826
Ludolph's Number 1610
lunar observation orbiter 1959
lunar module 1969
lunar vehicle, remote-controlled 1970
LUNIK (lunar observation orbiter) 1959
lutetium 1906
Lutetia (asteroid) 1852
Lutheran blood group system 1946
lymphatic glands 1650
lyre 673 BC
  enchanted 1821
lysine 1889

M

6-MP (drug) 1962
M & B 693 1938
macadamising 1819
Mach scale and angle 1887
machine-gun
  air-cooled 1911
  flintlock 1718
  gas-operated 1890
  Gatling 1862
  Maxim 1884
  propeller-synchronised 1915
  water-cooled/recoil-operated 1917
macromolecular chemistry 1926
Mae West lifejacket 1940
magic lantern 1640
MAGLEV (propulsion system) 1983
magnesium 1808
magnet
  ticonal G 1937
  traction 1985
magnetic disc storage system 1957
magnetic field 1819
magnetic ink character recognition 1961
magnetic-motor train 1967
magnetic needle dip 1576
magnetic North Pole 1831
magnetic pull 1824
magnetic sound-recording tape (plastic) 1929
magnetic variometer 1798
magneto, low-tension 1897
magneto-electric machine 1832
magnetron 1921
  cavity 1940
magnifying glass 1250
Magnox nuclear reactor 1962
mail, mechanised sorting 1957
mail coach 1784
'Mailgram' 1974
'Makrolan' 1958
malaria-carrying parasite 1897
malted milk 1887

manganese 1774
manganese steel 1839, 1883
Manhattan Project (atomic pile reactor) 1942
Manned Manoeuvring Units (MMU) 1984
map
    clay tablet 3800 BC
    manufacture 1617
    printed
        Bologna 1477
        China 1155
        set 1478
        world 220 BC, 150
map atlas 1595
map-maker, recording 1596
margarine 1868
marine engine drive 1899
marine outboard engine 1906, 1928
marmalade, orange 1797
Mars bar 1932
Martello tower 1804
maser
    atomic clock 1985
    lunar-reflected optical beam 1962
    optical-wavelength 1958
    prototype 1954
Masonite 1924
mass production 1798
    moving assembly line 1913
mass spectrograph, double-focusing 1919
mass spectrometer, tandem accelerator 1981
mass storage system 1975
Masuda wave power barge 1979
masurium 1925
matches
    friction 1826
    phosphoric 1786
    phosphorus 1830
    safety 1855
match-making machine 1842
mathemetical signs 1489, 1557, 1631
mausoleum 350 BC
Mavica video card system 1974
Maxim machine gun 1884
maximum and minimum thermometer 1794
McDonald hamburger 1948
MCPA (herbicide) 1941
measles vaccine 1953
measurement dial 550 BC
meat biscuit 1843
Meccano 1901
medical diagnosis 460 BC
medical stitching wire 1820
Mega-globe 1892
megaphone 1670, 1878
mega-submarine, nuclear-powered 1979
mellotron 1964
memory devices
    chips
        64k-bit 1978
        20-ns dynamic access 1984
        512k-bit 1983
        1,048k-bit 1984
    magnetic bubble binary 1982
memory-metal 1960, 1985
mendelevium 1955
Mercalli Scale (earthquakes) 1902
Mercator Projection 1568
merchant ship
    rigid folding sails 1980
    gas-turbine/electric 1956
    gas-turbine-engined 1951
meridional instrument 1540

'Merlon' 1958
merry-go-round 1620
mesmerism 1766
meson 1935
    D° 1976
mesotron 1935
metal polish, liquid 1905
metal tube, collapsible 1840
metallurgical blowing engine 31
metals, continuous casting 1927
methane 1785
methyl alcohol 1661
methyl group 1849
methyl methacrylate polymer 1934
metric system 1799
metronome 1810
mezzotint 1642
micro-analysis, organic 1916
micro-chip, ultra-fast gallium arsenide 1985
    (see also silicon chip and memory chip)
microcinematography 1903
microcomputer 1973
microfilm 1852
microfilm camera 1939
microfilm-reading machine 1919
microgroove record 1948
micrometer 1638
    engineering 1801
microphone 1827
    condenser-type 1916
    crystal-type 1931
    double-button type 1920
    dynamic 1877
    non-directional 1935
    ribbon-type 1923
micro-processor 1971
microscope
    atomic resolution 1983
    compound 1590
    contact X-ray electron 1984
    dioptric-achromatic 1827
    double convex compound 1611
    electron 1929
    miniature 1933
    pocket-sized 1740
    scanning acoustic 1978
    scanning electron 1965
    scanning tunnelling 1985
    super-scanning electron 1985
    vibration 1857
microscope condenser, dark-field 1853
micro-television 1962
microtones 1895
microwave oven 1945
microwave thermography 1980
mileometer, motor car 1901
milk
    condensed 1849
    evaporated 1849, 1884
    powdered 1855
milk bottles 1879
milking machine 1878
    pulsator model 1895
mill
    edge-runner 170, 400
    flour 1623
    trip-hammer 300–200 BC
    waterless 1444
millboard, impermeable 1898
milling machine 1848, 1862
    universal 1861
mimeograph 1875

mine, static 1842
mine drainage system 1484
mine railway, electric 1882
minehunter ship, glass-fibre 1972
mineral water 1741
miner's safety lamp 1815
minicomputer 1962
    mass-produced 1965
    sixteen-bit 1970
mining pump 1698
mini-skirt 1965
mirror
    rear-view 1896
    silvered glass process 1840
mischmetal 1902
missile
    air-to-air 1943
    air-to-air guided 1955
    air-to-surface 1957
    anti-missile 1959
    ground-to-air 1944
    intercontinental ballistic 1957, 1958, 1960
    surface-to-air 1945
    surface-to-surface 1952
missile launch warning satellite 1960
mnemonics 477 BC
Möbius strip 1858
molecular beam 1911
molecular weights, true 1858
molecules, active 1889
molybdenum 1778
money
    paper 910
    see also coinage
monoclonal antibodies 1975, 1982
monopak-still colour film 1935
monoplane 1907
    monocoque construction 1912
Monopoly (game) 1933
monorail 1888
    passenger 1872
monorail trolley system 1869
monotype 1888
moon map 1647
moon rock, crystal 1971
Moray car 1885
morphine 1805
Morse code 1838
mortar, giant 1914
mortar and pestle 10000 BC
mortice lock 1778
mosandrum 1879
Mössbauer effect 1957
motion pictures
    camera device 1895
    camera electric drive 1911
    feature-length sound 1927
    film show 1895
    Kinemacolour 1906
    louma crane 1967
    primitive machine 1832
    sound-on-film 1906
    stereophonic sound 1932
    technicolor 1926
    three-dimensional 1922
    tri-ergon sound 1922
    wide-screen 1925
motor ambulance 1895
motor barge 1894
motor-boat
    diesel-engined 1903
    gas-engined 1865

gas-turbine-engined 1947
Kevlar construction 1974
paraffin vapour 1886
rocket-powered 1927
transatlantic 1902
motor bus, petrol-engined 1895
motor cab taximeter 1897
motor car
    aero-engined 1910
    air-cooled 1902
    airscrew-propelled 1920
    all-metal body 1902
    all-steel chassis 1925
    automated manufacture 1952
    direct-drive 1898
    driverless six-wheeled 1985
    electric 1874, 1891
    forced lubrication 1904
    Ford Model T 1913
    fuel-injection system 1967
    fully aerodynamic 1981
    gas-driven 1807
    gas-turbine 1950
    high-speed electric 1899
    integral construction 1899
    key starting 1949
    linear-motor 1985
    magnetic-lift linear-motor 1978
    Morris Minor Mini 1959
    petrol-engined 1885
    power-steered 1951
    private diesel-engined 1936
    production model 1896
    rocket-propelled 1927
    saloon 1903
    solar-powered 1982
    streamlined production 1934
    supercharged 1907
    turbocharged 1973
    turbojet-engined 1960
    unitary fibreglass 1958
    'voice advice' 1982
motor car bodywork
    cleaner 1926
    enclosed 1898
    glass-fibre 1953
    polypropylene 1985
motor car brakes
    anti-skid 1978
    disc-type 1902
    hydraulic 1920
    servo-assisted 1919
motor car bumpers
    integrated 1978
    pneumatic 1906
    polypropylene 1972
motor car engine
    four-cylinder 1896
    lean-burn 1985
    90% plastic 1984
    straight-eight 1902, 1919
    straight-six 1903
    three-rotor rotary 1985
    V-8 1903
    Wankel 1963
motor car gear, direct top 1896
motor car headlamp wipers 1973
motor car mileometer 1901
motor car power steering 1926
motor car radio 1901
    commercial manufacture 1929

five-valve 1932
FM 1952
motor car reversing light 1921
motor car safety belt 1959
motor car self-starter 1911
motor car speedometer 1901
motor car tyres
  cord-type 1900
  Denloc 1982
  logging sensor 1985
  pneumatic 1895
  radial-bodied 1937
  radial-ply 1953
  snow-chains 1904
  tubeless 1947
motor car wheels, detachable wire 1905
motor car ventilation, air-flow system 1964
motor car windows,
  electric lift 1946
  rear-window heating 1948
motor car windscreen
  curved 1914
  toughened float-glass 1959
  wipers 1916, 1917, 1920
  zone-toughened 1961
motor caravan 1902
motor carriage, coal gas 1863
motor crash helmet 1902
motor-cruiser, glass-fibre 1951
motor-cycle
  anti-dive suspension 1979
  carbon-fibre chassis 1983
  centre-frame engine 1901
  desmodromic valves 1956
  electric self-starter 1913
  four-cylinder 1895
  in-line, air-cooled, four-cylinder engine 1903
  parallel vertical twin-engine 1904
  petrol-engined 1885
  pneumatic-tyred 1894
  quantity-produced 1895
  rotary-engined 1892, 1974
  telescopic fork 1904
  turbocharged 1980
  V-Eight 1908
  V-twin two-stroke engine 1902
  Wankel-engined 1974
motor horn, electric 1905, 1906
  tune-playing 1929
motor-racing circuit, banked 1907
motor-scooter 1915
  snow 1959
motor speech centre (brain) 1861
motor vehicles, power steering 1926
Motorola (radio) 1929
motorway 1924
Moulton bicycle 1960
mouse-trap, spring-operated 1910
mouth organ 1821
mower, circular-bladed 1805
mowing machine 1810
  see also lawn mower
mule (spinning machine), high-speed self-
  acting 1824
'multics' (computer system) 1965
multiplane, Venetian-blind 1907
multiplication sign 1631
mummy 2400 BC
muon neutrino 1962
mural arc 1688
mural circle 1789

music
  computerised algorithmic 1961
  printed 1472
music synthesiser
  electronic 1954
  voltage control 1964
musical instrument, super-computerised 1981
musical notation 1300 BC, 1026
  rhythmic 1200
musk xyol and ambrette 1891
musket, breech-loading 1786
mustard gas 1917
mustard mill 1845
mustard powder 1850
myographion 1850

N

nail violin 1770
napalm (incendiary) 1942
naphthalene 1819
narrow boat
  motor 1912
  steam-powered 1864
natural gas, North Sea 1965
navigation system, laser gyro inertial 1982
needle gun 1837
neodymium 1885
neon 1898
neon lighting 1910
neoprene 1932
nephoscope 1868
Neptune (planet) 1846
neptunium 1940
'Neracar' 1921
nerve
  depressor 1866
  optic 1538
nerve gases 1944
nervous reflex 1833
net-weaving machine 1830
neutrino 1931
Neutrodyne receiver 1919
neutron 1932
neutron bomb 1977
'New Science, The' 1537
newspaper 1609
  daily 1650
newspaper printing plate 1983
niacin 1937
nichrome 1906
nickel 1751
  process for pure 1890
nickel-cadmium storage cells 1908
nickel-ion alkaline storage battery 1908
nickel-plating 1869
Nike-Zeus (missile) 1959
niobium 1801
Nipkow scanning disc 1884
'nipple'-gun, grease-filled 1919
Nissen hut 1916
Nitonol memory metal 1960
nitric acid 1287
nitro-cellulose 1833
nitrogen 1772
nitrogen-from-air process 1903
nitroglycerine 1847, 1864
nitrous oxide anaesthetic 1844
Nivea cream 1911
nobelium 1958
nocturne (musical piece) 1834

'nodding boom' (wave-power device) 1974
noise reduction systems 1966, 1970, 1980
non-magnetic engine 1933
non-stick pan 1954
norwegium 1879
novocaine 1904
'N-rays' 1903
NRDC 143 and 161 (pesticides) 1974
nuclear bomb 1961
nuclear fission 1938
nuclear magnetic resonance (NMR) scanner 1981
nuclear power station *see* atomic power station
nuclear-powered ship
    cargo and passenger 1961
    icebreaker 1959
    mega-submarine 1979
nuts and bolts, standardised 1835, 1900
nylon 1938

O

Oakey glasspaper 1833
observatory 505 BC
    airborne 1982
    astronomical 640
    Uraniborg 1576
ocean liner, steam-turbine-engined 1907
oceanographic research ship 1950
octant 1731
'odic force' 1852
odometer, carriage 1851
odontology 1839
oil
    catalytic cracking 1928
    de-rusting 1617
    production vessel 1986
    transatlantic cargo 1861
    UK commercial quantities 1939
    underwater drilling for 1911
oil pipeline, American 1865
oil platform, tension leg 1983
oil refining process 1888
oil rig, all-steel 1890
oil spill solidification process 1985
oilfield
    Hibernia, Newfoundland 1980
    North Sea Forties 1970
    Mexican offshore 1937
    Persian 1908
    Persian Gulf 1951
    Russian 1873
    Sumatran 1884
    USA 1859
Ominimeter 1869
omnibus 1823
    Adams patent 1846
    double-decker 1847
        enclosed 1909
    eight-seater 1662
    electric 1897
    oil-engined 1899
    petrol-engined 1895
    steam-powered 1831
omnigripper 1985
ondes musicales (musical instrument) 1928
'Oorgat' (bridge slot) 1596
open hearth furnace 1864
opera hat, collapsible silk 1832
ophicleide (musical instrument) 1791
ophthalmoscope 1851

optic nerve 1538
optical fibre telephone cable 1977
orbital engine 1971
Ordnance Survey (of UK) 1819
organ
    radio-synthetic 1934
    steam 1859
    tone-wheel electric 1934
organic micro-analysis 1916
organic waste earthworm exploitation process
    1982
'orgone' accumulator 1940
Origin of Species, The (Darwin) 1859
Ormulum 1200
oscillator
    phase-shift 1959
    Wein Bridge (stabilised version) 1938
oscillograph 1897
osmium 1804
osmium glühlampe (light bulb) 1898
otheoscope 1877
Otto cycle 1877
Otto dicycle 1881
outboard engine
    four-cylinder two-stroke 1928
    petrol 1892
oven
    commercial electric 1891
    experimental electric 1889
    microwave 1945
    thermostatic regulator 1915
oxidation enzymes 1934
oximes 1882
oxygen 1777
    liquid 1877
oxygine 1777
oxyhydrogen blowpipe 1816
oyster-picking machine, automatic 1896
ozone 1840

P

pacemaker, internal 1957
paddle boat 1783
    electric 1838
paddle-wheel, cycloidal 1833
padlock 1381
page-turner, automatic 1887
paints
    acrylic 1964
    alkyd 1977
    cellulose 1924
    polyurethane 1939
palladium 1804
Panama Canal 1914
panharmonium (musical instrument) 1810
panomonico (musical instrument) 1810
Panopticon of Science and Art (building) 1853
panorama, Edinburgh 1788
pantaleon (drum) 1697
pantelegraph 1843
pantheon (temple) 124
pantograph 1603, 1869
pantothenic acid 1933
paper
    mulberry-based 105
    rot-proof 1878
    vegetable fibre 1800
    woven 1750
paper clip 1900

paper cup 1908
paper-cutting machine 1844
paper handkerchief, disposable 1924
paper-making machine 1799
  cylindrical 1809
  Fourdrinier 1803
paper-mill 1150
paper money 1910
paper-tape recorder 1575
paperback book 1935
papier-mâché 1759
papyrus
  gynaecological/veterinarian 1900–1800 BC
  map 1320 BC
  mathematical 1650–1600 BC
  unwritten 1950 BC
  written 2400 BC
parabellum (pistol) 1898
parabolic reflector 1752
parabolograph 1904
parachute 1797
  manually operated 1919
paraffin 1830
paralysis, general treatment for 1917
paranormal studies, isolation cage for 1871
parchment paper 1846
Parkesine (celluloid) 1856
parking meter 1932
  credit card operated 1984
Parkinson's disease 1817
parthenogenesis, artificial 1899
particle accelerator
  electrostatic 1933
  high-energy 1959
particleboard 1941
PASCAL (computer language) 1969
pasigraphy 1870
Pasini cipher 1411
pasta, mechanically produced 1800
pasta factory 1827
pasteurisation 1864
patent law 1474
Pauli exclusion principle 1924
pavement, rolling 1900
pay-phone, solar-powered 1985
PDP-1 (processor) 1959
pedal-harp, double-action 1825
pedometer 1799
  miniature pocket 1831
pen
  ballpoint 1938
  drawing 1954
  fibre-tip 1960
  fountain 1884
  geometrical 1752
  iron 1685
  kryptographic 1833
  plastic-tipped 1980
  twin-quill 1660
pencil 1792
  eraser-attached 1858
pendulum
  ballistic 1742
  compensated mercury 1726
  Foucault's 1850
pendulum clock 1657
pendulum motion 1581
penicillin 1928
  semi-synthetic 1957
  stabilised 1940
penny post, local 1663
pentode 1927

Pepper's ghost 1863
pepsin 1930
perambulator 1733
  motorised 1922
percussion (medical technique) 1761
percussion cap 1807
  copper 1816
perfume
  Chanel No. 5 1924
  synthetic 1891, 1893, 1926
Periodic Law 1869
periscope, submarine 1896
permanent waving 1906
pernicious anaemia, vitamin treatment for 1947
perpetual log 1802
perpetual motion 1598
  drum 1717
  machine 1245
Persil (detergent) 1907
perspective views (pinhole demonstration) 1437
Perspex 1934
Perspex lens 1936
petrol pump 1885
  clock dial 1925
petroleum see oil
phaeton (carriage) 1788
  crane-neck 1790
  equirotal 1838
phagocytes 1892
Phaistos disk 1700 BC
phase rule 1876
phase shift distortion 1922
Phenakistoscope (motion-picture machine)
  1832
phenolphthalein dye 1871
phenophthalmoscope 1870
phenylethyl alcohol 1876
phenylhydrazine 1875
philippium 1878
phloxin 1876
phonautograph 1857
phonofilm 1923
phonograph 1878
  commercial model 1888
  piano-attachment 1863
phonograph doll 1887
phonometer 1878
phonoscope 1862
phosphorus 1669
phosphorus matches 1830
photo-electric cell 1893
  practical 1900
photo-flash bulb 1930
photographic hat 1886
photographic pistol 1882
photography
  colour 1869
  daguerreotype 1837
  dye-coupler colour process 1914
  electrically lit 1877
  enlarger 1843
  fixative 1842
  half-tone process 1871
  high-speed flash 1851
  lens 1840
  instant printing 1903
  magnesium flash 1864
  medical 1863
  multiple-flash 1933
  portrait machine 1895
  prints 1838
  roll film 1889

serial 1877
slides 1850
underwater 1861
see also camera; cine camera; motion pictures
photoheliograph 1857
photophone 1880
photo-typesetting machine 1946
laser-driven 1980
phrenology 1770
pianoforte
double-escapement 1823
iron-frame 1800
with sustaining and damper pedals 1783
with sostenuto pedal 1874
pianola 1897
PIAT (grenade projector) 1941
picric acid 1771
pictographic accounts 3500 BC
pictographic inscription 3100 BC
pig-boiling, iron 1830
'pig, intelligent' (pipeline inspection) 1979
pile
atomic 1947
continuous current 1800
pile-driving machine 1738
steam, 1845
Pimms Number One (drink) 1840
pinacyanol 1904
'ping pong' 1889
pinhole demonstration (perspective) 1437
pin-manufacturing machine 1817
pion 1947
pipeline, cast-iron 1668
pipeline inspection 1979
pipes, cast-iron 1455
pistol 1540
automatic 1898
Derringer 1843
Luger *see* parabellum
pistolgraph 1858
pitchometer, propeller 1896
pitot tube 1732
planetarium 214 BC
geocentric type 1923
planimeter 1814, 1854
hatchet 1886
planing-machine, steam-powered 1842
plankton 1889
plant cells 1838
plant growth regulators 1978
plasticine 1897
plastics
bakelite 1908
Beetle 1924
casein 1897
nylon 1938
perspex 1934
polyamides 1938
polycarbonates 1958
polyester resin 1948
polypropylene 1954, 1964
polystyrene 1930
polytetrafluorethylene 1938
polythene 1933
polyvinyl chloride 1928
thermosetting 1924
platinum 1735
plastids-from-plastids 1883
playing cards 1440
plexiglass 1934
Plimsoll mark and line 1876
plough
all-iron 1808

copper parts 4000 BC
mould board 900
ox-pulled 2500 BC
replaceable parts 1814
Rotherham 1730
steam 1858
steel 1837
stump-jump 1876
three-furrow 1803
two-horse swing 1780
ploughing machine 1618
Pluto (planet) 1930
satellite 1978
plutonium 1940
plywood, applications for 1868
plywood factory 1885
pneumatic bumpers 1905
pneumatic despatch system 1859
pneumatic drill 1861
pneumatic underground railway 1870
Pneumatica c100
pneumatic tyres – *see* tyres
poetry, written 730 BC
poetry-producing machine 1846
poison gas
chlorine 1915
mustard 1917
nerve gases 1944
polar clock 1849
'Polaroid' (polarising material) 1936
Polaroid Land camera 1947
Polaris (submarine) 1959
poliomyelitis 1908
vaccine 1957
pollen 1694
polonium 1898
polyamides 1938
polycarbonates 1958
polyester resin 1948
polyethylene 1933
polygraph 1902
polymers
electrically conducting 1985
theory of 1922
see also plastics
polypropylene
fibrillated 1964
in sailing dinghy 1973
isotactic 1954
polystyrene 1930
polytetrafluorethylene 1938
polythene 1933
polyurethane paint 1939
polyvinyl chloride 1928
porcelain
hard, French system 1765
original 700
true 1716
'Portakabin' *see* Re-locatable building module
Portland cement 1824
positron 1932
post-coding 1957
post office, automatic 1924
postage-franking meter 1903
postage, prepaid ('Penny Post') 1840
postage stamp
adhesive 1834
perforated 1854
postcard 1861
poster, printed 1477
potassium 1807
potato, Burbank 1873

potato-digger, mechanical 1886
potentiometer 1904
potter's wheel 6500 BC
pottery
    coiled 7000 BC
    glazed 1500 BC
    protogeometric style 1000 BC
    repairs to 5000 BC
power looms 1787, 1789
power stations
    atomic 1956, 1962, 1976
    solar-powered 1977, 1982
    tidal 1967
powerboat
    aero-engined 1911
    catamaran 1985
    turbojet-propelled 1947, 1952
Power's accounting machine 1910
Practica Geometrica 1220
praseodymium 1885
praxinoscope theatre 1878
prayer beads 366
prefabricated construction
    for building 1578
    glass and iron 1840
Presdwood (hardboard) 1924
pre-selector gear box 1901
press-stud fastener 1860
pre-stressed concrete 1880
    strand for 1965
pressure cooker 1680
    aluminium 1905
    automatic 1978
    self-sealing 1938
'Prestel' viewdata system 1979
print roller 1816
printing
    background effects 1879
    clay characters 1049
    full-colour process 1719
    intertype process 1911
printing ink, ultra-violet 1975
printing press 1451
    cast-iron 1772
    curved printing plates 1851
    improved 1601
    rotary 1845
    rotary steam-powered 1812
    rounded cheek frames 1806
    solar-powered 1878
    web-offset 1865
printing system
    intertype 1911
    laser engraving of cylinders 1983
    multi-function bubblejet 1981
    non-impact 1974
    rotogravure 1895
processor, programmed data 1959
production line, computer-controlled 1960
progesterone 1934
programming languages *see* computer
    languages
proline 1901
PROLOG (computer language) 1971
promethium 1947
Prontosil dye, synthesised 1934
Prony brake 1750
proof-reader, computerised 1984
propeller
    adjustable-pitch 1844
    aerial screw 1859
    Arneson 1978

    folding 1848
    marine 1837
    nylon 1953
    tip-vortex-free 1976
    wave power 1980
propeller-rudder, marine 1926
prospector, portable 1855
protactinium 1913
protein, single-cell process 1979
proton-antiproton collider 1985
proton bombardment 1932
protoplasm
    full investigation 1846
    name coined 1839
pruning shears 1815
Prussian blue pigment 1704
'Pruteen' (feedstuff) 1979
psi (atomic particle) 1974
psychic pressometer 1854
psychrometer 1816
PTFE 1938
public address system, open-air 1916
puddling furnace 1784
pulley 400 BC
pulley-block-making machinery 1801
Pullman cars 1859
pulsars 1968
pulse ranging 1925
pulsed ruby laser 1960
pump
    annular-sail wind 1854
    fire 1699
    mining 1698
    solar-powered 1880
    tidal 1702
    water ejector 1852
pumping station, water 1582
    steam-powered 1801
punched cards
    data handling 1924
    processing 1889
pupin coil 1899
PVC 1928
pyramid
    smooth-sided 2575 BC
    stepped 2650 BC
pyridoxine 1936
pyrometer 1731, 1780
pyrophone 1873
Pyrosil (glass ceramic) 1957
pyrphoros (flamethrower) 190 BC

Q

quadrant,
    giant 995
    sea 1748
quadricycle 1851
quantum theory 1900, 1925
    subatomic 1913
quark 1963
quartz crystal clock 1929
quartz digital watch 1971
quasars 1963
    triple 1980
quaternions 1843
Queen's Ware (pottery) 1763
quern 8050 BC
    rotary 6000 BC

quinine 1820
quinquat lamp 1804
quinquereme (galley) 250 BC
'quiteron' (transistor) 1983

R

racing car, with aerofoils 1966
radar 1940
  microwave 1941
radar valve, centimetre 1940
radiation belts *see* Van Allen belts
radio
  all-transistor FM 1958
  integrated circuit 1966
  miniature 1985
  neutrodyne 1919
  portable two-way FM 1941
  superheterodyne 1917
  transistor 1954
radio advertisement 1922
radio astronomy 1931
radio beam navigation aid 1934
radio broadcasts
  public 1906
  regular service 1920
  short wave 1923
  stereophonic 1961
  UK station 1922
radio conducteur 1890
radio detection and ranging (radar) 1940
radio direction-finder 1910
radio pager, shirt-pocket 1971
radio-pulse detection system 1930
radio sources, quasi-stellar 1963
radio system, microwave point-to-point 1959
radio telescope 1937
  large 1955
radio tuning device (selective) 1916
radio valve
  five-electrode 1927
  miniature, two-volt multi-electrode type
    1935
radio waves, phase shift distortion 1922
radioactivity 1902
  measurement 1928
radiocarbon dating 1947
radiographic enlarger 1930
radio-immunoassay 1950
radiolocation 1935
radiology, bismuth effect 1897
radiometer 1876
radio-tracking, animal 1959
radiotelephone service
  cellular network 1985
  commercial 1946
  long-distance 1915
  on steamship 1929
  VHF 1932
radium 1902
radon 1900
railcar
  diesel-electric 1913
  double-decker 1867
  pneumatic-tyred 1931
railroad system, elevated and surface 1890
railway
  Advanced Passenger Train 1972
  amphibious passenger 1895
  atmospheric 1840

automatic air brake 1886
automatic electric block 1867
circular track 1808
diesel-electric 1934
electric 1881
electric mine 1882
high speed 1975
narrow-gauge electric 1883
route-setting and signalling system 1983
track-circuit-controlled 1902
underground electric 1890
underground pneumatic 1870
underground steam 1863
railway car, sailing 1878
railway carriage
  air-conditioned 1930
  corridor 1853
  gas-lit 1858
  Pullman 1859
railway gauge 1814
railway locomotive *see* locomotive
railway points, safety 1856
railway signals, safety 1856
rainbow theory 1611
rainfall, artificially-induced 1946, 1984
raised letters (for the blind) 1786
rake, mechanical 1852
305 RAMAC (control method) 1957
Raman effect 1928
Ramsden's lathe 1770
random access method 1957
'rapid bursters' (astronomy) 1976
'raudive voices' (from the dead) 1967
rayon
  acetate 1869
  cuprammonium 1890
razor
  electric 1931
  safety 1901
RDX (explosive) 1899
reactors, nuclear
  advanced gas-cooled (AGR) 1976
  experimental breeder No.1 1951
  fast breeder experimental (FBR) 1962
  heavy-water-moderated boiling water (BWR)
    1959
  Magnox 1962
  plutonium, power 1956
  pressurised water-cooled (PWR) 1957
  steam-generating heavy water (SGHWR) 1968
  uranium graphite-moderated gas-cooled
    1954
reading machine 1588
Réamur scale 1730
reaping machine 1828, 1831
rear-window heating 1948
record
  gramophone, 1887, 1904
  microgroove 1948
  vinylite disc 1946
  see also compact disc
record-player in automobile 1956
recording process, electrical 1920
recording tape (sound) *see* magnetic sound-
    recording tape
red blood corpuscles 1658
refinery, computer-controlled 1959
reflector, parabolic 1752
reflex, nervous 1833
refrigeration ship 1880
refrigerator 1850
  absorption system 1922

domestic 1918
freight wagon 1875
steam-pumped 1879
relascope 1948
relational data base 1970
relativity
  special theory of 1905
  general theory of 1916
re-locatable building module
  (loda-strut-leg system) 1960
remote manipulator system (RMS) 1983
reserpine 1953
resistance welding 1877
respirator
  charcoal-based 1854
  fireman's 1870
  war gas-mask 1915
revolver
  Colt 1835
  Smith and Wesson 1855
revolving door 1888
rhenium 1925
rhesus blood group system 1940
rhesus haemolytic disease treatment 1964
rhodamine 1887
rhodium 1805
riboflavin 1933
ribonucleic acid (RNA) 1955
Ricardo Comet catalytic unit 1979
rice cultivation 6000 BC
Richter scale (earthquakes) 1935
rickshaw 1869
rifle
  0.22 1847
  bolt action 1837
  breech-loading 1776
  chassepot 1866
  magazine 1875
  periscope 1915
  Winchester repeating 1860
rifle barrel 1520
rifle magazine box 1879
ring armature generator 1861
ring-spinning frame 1828
rising sector gate 1982
RNA 1955
road-making system 1764
road network 2100 BC, 300–200 BC
road-roller, steam 1859, 1865
road-stud, reflecting 1934
road transport, mechanical, for army use 1861
road vehicle
  electromagnetic 1866
  see also carriage; motor car
Roberval balance 1670
'Robin starch' 1899
Robinson's patent barley water 1823
robot 1928
  arc-welding 1984
  'Elmer and Elsie' 1950
  hand device 1985
  integrated 1984
  manufacture by robots 1983
  moulding and plating machines 1982
  name derivation 1921
  primitive industrial 1962
  shipbuilding arc-welding 1986
rock-trenching machine (RTM III) 1984
Rocket (locomotive) 1829
rockets
  and firelances 1100
  Congreve 1805

experimental 1688
high-altitude liquid-fuel 1949
liquid-fuelled 1926
spinning 1844
supersonic long-range war 1942
war 1780
rocket boat, liquid-fuelled 1980
rocket car, liquid-fuelled 1967
rocket plane 1847
rocking chair 1853
roll film, celluloid 1889
roll-towel cabinet 1917
roller bearings 1787
rollercoaster 1884
roller-conveyor, powered 1834
roller-skates 1760, 1823, 1863
rollers, use as wheel replacements 1691
rolling machinery, metal-plate 1734
Röntgen radiation 1906
roof tiles 640 BC
rope-spinning machine 1784, 1792
rope, reed 4000 BC
ropeway, monocable 1644
rosary 1202
Rosetta Stone 196 BC
rotary aero engine 1907
rotary hoe, steam-powered 1920
rotary hook (sewing device) 1852
rotary press, punched cards 1924
rotary stencil machine, automatic 1896
Rotherham plough 1730
rotogravure 1895
rotorship, Flettner 1922
rotovator 1920
rubella vaccine 1965
rubber
  foam 1929
  sponge 1846
  synthetic 1928
    Buna 1926
    GR-5 1942
    neoprene 1932
  vulcanised 1841
  waterproof 1823
rubber bands 1845
rubber engine mounting 1904
rubber solvent 1762
rubber stamps 1862
rubidium 1861
Rubik's cube 1979
ruby, artificial 1860
rudder
  Kort nozzle 1932
  ship's 700
rust preventative 1952
ruthenium 1844

S

saccharimeter, using polarised light 1847
saccharin 1879
saccharometer, brewing 1784
safety belt, motor car 1959
safety curtain, theatre 1794
safety film 1908
safety glass 1905
safety lamp, miner's 1815
safety pin 1849
safety razor, double-edged 1901

sail
  and rigging 100 BC
  Genoa jib 1927
  spinnaker 1865
  windmill-spring 1772
sail material
  nylon 1945
  rayon 1937
  terylene 1952
sailboard
  commercial 1970
  fibreglass 1964
  prototype wooden 1958
sailing dinghy, polypropylene 1973
salad cream 1927
saloon, swinging 1874
saloon car 1903
Saltaire (model village) 1853
Salvarsan 1909
samarium 1879
sandwich 1762
sandwich toaster 1922
sarrusphone 1856
satellite
  balloon communications 1960
  communications 1958
  Earth Resources Technology (LANDSAT) 1972
  gamma-ray hunting 1973
  geosynchronous 1963
  global communications 1965
  infra-red astronomical 1983
  international ultra-violet explorer 1978
  missile launch warning 1960
  navigation 1963, 1964
  observation/spy 1961
  solar maximum mission 1980
  Sputnik 1957
  tracking and data-relay (TDRS) 1983
  weather 1960
satellite docking 1967
satellite navigator, miniature 1979
satellite tracking of wildlife 1972
satellite transmission, transatlantic 1962
Saturn, observation 1610
saucepan, aluminium 1890
sawing engine 1683
saxophone 1844
scandium 1879
scanner, NMR 1981
scanning disc, Nipkow 1884
'Schaphandre' (diving apparatus) 1855
'Schizophone' 1890
Schottel-Rudder propeller 1950
scissors, mass-produced 1761
scooter 1897
Scotch Tape 1929
Scrabble (game) 1948
screw-cutting lathe 1569
screwdriver 1550
screw-manufacturing machine 1760
screw top (for bottles), external 1852
scribe's outfit 2600 BC
script
  Carolingian miniscule 800
  Demotic 700 BC
sculpture, stone 22000 BC
sea quadrant 1748
seal
  cylinder 3500 BC
  engraved 2400 BC
sealing wax 1556
seamless bust support 1903

seaplane 1910
  flying-boat-launched 1938
  supersonic delta-wing 1954
searchlight, naval 1868
Sears Tower 1973
seasickness preventative 1874
secateurs 1815
secretin 1902
sedan chair 1634
sedimentation equilibrium 1908
seed drill machine 1701
seed-drill plough 85 BC
seismograph
  electromagnetic 1905
  strain 1935
seismometer 1858
selenium 1818
self-raising flour 1871
self-service grocery store 1912
self-starter, automobile 1911
Sellotape 1929
semaphore 1801
  indicator arms 1905
  telegraph 1816
senet (game) 2450 BC
'Sensurround' (cinema) 1974
SEPADS device 1985
Seppings blocks 1800
septic tank 1896
servo-assisted brakes 1919
sewage treatment
  activated sludge 1912
  septic tank 1896
sewage sludge, polymer thickening process
    1980
sewers 600 BC
sewing machine 1830
  domestic 1857
  electronic 1975
  four-motion feed 1854
  lock-stitch 1833
sewing needles, bone 17000 BC
sex hormone 1927
sextant 1757
  equatorial 1856
  giant 990, 1425
SGHWR (nuclear reactor) 1968
shaduf (well-sweep) 1400 BC
sheep-shearing machine 1868
Sheffield plate 1743
shell-moulding 1941
shellac mouldings 1868
ship
  all-welded 1919
  armoured 1591
  caloric-engined 1852
  circular 1870
  cupola 1855
  diesel-engined 1904, 1911
  early 8000 BC
  gun-carrying 1339
  ironclad 1787, 1862
  jet-propelled 1839
  merchant 1859, 1862, 1874, 1980
  ocean liner 1881, 1907
  passenger excursion steam-turbined 1902
  rectangular sail 3200 BC
  refrigeration 1880
  research 1950
  steamroller 1896
  steam-turbined 1903
  steel-hulled 1862

tanker 1910
teak-built 1813
transatlantic 1819, 1903
see also boat; steamship
ship salvage engine 1635
ship's log, Chernikeef 1917
shock absorber, self-regulating hydraulic 1930
shoe polish 1906
shoe-sole-sewing machine 1858
shopping cart 1937
shorthand system, phonetic 1837
shorthand-writing machine 1827
shrapnel shell 1784, 1849
SHRDLU (computer program) 1970
'Shredded Wheat' (cereal) 1893
sialon (ceramic) 1970
side-saddle 1388
siege engine 865 BC
signalling, electro-pneumatic 1899
signature verification, electronic 1968
silencer, firearm 1908
silicon 1823
silicon carbide 1891
silicon chip
    288k 1983
    32-bit CPU 1981
    see also memory chip; microprocessor
silicon nitride 1857
silicones 1905
silk
    artificial 1889
    spun 1671
silk draw loom 1567
silk-flyer 1090
silk-knitting machine 1599
silk loom, mechanical 1805
silk machinery 1310
silk production 2640 BC
silk-reeling machinery 100 BC
silk-spinning machine, long-haired 1836
silo 1873
Silurian system (geological) 1839
silver coinage 550 BC
silver polish, liquid 1912
silvered glass process 1840
'Silvo' (liquid polish) 1912
Simms-Bosch magneto 1837
Sinclair C5 (tri-cyclecar) 1984
single-cell protein process 1979
siren 1819
sistrum (rattle) 2400 BC
skateboard 1966
    urethane wheels 1973
ski 2000 BC
'Ski-Doo' (motor scooter) 1959
ski slope, artificial 1946, 1960
skyscraper 1882
    reinforced concrete 1903
    riveted skeleton 1889
    Sears Tower 1973
skywriting 1922
sleeping sickness carrier 1895
slide rule 1621, 1630
slides, photographic 1850
SLURP (pollution removal) 1972
smallpox vaccination 1796
smelting oven, regenerative 1863
smokeless powder 1864
smoke-trails 1922
snooker 1875
snow chains 1904
snow skis 2000 BC
soap, transparent 1798

soda-water-making apparatus 1820
sodium 1807
sodium carbonate production process 1863
solar battery 1954
solar-collector 1875
solar distillation system 1872
solar energy mirrors 213 BC
solar engine
    electric 1893
    flat plate/two-fluid system 1902
solar furnace 1747
solar house-heating 1935
solar maximum mission satellite 1980
solar motor 1615
solar polar mission, international 1986
solar-power plant 1977, 1982
solar-powered pay-phone 1985
solar-powered steam engine 1864
solar spectrum 1821
    map 1880
solar thermoelectric device 1908
solid logic technology (SLT) 1964
solitons 1874
Solomon's Temple 950 BC
soluble aspirin 1948
SONAR (echo-sounding) 1917
sonic destructor 1270 BC
sonnet 1024
SOS signal 1906
sound, velocity of 1826
sound-on-film motion pictures 1906, 1923
    (*see also* phonofilm)
Space Center (US) 1961
space craft
    animal-carrying 1957
    comet-intercepting 1985
    computer-carrying 1965
    crew-carrying 1964
    docking module assembly 1975
    man-carrying 1961
space probes
    'Hipparchus' 1986
    Jupiter 1986
    Mars 1976
    outer solar system 1972
    planet-to-planet unmanned 1977
space shuttle orbiter 1977
space station 1982
    earth orbiting 1971, 1973
space telescope 1986
spaceflight docking, manned 1969
Spacelab 1983
spanner 1550
spark plug 1883
'Speak and Spell' (machine) 1978
speaking clock 1933
Species, On The Origin of (Darwin) 1859
spectacle lens *see* lens
spectacles
    for myopia 1285
    cemented bifocals 1884
    manufacture of 1478
    nylon supra 1939
    pantascopic 1834
    Temple 1728
'Spectra 900' (fibre) 1985
spectroheliograph 1891
spectroscope 1859
    atmosphere trace molecule 1985
speech-simulating machine 1939
speedometer
    centrifugal-type 1896
    locomotive 1855

motor car 1901
spheres, celestial/terrestrial 552 BC
sphygmograph 1863
sphygmomanometer 1896
spinal meningitis vaccine 1967
spinnaker (yachting sail) 1865
spinning frame mill 1769
spinning jenny 1764
spinning mule 1779
spinning wheel, belt-driven 1280
spinthariscope 1877
spirit duplicating 1923
spirit level 1662
spirit stove 1902
Spitfire (aircraft) 1936
sponge rubber 1846
spray gun 1803
spring and damper 1984
sprinkler system 1874
Sputnik 1957
stage, revolving 1758, 1793
stainless steel, chromium alloys 1912
stamp mill 1598
stapler 1868
statistics 1687
statues, chryselephantine 447
steam car 1770
    high-speed 1906
steam coach 1829
steam engine
    compound 1781
    double cylinder expansion 1804
    experimental 1690
    rotative 1779
    solar-powered 1864
    Z crank 1855
steam engine condenser 1765
steam engine cylinder borer and polisher 1775
steam engine piston 1712
steam engine valve gear 1834
steam gun 1824
steam hammer 1838
steam man 1868
steam ram 1836
steam shovel 1836
steam tram 1837
steamboat 1786
    commercial 1802
    long-distance 1807
    multi-hulled 1788
    offshore 1808
    water-jet-propelled 1787
steamship
    cigar-shaped 1866
    commercial 1812
    iron screw-propelled 1843
    steel-construction 1879
    submarine-hulled 1880
stearine candles 1850
steel
    air-hardening tool 1868
    high-speed 1899
    manganese 1883
    manganese crude 1839
    stainless 1912
steel-making process
    Bessemer converter 1856
    oxygen-lance 1929
steel rolling mill
    automatic thickness control 1957
    continuous hot strip 1922
stereo, personal 1980

stereophonic radio broadcast 1961
stereophonic tape recording 1955
stereoscope 1838
stereoscopic photography 1861
stereotyping 1725
stethoscope
    binaural 1850
    with earpiece 1828
    wood-turned cylinder 1816
still, alcohol 1832
stimulation of paralysed limbs 1982
Stirling-cycle engine 1816
stirrup 850 BC
stock-ticker 1870
stocking-frame machine 1758
stone, artificial glazed 2300 BC
stopwatch 1855
storage cells
    alkaline 1900
    nickel-cadmium 1908
    see also battery
strain seismograph 1935
street-cleaning machine 1841
street-crossing signals 1868
street-lighting, electric 1857
street railway, elevated 1868
streetcar, electric 1879
streptomycin 1943
'stretch' (computer system) 1961
string galvanometer 1903
strip lighting, electric 1905
stroboscope 1836
stroboscopic disc 1832
strontium 1793
strychnine 1818
stylometric analysis, computerised 1963
sub-aqua breathing device 1982
submarine 1776
    ballistic-missile 1962
    battery-electric 1888
    cigar-shaped 1800
    diesel-engined 1906
    electric 1886
    explosively charged 1863
    nuclear-powered 1955
    petrol-electric 1898, 1900
    Polaris 1959
    steam-powered 1879
    wooden boat 1620
submarine cable 1842
submarine decompression chamber 1929
submarine-hunting device 1985
submarine lamp 1850
submarine periscope 1896
submarine pipeline 1975
submersible, glass-fibre 1968
Suez Canal 1869
sugar, filter for 1811
sugar beet factory 1801
sugar evaporation, multiple effect 1832
sugar-refining vacuum pump 1812
sulphonamides 1934, 1938
sulphur, extraction process 1902
sulphur dioxide 1770
sun-and-planet gearing system 1781
sundial of Ahaz 730 BC
'Super Amp' (amplifier) 1949
supercocotte 1953
super-computer 1976
superheterodyne radio receiver 1917
supermarket 1930
    holographic terminal 1980

trolley for 1937
superphosphate fertiliser 1842
supersonic aircraft 1947
supertanker 1966
'supral' (alloy) 1975
surgical thread 1920
surveying chain 1606
surveying (instrumental triangulation) 1615
suspension bridge 1883
swell pedal, organ 1712
swimming aid, inflatable 880 BC
swimming pool, indoor 1742
swinging axles 1903
switch socket, electric 1888
Sydney Opera House 1973
synchromesh gearbox 1929
synchronous motor 1934
synchrotron
    AG electron 1962
    AG proton 1959
synthesiser, programmable 1977
syringe, hypodermic 1853
SYSTRAN (translation system) 1970

T

table-tennis 1889
tactel (fibres) 1983
talking beacon 1928
talking card system 1976
tampon, hygienic 1930
tandem-aerofoil boat 1976
tank (tracked vehicle)
    amphibious 1931
    English design 1915
    French design 1915
    radio-controlled 1940
    revolving turret 1922
tanker ship, diesel-engined 1910
Tannoy (loudhailer) 1934
tantalum 1802
tantalum light filament (incandescent lighting)
    1904
tap, screw-down 1800
tape-recorder,
    commercially viable 1935
    magnetisable steel tape 1899
    personal 'Walkman' player 1979
tape-recordings, stereophonic 1955
tartaric acid 1769
taxi 1640, 1847
taximeter 1847, 1897
tea, packeted 1826
tea-bags 1920
tea fountain 1774
tea-maker, alarm-clock 1902
teaching machine 1925
tear gas 1914
technetium 1937
Technicolor 1926
teddy bear 1902
teeth, false 1770, 1788, 1845
Teflon 1943
    used as non-stick coating 1954
telautograph 1881
teledynamic transmitter 1850
telegraph
    automatic printing 1867
    computing 1846
    electric 1816

electrolytic 1810
letter-printing 1854
optical 1792
semaphore 1816
ship's 1875
speaking 1848
telegraph cable
    Atlantic 1866
    cross-Channel 1851
telegraphy
    duplex 1853
    quadruplex 1874
    multiplex 1853
    synchronous multiplex 1884
    vocal 500 BC
    wireless 1894
    wireless long-distance 1901
telekouphonon (speaking telegraph) 1848
telepen 1972
telephone 1861
    amplifier valve 1906
    articulating 1876
    bellowing 1898
    dial 1896
    experimental 1849, 1854
    wireless 1899
telephone cable
    optical-fibre 1966
    transoceanic 1955
telephone call-box
    coin-operated 1889
    solar-powered 1985
telephone call maker, store and press 1978
telephone exchange/switchboard 1877
    automatic 1889
    electro-mechanical 1926
telephone speaking clock 1933
telephotography 1881
telescope 1608
    astronomical 1609
    binocular 1825
    double convex 1611
    giant reflector 1845
    millimetre-wavelength 1985
    radio 1937
    reflecting 1668
    refractor 1758
    space 1986
telescope lens, Pyrex 1944
teletext 1974
télétroscope 1881
teletype 1928
teletypewriter exchange service (TWX) 1931
television
    cable 1949
    colour 1928
        experimental 1941
        high-definition 1938
        regular broadcasts 1954
    echo cancellation chip 1982
    pocket receiver 1977
    recording system 1928
    three-dimensional 1985
    VHF electronic 1935
television advertisement scheme 1941
television camera
    as viewfinder 1960
    iconoscope electronic 1931
    outside broadcast 1938
    Plumbicon tube 1957
television transmitting studio 1929

television tube/screen
  colour 1949
  image dissector 1928
  largest 1985
  pocket flat 1984
  Trinitron 1968
televisual telephone 1930
Telex system 1931
tellurium 1782
temnograph 1879
temperature control, automatic 1883
temperature scales
  Celsius 1742
  Fahrenheit 1714
  Réamur 1730
temple,
  prefabricated 1300 BC
  Solomon's 950 BC
ten-gallon hat 1865
tennis racket, carbon-fibre 1980
tensile-testing machine 1729
terbium 1843
terylene 1941
test-tube baby 1978
tetra-alkyl lead 1921
text-to-speech system 1985
textile mill, water-turbine-powered 1843
texts, clay geometrical 1800 BC
thallium 1861
Thames Flood Barrier 1982
thaumatrope 1825
theatre, electrically lit 1878
theatre curtains 1664
  safety 1794
Theatrum Orbis Terrarum (atlas) 1570
theodolite 1571, 1787
thermionic valve 1904
thermit process 1898
thermocouple 1822
  solar 1888
thermodynamics, third law 1906
thermographic duplicating machine 1973
thermography, microwave 1980
thermolampe 1799
thermometer
  maximum and minimum 1794
  medical/clinical 1616
  spirit 1730
thermo-multiplier 1831
Thermos flask 1892
thermostat, bi-metal 1830
thesaurus 1852
Thetis (asteroid) 1852
thiamine 1912
thimble 1684
thiokol A (synthetic rubber) 1928
thiophene 1883
thiram 1932
thoracic duct 1647
thorium 1828
thoron isotope 1899
thought machine 1908
three-dimensional images 1947
three-dimensional laser-holography 1964
three-dimensional television 1985
threshing machine 1732, 1776
threshing/fanning machine 1834
thulium 1879
thyroxine 1919
ticket machine, coin-operated 1886
ticonal G (magnet) 1937
tidal-electric power project 1967
tide-calculating machine 1875

ticlocken coat 1899
Tilbury gig 1820
tile, machine-made 1334
timber-bending machine 1855
timber-sawing machine 1592
time bomb 1585
Time Machine (story) 1895
time measurement candle 886
time-recorder, employee's 1894
time-signature, musical 1674
time-switch 1867
tinlet paint pot 1937
tinplated food cans 1865, 1875
tintometer 1886
titanium 1937
  in ilmenite 1791
TNT (explosive) 1863
toaster, electric 1893
  micro-chip 1981
  pop-up 1927
tobiscope 1967
toilet roll 1871
tomato ketchup, bottled 1876
tomography 1928
  positron emission 1985
tonic sol-fa 1835
tonic water 1840
tonometer 1834
tool steel, air-hardening 1868
toothbrush 1498
  electric 1885, 1961
  nylon 1938
toothpaste tube 1892
top quark (particle) 1984
torch, electric 1891, 1898
torpedo 1842
  self-propelling 1866
Torricellian Tube 1644
torsion bar suspension 1945
tourniquet 1674
toxicology 1813
toy tiger 1000 BC
toy train, clockwork 1856
tractor
  caterpillar diesel-engined 1931
  crawler 1904
  diesel-engined 1922
  hydrostatic steering 1968
  mass-produced 1917
  petrol-engined 1889
  rear-wheel power adjustment 1947
  rubber-tyred 1931
  three-point hitching system 1933
traffic lights, electric 1914
  automatic 1926
  red-green-amber 1918
train
  see locomotive; railway
train shed, giant 1854
tram
  compressed air 1888
  electric 1888
  passenger 1832
  steam 1837
tram-roads 1800
tranquilliser drug 1950
transformer, electrical 1884
  principle of 1831
transistor
  high-speed switching surface barrier 1954
  junction 1951
  silicon-grown 1957

memory 1971
point contact 1948
superconducting 1983
transistor radio 1954
translation machine, computerised 1984
traveller's cheque 1891
treacle, canned 1885
treadmill 1817
tree-measuring instrument 1948
tricycle
electric 1889
Humber tandem 1885
motor 1888
Salvo quad 1878
steam-powered 1881
streamlined cabin 1980
tri-cyclecar, pedal-electric (Sinclair C5) 1984
tri-ergon sound film 1922
trifocal lens 1827
'trigger relay' (electronic valve) 1919
trigonometrical abbreviations 1621
trigonometry 130 BC
trimaran
sailing 1943
steam-engined/aerially propelled 1903
Trinitron television tube 1968
triode, regenerative and oscillating features 1912
triode amplification tube 1906
trip-hammer mill 20
triplane, triple-hydro- 1921
Triplex glass 1910
tripod, aquatic 1822
trireme (galley) 750 BC
Trojan Horse 1250 BC
trolleybus, prototype 1881
trombone, tenor-bass 1839
'true water cure' 1849
trunk, life-saving 1899
truss, railroad bridge 1847
tryptophan 1900
tse-tse fly 1895
tuberculosis germ 1882
tuberculosis vaccine 1908
tumour-inducing virus 1910
tungsten 1783
tungsten carbide 1926
tungsten filament 1908
tungsten filament lamp, gas-filled 1913
tuning fork 1711
tunnel, canal 1777
tunnelling shield 1835
Tupperware boxes 1945
turbine
axial 1837
steam impact 1887
steam-powered 1884
vortex 1850
water see water turbine
tuxedo 1886
twine-binder 1876
TWX exchange service 1931
type, cast 1720
type-composing machine 1840
type foundry 1392
typeface
italic 1476
Jensen 1470
typesetting
by casting 1720
by computer 1960
photographic 1895

typewriter
automatic 1908
electric 1901
electric mass-produced 1935
electronic 1978
for the blind 1808
golf-ball 1961
intelligent see word processor
mass-produced 1873
noiseless 1908
pneumatic 1891
pocket 1887
portable 1892
portable electronic 1982
simulated listening 1981
thermal electronic 1985
visible 1896
typhus fever body louse 1909
tyre
cord 1900
cross-ply 1903
low-pressure 1923
tractor 1932
radial-ply (steel-reinforced) 1946
radial (asymmetric tread) 1965
tubeless 1947
tyre valve (pneumatic) 1898

U

UK Patent No. 1,000,000 1965
ultramarine blue, artificial 1828
ultramicroscope 1903
ultrasonic scanner 1979
ultrasound 1955
ultra-violet radiation 1802
umbrella
alpaca covering 1848
collapsible pocket 1887
curved steel rib frame 1874
lightning-conducting 1786
telescopic 1930
underground railway 1863
automatic doors 1922
electric 1890
pneumatic 1870
undulatory theory of matter 1924
UNIVAC 1 (computer) 1952
universal joint 1676
unnilenium 1982
unnilhexium 1974
unniloktium 1984
unnilpentium 1970
unnilquadium 1964
unnilseptium 1976
Uraniborg observatory 1576
uranium 1789
Uranus 1781
urea, synthesis of 1827
ureas, compound 1851
urease 1926
US Patent No. 1,000,000 1911
US Patent No. 2,000,000 1935
US Patent No. 3,000,000 1961
US Patent No. 4,000,000 1976

V

V2 rocket 1942
vaccination (smallpox) 1796

vaccine
  cholera 1892
  foot-and-mouth disease 1899
  measles 1953
  polio 1957
  rubella 1965
  17-D 1939
  spinal meningitis 1967
  tuberculosis 1908
  whooping cough 1923
  yellow fever 1939
vacuum cleaner
  electrically powered 1901
  Hoover 1907
  manually operated 1869
  upright 1907
vacuum flask 1892
vacuum tube thermionic valve 1904
valine 1901
Valium 1961
valve, electronic
  as switch 1919
  thermionic 1904
valve gear, steam engine 1834
Van Allen radiation belts 1958
vanadium 1830
vapour-powered machinery 1794
varnish 1772
vaulting horse 1761
vegetable oils, catalytic hydrogenation of 1899
Velcro (fastening device) 1948
velocipede
  amphibious 1867
  chain-driven 1869
vending machine, automatic 1883
veneer-cutting machine 1794
Venetian blind 1769
venticulography 1918
ventilators 1741
Venus (planet), colour TV pictures 1982
vernier scale 1631
Very High Frequency (VHF) 1917
VHF radio-telephony 1932
vibraphone 1916
video camera, underwater 1980
video 8 camera 1985
video card magnetic 1974
video disc 1970
  arcade game 1983
  laser-read 1972
video game 1972
  mass-produced 1975
videophone service 1936 (see also televisual
  telephone)
video-tape
  cassette system 1969
  commercial use 1956
  experimental 1952
  home system 1975
  transistorised system 1961
video-tape recorder, home-use all-transistor
  1964
Viewdata system 1979
village 6000 BC
vinyl floor covering 1945
vinyl polymers 1935
vinylite disc (gramophone record) 1946
violin, nail 1770
virus 1892
viscose process 1892
visible speech 1866
vitamin A 1913
  synthetic 1947
vitamin B1 1912

vitamin B2 1933
vitamin B6 1936
vitamin B12 1956
  as anaemia treatment 1947
vitamin C, isolated from citrus fruits 1928
vitamin C, synthetic 1932
vitamin D 1898
  antirachitic factor 1918
vitamin E 1922
vitamin H 1940
vitamin K 1934
vitamins, growth-stimulating 1906
Vitrifix process (asbestos waste re-cycling) 1984
Volkswagen 1936
Volkswagen, amphibious 1940
'Volontas' 1822
Vostok I (spacecraft) 1961
VTOL aircraft 1960
  fighter-piane 1969
vulcanised rubber 1841
vulcanite 1851

W

wagonette 1842
wagon-fort 1420
wagon-mill 340
'walkie-talkie' radio 1941
'Walkman' (tape-replayer) 1979
walled town 8350–7350 BC
wallpaper 1481
  block-printed 1560
  continuous pattern 1688
  cylinder-printed 1835
Waltham watch 1850
Wankel engine 1963
  motorcycle 1974
warehouse design 193 BC
warship, armoured 1813
washing apparatus (for ore) 1512
washing-machine
  electric 1906
  hand-rotated 1884
  spin-dry 1924
  steam-operated 1862
  ultrasonic 1985
washtub, mechanised 1858
watch
  quartz digital 1971
  jewelled 1704
  machine-made 1850
  pocket 1510
  talking 1895
  Waltham 1850
  waterproof 1926
  see also wristwatch
water bed 1813
water closet 1589
  valve 1775
  washdown flushing 1889
  washout 1870
water-displacer (WD-40) 1952
water-filtration system 1675 1829
water frame (for spinning) 1769
water-skiing, powered 1922
water turbine 1827
  bucket-wheel 1870
  cased-in 1855
  variable propeller 1924
watercolour pigment cakes 1766, 1832
watermark 1286

watermill 555
waterproof rubber 1823
watertight compartments, ship's 400
waterwheel shaft, cast-iron 1769
wave power
  boat propulsion 1980
  device 1974
  energy absorber and diverter 1944
  measurement of 1875
wave theory of light 1801
WD-40 1952
weather barometer 1672
weather satellite 1960
weaving loom, *see* loom
weaving mill, ribbon 1605
weighbridge, cart 1743
weighing machine
  personal 1928
  with dial 1772
welding
  arc 1865
  resistance 1877
Wells hive 1890
Western hat 1865
wheat-stripper 1843
wheel
  detachable wire 1905
  overbalancing 1245
  tangent-spoke 1874
  threaded spoke 1870
  toothed 200 BC
'Wheel of Life' (optical toy) 1832
wheelbarrow 231
wheel-bearings 100 BC
wheelchair 1650
wheel-cutting machine 1670
'wheels, aerial' (pneumatic) 1844
Whirling Aeoliphile, steam-powered 100
whisky 450
whist (game) 1743
Whit classification 1835
'white coal' (hydro-electric power) 1867
whole-body X-ray scanner 1973
wildlife-preserving park 1897
wildlife satellite-tracking 1972
Winchester rifle 1860
wind gauge 1667
  Beaufort Scale 1806
wind tunnel 1871
wind turbine
  megawatt farms 1982
  2-megawatt 1979
  variable-geometry vertical-axis 1985
winding colliery engine 1784
windmill
  automatic fantail 1745
  for electricity generation 1890, 1931, 1941
  for grinding corn 644
  for shammy dressing 1592
  patent sails 1807
  roller reefing sail 1789
Windolene (polish) 1923
window blinds, parallelogram 1760
window polish 1923
windows, stained glass 535
windscreen, toughened plate-glass 1929
windsurfer 1970
wine-pressing machine 1824
wing, flexible 1956
winnowing machine, rotary 40 BC
wire-drawing 1410, 1563

wireless
  communication to an aeroplane 1910
  pedal 1928
  see also radio
wirephoto 1921
'*Wisdom Text*' 2380 BC
wood glue, synthetic 1912
wood-planing machine 1802
wood pulp
  alkaline process 1857
  chemical 1854
  mechanically ground 1843
  prototype sulphite process 1866
  sulphite process 1872
wool-combing machine 1790
Woolworth Building 1913
Woolworth's store 1879
wolfram 1783
Worcester sauce 1837
word processor 1965
word-teaching game 1982
'*World, History of the*' 1311
worm gear 1897
wristwatch 1790
  self-winding 1922
'writing ball' (mechanical device) 1867
writing machine 1714, 1753
  swinging sector 1829
wrought iron
  beam 1824
  manufacture 1851

X

xenon 1898
xenon short-arc lamp 1954
xerographic copying machine
  prototype 1938
  production office model 1950
X-ray analysis 1920
X-ray astronomy 1962
X-ray scanner (computerised axial
  tomography) 1973
X-ray spectroscopy 1916
X-ray system, filmless 1985
X-rays 1895
xylonite 1869
xylophone 1840
xylotechnographica (wood-staining) 1875
xylyl bromide (tear gas) 1914

Y

yacht
  Bermuda sail 1808
  carbon-fibre 1985
  catamaran 1662
  ice 1790
  motor (glass-fibre construction) 1954
  pneumatic 1889
  swift 1595
  turbine-driven 1897
  winged keel design 1983
Yale cylinder lock 1865
yarn spinner, self-twist 1961
yeast growth 1836
yellow fever vaccine 1939
ytterbium 1878
yttrium 1794

Z

Zeppelin raid 1915
zero (mathematics) 10000 BC
Ziggurat, Ur-Nammu 2100 BC
zinc 1541
zinc white 1790
zinc-coated steel 1840

zip fastener 1891
  hook-less 1906
zirconium 1789, 1937
zither, bowed 1823
Zoetrope 180
zoo, public 1793
zoom lens 1932, 1945